T5-BBB-683

Magika Hiera

Magika Hiera

Ancient Greek Magic and Religion

EDITED BY

Christopher A. Faraone
Dirk Obbink

New York Oxford
OXFORD UNIVERSITY PRESS
1991

Oxford University Press

Oxford New York Toronto
Delhi Bombay Calcutta Madras Karachi
Petaling Jaya Singapore Hong Kong Tokyo
Nairobi Dar es Salaam Cape Town
Melbourne Auckland
and associated companies in
Berlin Ibadan

Published by Oxford University Press, Inc.,
200 Madison Avenue, New York, New York 10016

Library of Congress Cataloging-in-Publication Data
Magika hiera : ancient Greek magic and religion / [edited by] Christopher A.
Faraone and Dirk Obbink.
p. cm. Bibliography: p. Includes indexes. ISBN 0-19-504450-9
1. Magic, Greek. 2. Greece--Religion. I. Faraone, Christopher A.
II. Obbink, Dirk.
BF1591.M35 1991 88-37685
133.4'3'0938--dc19

9 8 7 6 5 4 3 2 1
Printed in the United States of America
on acid-free paper

Preface

Individuals in antiquity who did not particularly like their neighbors or colleagues or became enamored of others, who wanted to win big at the races or guard against a life-threatening disease or forecast a rise in personal income or come to terms with a forgotten past, had a variety of methods at their disposal to attain their goals, or at least express their desires. Some are still familiar; others have fallen into obscurity (at least in Western society) together with their origin, operation, and social function. The practice of writing curses (*defixiones, κατάδεσμοι*, i.e., "binding spells") on lead tablets, for instance, and dropping these tablets into a well, spring, or grave has been documented wherever Greeks or Romans lived and exercised their influence.* The practice is attested as early as the fifth century B.C. in places as far-flung as Sicily, Attica, and the shores of the Black Sea. Lead "voodoo dolls" have been unearthed even in Attica, inscribed with the names of famous fourth-century orators and then pierced through with iron nails or bronze pins. Papyrus finds have brought to light large sections of magical handbooks in which various professionals purveyed their selections of detailed prescriptions or recipes for acquiring a lover, curing disease, prevailing in court, securing the tutelage of a particular deity, or protecting an individual's home or workplace against others or against potential threats in the community at large.

Individuals in antiquity turned to such rituals in the hope of bettering their fortunes in a natural world that seemed hostile and unpredictable, in a society that competed fiercely for the use and control of limited resources and advantages. Farmers ensured a bountiful harvest by encouraging rainfall and inhibiting the attacks of noxious insects and other agricultural blights. In the cities and towns merchants, artisans, and politicians attempted to increase their profits and personal prestige by cursing the activities of their rivals in the agora or in the popular assembly. At some point in their lives virtually every man and woman would have had the option of recourse to these traditional rites to learn about the future, turn the head of a potential lover, or prevent plague and other diseases from falling on their families and flocks. Indeed a close reading of the extant sources for daily life in the ancient world reveals many such common fears and persistent uncertainties that

* Many of the phrases used in this opening paragraph came from an unpublished essay by H. S. Versnel.

daily beset all men and women, rich and poor, slave or free. Unrequited love, sterility, impotence, gout, eye disease, bad luck at the races, or an unexpected setback in a legal case—all these and a multitude of other distresses are revealed in the texts of the magical inscriptions and papyri.

All these practices border ostensibly on the sphere of *religion* (perhaps of a private or familial sort) insofar as they document attempts on the part of individuals to influence factors in their environment that are beyond their immediate control. In many cases these private, "magical" rites have clear parallels with well-known forms of corporate and civic cult. Yet the relationship between magic and religion with respect to such practices has historically been, and continues to be, a very problematic one. Students of ancient religion have treated such practices in turn as superstitious religion, vestiges of primitive religion, perverse or corrupt forms of religion, and the very inverse of religion. Others, having introduced theories of the development of science and scientific thinking, have claimed that the relationship between men and the gods exhibited in magical practices was fundamentally different from that in religious rites; or that magic involved manipulation, religion supplication; or that magic presupposes principles of cosmic sympathy and antipathy, whereas religion does not. More recent work (particularly, though not exclusively, influenced by developments in sociology and anthropology) has brought trenchant challenges against these distinctions. Many now view magic as a type of religious deviance and treat magical practices as nondichotomous variations in ritual procedure, arguing that the antithesis between magic and religion arbitrarily separates a continuous spectrum of interlocking religious phenomena. There cannot at present be said to exist anything approaching a consensus over the deployment and definition of terms (especially with regard to theories of historical development). Many continue to cling, consciously or not, to the standard dichotomy. The situation "resembles nothing so much as the endless shuffling and redealing of a deck of but three cards."*

In any field of inquiry progress is achieved by two developments: either the existing pool of data is significantly enlarged or otherwise improved so as to prompt new investigations according to existing approaches; or refinements or (r)evolutions in methodology prompt investigators to look at the existing data "through different-colored lenses." In the study of ancient Greek magic and religion both developments have occurred. The first four chapters in the present volume are devoted to newly found or reedited inscriptional material and to the subsequent refinement of categories and theories of historical development attendant on the incorporation of such new data. The remaining essays in the book deal with changes in the study of this evidence—with particular attention to *specific ritual practices and procedures*—and with new definitions of the fields of religion and magic or science that have been prompted by refinements or changes in methodology.

In each case authors were urged to consider in detail a specific area of ritual activity and to ask *whether the traditional dichotomy between magic and religion*

* C. R. Phillips, *ANRW* 16.3 (1986):2723, with reference to the terms "religion," "magic," and "science."

helped in any way to conceptualize the objective features of the evidence examined. This volume arises out of our conviction that a case-by-case examination of specific rituals and their contexts will eventually yield a comprehensive account of the areas of convergence and divergence between ancient magic and religion and establish the study of magic as an area to be ignored by students of ancient religion and society only at their peril.

Blacksburg, Virginia C. A. F.
New York City D. O.
May 1990

Acknowledgments

We acknowledge a special debt of gratitude to John J. Winkler who offered encouragement, advice, and criticism at every juncture. E. Courtney, A. Henrichs, M. H. Jameson, D. Jordan, L. Koenen, R. Kotansky, M. McCall, P. Parsons, A. E. Raubitschek, S. Stephens, S. Treggiari, and H. S. Versnel all offered valuable advice in planning and producing the book and in many instances read and commented on earlier drafts of individual essays. D. Obbink translated the essay of S. Eitrem from the original German. J. Ringgold translated H. S. Versnel's contribution from the Dutch original. Thanks are owed to the Department of Classics and the Dean of Graduate Studies of Stanford University for providing funds for the costs of translation, typing, copying and mailing. Finally, we are grateful to the Louvre for permission to publish the photograph used on the dust jacket.

Contents

List of Abbreviations

1 The Agonistic Context of Early Greek Binding Spells, 3
CHRISTOPHER A. FARAONE
Virginia Polytechnic Institute and State University

2 "Cursed be he that moves my bones", 33
J. H. M. STRUBBE
University of Leiden

3 Beyond Cursing: The Appeal to Justice in Judicial Prayers, 60
H. S. VERSNEL
University of Leiden

4 Incantations and Prayers for Salvation on Inscribed Greek Amulets, 107
ROY KOTANSKY
Los Angeles, California

5 The Pharmacology of Sacred Plants, Herbs, and Roots, 138
JOHN SCARBOROUGH
University of Wisconsin

6 Dreams and Divination in Magical Ritual, 175
† SAMSON EITREM
formerly Oslo University

7 Prayer in Magical and Religious Ritual, 188
FRITZ GRAF
University of Basel

8 The Constraints of Eros, 214
JOHN J. WINKLER
Stanford University

9 Magic and Mystery in the Greek Magical Papyri, 244
HANS DIETER BETZ
University of Chicago

10 *Nullum Crimen sine Lege:* Socioreligious Sanctions on Magic, 260
C. R. PHILLIPS III
Lehigh University

Selected Bibliography of Greek Magic and Religion, 277
Index of Greek Words, 285
Index of Latin Words, 289
General Index, 291

Abbreviations

Ancient authors and works are referred to by standard abbreviations. Abbreviations of frequently cited periodicals, dictionaries, book series, and epigraphical and papyrological collections are listed below.

AJA	*American Journal of Archaeology*
AJP	*American Journal of Philology*
AM	*Mitteilungen des deutschen archäologischen Instituts, Athenische Abteilung*
ANRW	*Aufstieg und Niedergang der römischen Welt.* Berlin, 1972–.
APF	*Archiv für Papyrusforschung*
ARW	*Archiv für Religionswissenschaft*
BASP	*Bulletin of the American Society of Papyrologists*
BCH	*Bulletin de correspondance hellénique*
BE	*Bulletin épigraphique,* published in *REG* (cited by year)
BIFAO	*Bulletin de l'Institut Français d'Archéologie Orientale*
BJ	*Bonner Jahrbücher*
BSA	*Annual of the British School at Athens*
CJ	*Classical Journal*
CPh	*Classical Philology*
CR	*Classical Review*
CRAI	*Comptes rendus de l'Académie des Inscriptions et Belles-Lettres*
DAGR	Daremberg, C. V., and E. Saglio, eds. *Dictionnaire des antiquités grecques et romaines*. Paris, 1877–1919.
DenkschrWien	Denkschriften der Akademie der Wissenschaft der Wien, Philos.-hist. Klasse
DT	Audollent, A. *Defixionum Tabellae*. Paris, 1904.
DTA	Wünsch, R. *Defixionum Tabellae Atticae. IG* 3.3. Berlin, 1897.
EA	*Epigraphica Anatolica*
EPRO	Études préliminaires aux religions orientales dans l'empire romain
GMPT	Betz, H. D., ed. *The Greek Magical Papyri in Translation*. Chicago, 1986.
GöttNachr.	Nachrichten von der Gesellschaft der Wissenschaften zu Göttingen, Philologische-historische Klasse

GRBS Greek, Roman, and Byzantine Studies
HR History of Religions
HSCP Harvard Studies in Classical Philology
HTR Harvard Theological Studies
IG Inscriptiones Graecae. Berlin, 1873–.
JHS *Journal of Hellenic Studies*
JOAI *Jahreshefte des österreichischen archäologischen Instituts*
JRS *Journal of Roman Studies*
JWCI *Journal of the Warburg and Courtauld Institute*
LSJ Liddell, H. G., R. Scott, and H. S. Jones. *A Greek English Lexicon*. 9th ed. Oxford, 1925–1940; Suppl. 1968.
MAMA *Monumenta Asia Minoris Antiqua*. Manchester, 1928–61.
OZ Hopfner, T. *Griechisch-ägyptischer Offenbarungszauber*. Studien zur Palaeographie und Papyruskunde 21, 23. Leipzig, 1973-1974.
PDM Egyptian Demotic Papyri translated in *GMPT*
PGM Preisendanz, K. et al., eds. *Papyri Graecae Magicae: Die griechischen Zauberpapyri*. 2d ed. Stuttgart, 1973–1974.
RAC Klauser, T., ed. *Reallexikon für Antike und Christentum*. Stuttgart, 1950-.
RE Pauly, A., and G. Wissowa, eds. *Real-Encyclopädie der classischen Altertumswissenschaft*. Stuttgart, 1894–1980.
REA *Revue des études anciennes*
REG *Revue des études grecques*
REL *Revue des études latines*
RGVV Religionsgeschichtliche Versuche und Vorarbeiten
RhM *Rheinisches Museum für Philologie*
RHR *Revue de l'histoire des religions*
RIB *Roman Inscriptions of Britain*
Roscher W. H. Roscher, ed. *Ausführliches Lexikon der griechischen und römischen Mythologie*. Leipzig, 1884–1937.
SBAW Sitzungsberichte der bayerischen Akademie der Wissenschaften, Munich
SBBerlin Sitzungsberichte der Akademie der Wissenschaften, Berlin, Philos.-hist. Klasse
SBHeidelberg Sitzungsberichte der Akademie der Wissenschaften, Heidelberg, Philos.-hist. Klasse
SBLeipzig Sitzungsberichte der Akademie der Wissenschaften, Leipzig, Philos.-hist. Klasse
SBWien Sitzungsberichte der Akademie der Wissenschaften, Wien, Philos.-hist. Klasse

SEG *Supplementum Epigraphicum Graecum*. Leiden, 1923-.

SGD Jordan, D. R. "A Survey of Greek Defixiones Not Included in the Special Corpora." *GRBS* 26 (1985): 151-97.

SIG Dittenberger, W., ed. *Sylloge Inscriptionum Graecorum*. Leipzig, 1898–1924.

SO *Symbolae Osloenses*

TAPA *Transactions of the American Philological Association*

TWNT Kittel, G., ed. *Theologisches Wörterbuch zum Neuen Testament*. Stuttgart, 1933-.

WS *Wiener Studien*

ZPE *Zeitschrift für Papyrologie und Epigraphik*

Magika Hiera

1

The Agonistic Context of Early Greek Binding Spells

Christopher A. Faraone

A flattened lead "gingerbread man" now in the Bibliothèque Nationale in Paris is deceptively benign at first glance; closer examination reveals two brief texts inscribed at different points on its surface:[1]

> *καταγράφω* Εἰσιάδα *τὴν* Α⟨ὐ⟩*τοκλέας*
> *πρὸς τὸν* Ἑρμῆ *τὸν κάτοχον.*
> *κάτεχε αὐτὴ⟨ν⟩ παρὰ σα⟨υ⟩τόν.*
>
> *καταδεσμεύω* Εἰσιάδα *πρὸς τὸν* Ἑρμῆ
> *τὸν κάτοχον·* [χ]ἔρες,
> *πόδες* Εἰσιάδος, *σῶμα ὅλον.*
>
> I register[2] Isias, the daughter of A(u)toclea, before Hermes the Restrainer. Restrain her by your side!
>
> I bind Isias before Hermes the Restrainer, the hands, the feet of Isias, the entire body.

Apparently found at Carystus on the island of Euboea, these two messages date to the fourth century B.C. and are good examples of a uniquely Greek form of cursing known as a *κατάδεσμος*, or *defixio,* terms that I shall use interchangeably to mean "binding spell."[3] Nearly six hundred Greek *defixiones* have been published to date and more than four hundred others have been unearthed and are awaiting study.[4] The earliest examples are found in Sicily, Olbia, and Attica and date to the fifth century B.C.; by the second century A.D. they begin turning up in every corner of the Greco-Roman world. In the classical period they are usually inscribed on small sheets of lead, which are folded up, pierced with a bronze or iron nail,[5] and then either buried with the corpse of one of the untimely dead (*ἄωροι*)[6] or placed in chthonic sanctuaries.[7] In later periods they are more often placed in underground bodies of water (e.g., wells, baths, fountains).[8] Sometimes the findspot indicates the target of the curse in use; *defixiones* aimed at charioteers, for example, have been discovered beneath the starting gates and amidst the ruins of the *spinae* in late Roman hippodromes.[9]

My aim is to provide an analysis of the function and social context of the *κατάδεσμοι* in early Greek society. The approach will be twofold: I shall (1) analyse the various formulas used in the binding curses and demonstrate that they originally aimed at binding but not destroying the victim and (2) suggest that an agonistic relationship was the traditional context for the use of *defixiones* and that they were not employed as after-the-fact measures of vengeful spite but rather as effective "preemptive strikes" against a formidable foe in anticipation of a possible or even probable future defeat.

THE BINDING FORMULAE

From the available evidence it is still somewhat unclear whether these binding rituals are a traditional form of self-help to which the ancient Greeks themselves turned in times of crisis or whether a professional "magician" was employed to perform the ritual in their stead. The act of flattening out a soft piece of lead and then scratching a name into it certainly did not require much more effort or technical skill than inscribing a potsherd for a vote of ostracism.[10] On the other hand, Plato refers clearly to peripatetic magicians who perform κατάδεσμοι for a price (*Resp*. 364c), and there are several well-documented instances (albeit dating to the Roman period or later) that involve caches of *defixiones* mass-produced by the same individual(s) working from a formulary.[11] Four bound lead "voodoo" dolls—each enclosed within a lead box inscribed with a binding curse—were recently discovered in two different graves in the Kerameikos; dating to circa 400 B.C., they seem to have been produced by the same person(s), perhaps providing the earliest extant material evidence for the professional magician at work in Greece.[12]

The actual layout of an inscribed κατάδεσμος occasionally bears close similarities to other written forms of public or private communication. A few Attic examples from the early fourth century, for example, consist solely of names designated in the formal manner (i.e., with patronymic and demotic) and laid out neatly in columns (e.g., *DTA* 55 or SGD 48); in two cases this imitation of contemporary public monuments is made explicit by the heading ἀγαθῇ τυχῇ · θεοί (SGD 19 and *DTA* 158). At other times the tablet is referred to as an epistle sent to a chthonic god or a restless *nekydaimon;* such a form would naturally suggest itself to a Greek, since lead seems to have been a common medium for letter writing in the earlier periods.[13] The efficacy of a few early Attic curses may hinge on the corpse's ability to read the tablet placed in his grave and act accordingly.[14] One curse from Piraeus (*DTA* 103) reads, "I send this letter to Hermes and Persephone"; while another (*DTA* 102) takes the form of a bill of lading: "I send a letter to the daemons and Persephone bearing NN (name of person to be supplied)."[15] In at least three cases, the inscription of the names of chthonic gods on the outside of the rolled-up tablet may be meant to imitate the method in which ordinary letter scrolls were regularly addressed.[16]

These parallels to other forms of written documents are infrequent and seem to be idiosyncratic inventions or variations that probably do not point to the origin of this uniquely Greek form of cursing.[17] Some scholars, in fact, have argued that the *defixio* was originally a purely verbal curse, although I prefer to think that both the spoken formula and the attendant gesture (i.e., the distortion of lead, wax, or some other pliable material) developed simultaneously.[18] A cache of some forty blank tablets, rolled up and pierced with nails, may suggest that the name of the victim and the cursing formulae could merely be recited over the tablet while it was being twisted and perforated.[19] Another clue is the fact that the earliest terms used to refer to *defixiones* do not allude to the act of writing;[20] a fifth-century Sicilian curse tablet (SGD 91) refers to itself as an *εὐχά* ("prayer" or "boast") and Plato (*Resp.* 364c) speaks generally of binding spells called *κατάδεσμοι*. The *ὕμνος δέσμιος* (lit., "binding song") of the Erinyes in Aeschylus' *Eumenides* seems to be a purely verbal form of *defixio;* it aims at binding the verbal and mental faculties of Orestes in hopes of inhibiting his performance at his forthcoming murder trial.[21] Indeed the very fact that the great majority of the earliest Sicilian and Attic *defixiones* consist solely of lists of names strongly suggests that a verb of binding was uttered aloud sometime during the ritual and that the later development of more-complex, written formulae reflects a desire to inscribe more and more of the spoken charm on the tablet, a process that was undoubtedly accelerated by the gradual spread of literacy in the classical period.

The Greek *κατάδεσμοι* that mention only the name of the intended victim steadily decrease in frequency from the classical age until their total disappearance in the first century A.D. The complex formulae become correspondingly more popular in the later periods[22] and can be divided roughly into four groups:[23]

1. *Direct binding formula.* The *defigens* (lit., "the one who binds") employs a first-person singular verb that acts directly upon the victims or specified parts of their bodies, for instance, *καταδῶ τὸν δεῖνα* ("I bind NN").
2. *Prayer formula.* Gods or daemons are invoked and urged by a second-person imperative to perform similar acts of binding, for instance, *κατέχετε τὸν δεῖνα* ("Restrain NN!").
3. *Wish formula.* The victim is the subject of a third person optative, for instance, *ὁ δεῖνα ἀτελὴς εἴη* ("May NN be unsuccessful!").
4. *Similia similibus formula.* This type employs a persuasive analogy, for instance, "As this corpse is cold and lifeless, in the same way may NN become cold and lifeless."

The direct binding formula (no. 1) is best described as a form of performative utterance that is accompanied by a ritually significant act, either the distortion and perforation of a lead tablet or (more rarely) the binding of the hands and legs of a small effigy. Often various bodily parts or personal possessions are listed alongside the person's name as more specific targets, for instance, *DTA* 52 (Attic, third century B.C.): "I bind Mnesithides and the tongue, work, and soul of Mnesithides." The most common elaboration of the direct binding formula is the addition of the name(s) of a deity or deities who appear as witnesses or overseers of the act, for instance *DTA* 91 (Attic, third-century B.C.): "I bind Ophelion and Katheris before (*πρός*) Hermes Chthonios and Hermes Katochos." Most of the other verbs used in the first-person (e.g. *κατατίθημι*, *παραδίδωμι* or *καταγράφω*), like *ἀπο-* and *ἐγγράφω* on the early Sicilian tablets, seem to be legal or technical terms that shift responsibility for the binding to the divine sphere;[24] thus the *defixio* that reads simply, *καταδίδωμι τὸν δεῖνα* ("I assign NN") is probably shorthand for the full expression that would include a "prayer formula" as well, such as on the text from

Carystus quoted above: "I register NN before Hermes the Restrainer. (You, Hermes,) restrain her by your side!"

The prayer formula (no. 2) brings the gods more directly into the action as the actual agents of the binding. It, too, was often accompanied by the manipulation of lead tablets and images. Limited evidence from Sicily seems to suggest that the prayer formula was used as early as the mid–fifth century B.C., but it is impossible to know this for certain.[25] By the early fourth century it appears frequently in Attica and elsewhere. Like any proper Greek prayer, these tablets begin with an invocation of a particular chthonic god who is addressed directly in the second person.[26] Hermes, Ge, Hecate, and Persephone are (in that order) the most common deities to appear on the tablets. As is true for most traditional forms of prayer, there is an unlimited opportunity for the *defigens* to expand the text at the invocation, either by multiplying the number of gods invoked or by increasing the number of epithets and powers of the god(s) already invoked. This opportunity is used with greater frequency in the Roman period when the addition of any and all foreign-sounding deities, epithets and *voces magicae* was thought to increase the efficacy of every "magical" operation. The verbal form that followed the invocation was usually a second-person imperative,[27] for instance, Ἑρμῆ κάτεχε τὸν δεῖνα ("O Hermes, restrain NN!"), but this is occasionally replaced by a third-person passive imperative, for instance, *DTA* 105a (Attic, third century B.C.) Ἑρμῆ χθόνιε ὁ δεῖνα καταδεδέσθω ("O Hermes Chthonios, NN must be bound!").[28]

It must be granted that Kagarow's designation of simple, unadorned imperatives as "prayer formulae" uses such general criteria for defining "prayer" that it could conceivably include "coercive" imperatives (e.g., with a verb like ὁρκίζω understood) as well as "submissive" imperatives (e.g., with a verb like ἱκετεύω understood). As such it might be argued that it is of little or no use as an analytical category in a taxonomy of κατάδεσμοι. I would suggest, however, that whenever such distinctions were important to the *defigens,* he or she took pains to make them clear; thus we do have a few examples of clearly submissive terminology in *defixiones* from the early period (see Versnel's treatment of the "borderline" *defixiones,* chap. 3). The most important fact is that the overwhelming majority of early Greek κατάδεσμοι that directly address the gods make no attempt whatsoever to inform the deity that the approach of the *defigens* is either coercive or submissive; I suggest that concern over such matters is related to later and perhaps non-Greek religious mentalities. The author of a text that simply reads, "O Hermes Restrainer, restrain NN!" has stripped his prayer down to the bare essentials (the invocation and the request) undoubtedly because he thinks that such a simple, blunt approach is effective exactly as it stands. Thus Kagarow's rather general definition of the "prayer formula"—although its usefulness is limited when analyzing later, more verbose texts—is particularly suited for the analysis of early Greek *defixiones* because it is ultimately dictated by the nature of the texts themselves.

The third category ("wish formula") is distinguished on formal grounds here, but in actual practice it is usually employed as the second part of the so-called *similia similibus* formula (category no. 4). This final category is potentially the most revealing type because it seems to give us insight into the rationale behind many of the details that constitute these binding rituals. A fourth-century Attic curse (*DTA*

67) appears to be the earliest extant example of this type of formula; it refers to the fact that the text of the tablet is inscribed in retrograde: "Just as these words are cold and backwards (lit., *ἐπαρίστερα*, "written right to left"), so too may the words of Krates be cold and backwards." A popular tactic is to allude to the fact that the *defixio* has been placed in a tomb, for instance, *DT* 68 (Attic, fourth century B.C.): "Just as this corpse lies useless, so too may everything be useless for NN." Another Attic tablet, *DTA* 106b (third century B.C., cf. *DTA* 105 and 107), uses the lead medium as a point of reference: "Just as this lead is useless (*ἄχρηστος*), so too may the words and deeds of those listed here be useless." As all three of these types of *similia similibus* formulae might be interpreted as proof of the homicidal intent of the *κατάδεσμοι*, a detailed discussion seems warranted here, especially in light of some recent anthropological work on modern misconceptions about so-called sympathetic magic.

Although Wünsch recognized from the very beginning that lead was probably used because it was a cheap writing medium in classical Attica and elsewhere (it is a by-product of silver mining), he suggested that the peculiar coldness and color of the metal ("like the pallor of a corpse") might have made its use for *defixiones* especially appealing.[29] Other scholars, such as Kagarow, went so far as to suggest that the *similia similibus* formulae can be used as proof that the use of lead, as well as the employment of retrograde writing and the deposit in tombs, was originally of great importance to the efficacy of this kind of curse. Some evidence, however, points to the use of media other than lead. One fourth-century Attic *defixio* suggests that *κατάδεσμοι* inscribed on wax, another cheap and popular writing material, may have been used as well: "I bind all of these people in lead *and in wax*" (*DTA* 55a); here we might imagine that two tablets, one wax and one lead, were inscribed with the same text in hope of doubled efficacy. Unfortunately, only the lead tablet has survived the passing of the centuries. Ovid (*Am.* 3.7.29) imagines a person performing a *κατάδεσμος* by writing a name on wax and then piercing it: *sagave poenicea defixit nomina cera*.[30] There are no extant wax *defixiones,* nor would anyone expect them to survive in the harsh climate of the Aegean. A similar state of affairs occurs with the material evidence for Greek voodoo dolls of which only lead and bronze examples survive.[31] Here, too, there is reason to suspect that these are only the more durable representatives of effigies fashioned from a wide variety of equally inexpensive materials.[32] In his discussion of the different forms of malevolent magic that he would prosecute under the laws pertaining to nonfatal injury Plato (*Leg*. 933a–c) mentions the fears that overtake some men when they see "molded images of wax" (presumably voodoo dolls) at the crossroads or in graveyards. Wax and clay images, moreover, regularly appear in the stock scenes of erotic magic in hellenistic and Augustan literature (e.g., Theoc. *Id*. 2.28–29; Verg. *Ecl*. 8.75–80; and Hor. *Sat*. 1.8.30–33 and *Epod*. 17.76). Examples of wax and clay dolls (sometimes employed in conjunction with lead *defixiones*) have actually survived in the more stable climate of Egypt.[33]

The popular habit of inscribing *defixiones* in retrograde during the late classical period provides a clue to the process by which lead came to be the preferred medium for written binding curses in the Roman and late antique periods. In classical times, Attic *κατάδεσμοι* are somewhat regularly inscribed in reverse on lead or (more rarely) in boustrophedon style, as were many inscriptions regardless of the medium into which thay were scratched or chiseled.[34] When in the later classical period the habit of writing retrograde was generally abandoned in other profane forms of inscription, the originally accidental direction of inscribing *defixiones* became "petrified" in the ritual and henceforth assumed greater significance—a significance that then came to be expressed in certain curse formulae (such as that quoted above) that rationalize their part in the ritual.[35] Similarly, we might imagine that when papyrus and other cheaper writing materials became more popular for day-to-day writing, the use of lead was similarly retained in the binding ritual and thereby was thought to be the preferred-if-not-necessary medium in the later periods.

The *similia similibus* formulae and the voodoo dolls employed in Greek binding spells are traditionally referred to as "sympathetic" or "homeopathic" rituals, but they can be more precisely described as "persuasively analogical" according to the definitions laid down by the anthropologist S. J. Tambiah.[36] He argues against the prevailing theory that "sympathetic" or "homeopathic" magic is based on (poor) observation of empirical analogies and differentiates instead between the operation of "empirical analogy" used in scientific inquiry to *predict* future action and "persuasive analogy" used in ritual to *encourage* future action. In order to increase their crop, the Azande prick the stalks of bananas with crocodile teeth while saying, "Teeth of crocodile are you. I prick bananas with them. May bananas be prolific like crocodile teeth."[37] This ritual act is not based on any (mistaken) empirical analogy between bananas and crocodile teeth but rather on the hope that correct performance of the ritual gesture and the incantation will "persuade" the bananas to become analogous to crocodile teeth with regard to their plenitude. The crucial difference is the use of the optative in the second part of the analogy, which (like the second part of a *similia similibus* formula) "urges" or "persuades" an object to become similar in a circumscribed way (i.e., in the case of a lead *defixio,* with respect to coldness or uselessness) to something that is obviously dissimilar. The rationale of this kind of ritual is not, therefore, based on poor science or a failure to observe empirical data but rather on a strong belief in the persuasive power of certain kinds of formulaic language.

It is important to note the limitations that are placed on the analogies in the *similia similibus* formulas, for it is all too easy to assume that the ultimate purpose of the *defigens* is to kill the victim when he encourages an analogy to a corpse or some inorganic material. In fact, in stark contrast to funerary imprecations or other traditional forms of cursing that specifically call for the utter destruction of the victim (with all his kith and kin), the ultimate goal of early Greek *defixiones* is very rarely the death of the victim.[38] Thus when a formula says, "Just as this corpse is unsuccessful, so may NN be unsuccessful" or "Just as this lead is cold and useless, so may NN be cold and useless," one must recognize that the analogy is limited to the corpse's lack of success or the frigidity and uselessness of the lead and does not imply that the lifelessness of both are wished upon the victim. Such an inference would be equivalent to supposing that the Azande believed that the incantation quoted above would cause their bananas to become like crocodile teeth in every aspect (i.e., hard, sharp, white, inedible, and not just plentiful). The limited scope of the persuasive analogies in the Greek *defixiones* clearly suggests that their main motivation was restraining or inhibiting, not destroying, the victim.

A comparison of these private binding spells (both the voodoo dolls and the inscribed lead tablets) with analogous public rituals used to protect entire Greek cities seems to confirm the circumscribed effect of Greek *κατάδεσμοι*. The Orchomenians, at the direction of the Delphian Apollo, erected a bronze statue of Actaeon and bound it to a rock with iron as a means of protecting their people against the attacks of his vengeful ghost.[39] In the first century B.C. the people of Syedra, plagued as they were by the incessant attacks of pirates, were advised by the Clarian Apollo to erect a statue of Ares bound in the "iron chains of Hermes" and supplicating a figure of Dike.[40] They were also advised to continue their own defensive military maneuvres, and it seems as if the binding ritual was meant to tip the scales in their favor. In addition, we know of three archaic silver statuettes buried (facing north) in an old sanctuary in Thrace, which were unearthed in the fourth century A.D.; each figurine has its arms bound behind its back like many of the extant Greek voodoo dolls employed against individuals. According to the local people these effigies had been so deployed to inhibit the incursions of three different barbarian tribes to the north.[41] In all three reports, there is no hint that the enemy is destroyed by the creation and binding of the civic voodoo dolls; they are employed defensively—almost like phylacteries—to restrain an enemy and prevent him from doing harm. Indeed, this sense of binding magic as a defensive act against an enemy or rival with an "unfair" advantage recurs below in the rare texts where we get some sense of the motivation behind these spells.

Sophronius, a late sixth-century A.D. Christian writer, gives us the only detailed narrative of a person's escape from a binding spell.[42] He reports how an Alexandrian paraplegic named Theophilos is visited in a dream by two saints and told to go down to the harbor and offer a large sum of money to the fishermen there to buy their next catch, sight unseen. He does as he is bidden, and when the nets are hauled up a small box is discovered amidst the struggling fish; some bystanders pry it open and find a bronze statuette with nails driven into its hands and feet. As each of the nails is withdrawn from the effigy, the paralysis in the corresponding limb of Theophilos ceases immediately. Although certainly no one would vouch for the historicity of this particular incident, the need for verisimilitude in the details of such miracle stories suggests that the underlying assumptions about the paralyzing but nonfatal effects of voodoo dolls were common knowledge, at least in the early Byzantine period.

The rationale behind the placement of the tablets in graves and chthonic sanctuaries has been similarly misinterpreted. It is true that contact with the coldness and inertia of corpses provides the motivation of some *similia similibus* formulae, but, like the formulae that refer to the lead medium or the retrograde writing, these formulae seem to rationalize what was originally a rather mundane procedure—the practice of communicating with the gods or the dead by means of sealed *πινάκια* or *libellae* (e.g., the oracle questions and petitions discussed in detail by Versnel, chap. 3). In fact, this technique also helps explain the two peculiar variants of the direct binding formula discussed above: the use of prepositional phrases to implicate the gods in the procedure, namely, "I bind NN before (*πρός*) or near (*παρά*) the god(s)" and the use of technical or legal terminology such as *καταγράφω* to "register" or "transfer" the responsibility of the binding to the divine sphere of activity. Such verbs are usually very loosely translated as "consign" or "devote," which often wrongly implies that by such an action the victim is literally sent to the underworld, that is, killed. Again, the evidence collected by Versnel suggests that these expressions ought not be taken in the literal sense, but rather in a legal or technical sense, that is, they refer in an abstract way to the domain of the god's jurisdiction and influence.[43] An inscribed lead tablet from Crete,[44] dated to the imperial period, provides an interesting parallel: "I hand over (παραδίδωμι) this gravestone to the gods of the underworld to guard." Two nearly identical statements appear directly on two Attic grave steles of comparable date (*IG* II^2 13209–10) and seem to confirm the usage; the gravestones are placed under the control of the underworld gods, not literally "sent to the underworld" or destroyed. What is so illustrative about the Cretan example is the fact that the transfer was inscribed on a lead document separate from the gravestone itself and then placed nearby or buried beneath it as a sort of "legal writ of cession."

Thus, there are three very different styles of binding spells in ancient Greece:

1. The direct binding formula, which is a performative utterance, that is, a form of incantation by which the *defigens* hopes to manipulate his victim in an automatic way
2. The prayer formula, which is exactly that—a prayer to underworld deities that they themselves accomplish the binding of the victim
3. The so-called *similia similibus* formula, which is better understood as a form of "persuasive analogy" (also an incantation), in which the binding is accomplished by a wish that the victim become similar to something to which he or she is manifestly dissimilar.

It is important to emphasize the heuristic purpose of these divisions and to point out that within fifty years of the first appearance of written κατάδεσμοι all three types of formulae are being employed side by side, sometimes on the very same tablet (see, e.g., the Carystus tablet quoted above, which contains both a direct binding formula and a prayer formula).

THE AGONISTIC CONTEXT

Defixiones, then, provide a means of binding or restraining enemies without killing them. I shall now investigate the social context in which such a powerful weapon was deployed. More than three quarters of the published Greek κατάδεσμοι are inscribed with names only or are so laconic that they give us no hint whatsoever of their specific purpose; the more discursive texts, however, contain details that allow us to place them into four[45] general categories according to social context: commercial curses (25 examples); curses against athletes or similar kinds of public performers (26 examples); amatory curses (38 examples) and judicial curses (67 examples). As will be discussed below in further detail, the amatory curses can be further subdivided into two rather different types: "separation" curses and "aphrodisiac" curses. Generally speaking, the judicial and commercial curses date to the classical and hellenistic periods, while those that bind public performers usually date to the late Roman period (from the second century A.D. onwards). The few separation curses that we have seem to be evenly distributed, while the aphrodisiac curses are exclusively a late phenomena. Keeping these important chronological limitations in mind, I propose to examine these four categories (excluding, for reasons that will become clear, the aphrodisiac *defixio*) and to argue that an essential feature of all four types is that they refer to agonistic relationships, that is, relationships between rival tradesmen, lovers, litigants, or athletes concerned with the outcome of some future event.

Hesiod begins the *Works and Days* with a eulogy of the beneficial kind of rivalry that existed among the craftsmen and artisans of his day. There is some evidence from the classical period that such tradesmen (and innkeepers as well), in their efforts to stay ahead of the competition, employed *defixiones* to inhibit the success and profit of their rivals. The popular habit of qualifying many of the victims' names on *defixiones* as tradesmen may or may not point to a context of commercial competition.[46] The likelihood, however, increases dramatically in the case of twenty-five tablets that explicitly bind the business, profits, or workshop of others.[47] The earliest example (c. 450 B.C.) comes from the necropolis at Camarina in Sicily (SGD 88): "These people are registered for (lit., "written down for") a downturn in profits (ἐπὶ δυσπραγί[αι τôν] κέρδōν)."[48] A third-century-B.C. Attic curse (SGD 52) binds two net weavers, as well as their ἐργαστήρια (workshops), while a contemporary Sicilian tablet (SGD 124) reads, "I bind the workshops of these men . . . so that they may not be productive but be idle and without luck." A rather long Athenian *defixio* (*DTA* 75, Attic, third century B.C.) curses two workshops, one inn, and one store. A fourth-century-B.C. κατάδεσμος found buried in the mud-brick wall of a house situated in the ancient Athenian industrial quarter binds a number of bronze workers (SGD 20). There is some later literary evidence that magic was popular among certain types of craftsmen, especially in those professions like bronze working where delicate heating and cooling processes were necessary to avoid breakage; Pollux (7.108), for example, refers to talismans that bronze workers placed in their foundries to ward off the evil eye, and Pliny (*HN* 28.4) discusses the incantations that potters were said to cast against the kilns of their rivals.

One of the most straightforward applications of *defixiones* (and one that most clearly shows their agonistic context) is in situations involving public performances in athletic or theatrical contests, where they are employed by or on behalf of one contestant to alter or impede the performance of an opponent. Pelops' prayer to Poseidon in Pindar's first *Olympian* may preserve the earliest example of this kind of curse (lines 75–78):

> Look you, Poseidon, . . . block (πέδασον, lit., "bind!")
> the brazen spear of Oenomaos, and give me the fleeter chariot by Elis' river,
> and clothe me about in strength. (trans. Lattimore)

Although it can be argued that this particular race was neither fair nor sportsman-like, it is interesting that Pelops demands both the restraint of his rival and at the same time an enhancement of his own ability to win. It is clear that Pelops wishes to protect himself from Oenomaos' sharpshooting (an unfair advantage), and once again we get the impression that, as with the bound effigy of Ares at Syedra, the binding aims at evening up the odds of a contest that from the perspective of the *defigens* seems rather lopsided.

There are four binding curses written against individuals involved in theatrical competitions, dating primarily to the classical and hellenistic periods. An Attic curse (*DTA* 45, third century B.C.) reads simply, "I bind NN, the actor (ὑποκριτής) in a leaden bond." As with other victims who are only qualified by vocation, one cannot be sure whether this man is being bound by a rival in the theater or whether the title *actor* only serves to differentiate him from other individuals with the same name. Two contemporary lead disks probably written by the same person (*DTA* 33 and 34) are inscribed in a confused serpentine pattern and carry binding curses against men described as διδάσκαλοι and ὑποδιδάσκαλοι; the second tablet leaves out the names of the men and identifies them only as "those who are with Theagenes," who is presumably the χορηγός.[49] The anonymity of the διδάσκαλοι on the second disk suggests that the real target here may be Theagenes or the rival theatrical group he heads. The *defigens* in a mid-fifth-century-B.C. Sicilian tablet (SGD 91) consigns (on behalf of his friend Eunikos, the χορηγός) all the other χορηγοί and their children and parents "to futility both in the contest (ἐν ἀγôνι) and outside the contests." On account of the obscurity of the term χορηγός in early non-Attic Greek, it is unclear whether the victims listed here are liturgists (as they would be in Athens) or actual performers.[50] The agonistic context of the curse is, however, unmistakable, and is echoed in a second-century-A.D *defixio* written in Latin that reads in part, "Sosio must never do better than the mime Eumolpos. . . . Sosio must not be able to surpass the mime Fotios."[51]

From the later periods there is much evidence for the continued use of binding spells against public performers, especially athletes and charioteers. As they are all clearly employed in very similar agonistic situations, I include them here in my survey. Five curses against wrestlers (SGD 24–28) were discovered in a well in the Athenian agora and date to the third century A.D., more than six hundred years later than the Attic *defixiones* discussed above. Three of these curses aim at binding and "chilling" the same wrestler, a man named Eutychian, with the expressed hope that he "fall down and disgrace himself." Each curse, however, seems to have been designed for a different match: the first refers to a bout that is to occur on "this coming Friday"; the second refers to a match against Secundus and the third to one against Hegoumenos.[52] All three of these binding curses contain *similia similibus* formulae of a general nature such as, "As these names (of the invoked daemons) grow cold (ψύχεται), so may Eutychian's name and breath, impulse, charm, knowledge, reckoning grow cold (ψυχέσθω). Let him be deaf, dumb, mindless, harmless, and not fighting against anyone" (trans. D. R. Jordan). Other athletic *defixiones* offer similarly general imperatives that aim at curtailing the various talents and strengths of the opponent; a *defixio* from Oxyrrhynchus is employed to bind the sinews and the limbs of two racers "in order that they should be neither powerful nor strong." The spell goes on to invoke a daemon who is to keep the victims up all night and prevent them from sleeping or eating anything in preparation for the race.[53] The only other published curse against a runner was found in the same Athenian well as the three wrestling curses against Eutychian. It contains a rather curious two-part request that suggests that the daemons did not always succeed on the first try: they are asked to prevent a certain Alkidamos from "getting past the starting lines of the Athenaia [a recurrent ephebic festival]" but add that if he manages to do so in spite of their efforts, they should then make him "veer off course and disgrace himself."[54]

Like the athletic binding spells discussed above, *defixiones* against charioteers begin to appear in the second or third century A.D. With the exception of the so-called *Sethianorum tabellae*[55] from Rome, they are to be found exclusively in North Africa and Syria.[56] In a well-known κατάδεσμος from Carthage (*DT* 242, third century A.D.) we find once again the mention of one particular race ("tomorrow morning") and a detailed listing of the parts of the body to be bound: the shoulders, arms, elbows, wrists of the charioteers, as well as their eyes and the eyes of their horses, who are all named alongside the names of their drivers.[57] Finally there are six Greek binding spells of similar date and provenance that aim at inhibiting the performance of *venatores,* a special class of gladiator who fought wild animals in the amphitheater. In these texts the daemons are exhorted to protect the animals from harm (certainly not out of any humanitarian spirit) and to bind and impede the actions of the *venator* so that he may fall an easy prey.[58]

The third area of intense personal competition in which the Greeks employed κατάδεσμοι was the battlefield of Eros, especially in the situation of a "lovers' triangle," where two individuals were competing for the affections of a third. Here, too, if a lover or would-be lover feared the outcome of a contest, he might turn to the use of a *defixio* in order to impede the advances, the flirting, and even the sexual performance of his or her rival. As I mentioned briefly above, it is helpful here to distinguish between two different types of amatory curses: (1) the so-called separation curse or *Trennungszauber,* which was usually aimed at the rival lover, and occasionally at the beloved (i.e., it sought to inhibit the conversation and contact between the two) and (2) the so-called aphrodisiac or erotic curse in which only the beloved is mentioned (i.e., it sought to torture the "beloved" victim with burnings, itchings, or insomnia that could only be assuaged by submitting to the desires of the *defigens*).

The first group, of which there are fifteen published Greek examples,[59] is attested in the classical and hellenistic periods and fits rather easily into the patterns of use that we have observed in the athletic and circus curses. A long Boeotian curse (*DT* 85, third or second century B.C.) is designed to restrain a rival lover named Zoilos. As the curse is rather lengthy, I translate only a few relevant sections:[60]

> Just as you, Theonnastos (the dead person with whom the tablet was buried) are powerless (ἀδύνατος) in the movement of your hands, your feet, your body . . . , so too may Zoilos be powerless (ἀδύνατος) to come to Antheira, and in the same way Antheira (to) Zoilos. (Side A 1–4)
>
> Just as this lead is in some place separate (ἔν τινι ⟨τόπωι⟩ χωριστῶι) from the haunts of mankind (i.e., a tomb), so too may Zoilos be separated (⟨κε⟩χωρισμένο⟨ς⟩) from the body and the touch of Antheira, and the endearments and the embraces of Zoilos and Antheira. (Side A 8–11)
>
> Just as this lead is buried (ὀρώρυχτ⟨αι⟩) . . . , so too may you (i.e., the corpse) utterly bury (κατορύχοις) the works, the household, the affections, and everything else of Zoilos. (Side B 16–21)

Here, as in the Attic athletic curses discussed above, a rival is being "persuaded" to be inert and kept away from the prize, in this case a woman named Antheira.

A rather simple Attic curse (*DTA* 78, fourth century B.C.) seems to have been written by a jealous wife or fiancée:

> Ἀρι[σ]τοκύδη καὶ τὰς φανο⟨υ⟩μένας
> αὐτῶι γυναῖκας.
> μήποτ' αὐτὸν γῆμαι ἄλλην γυναῖ⟨κα⟩ μηδὲ παῖδα.

> [I bind?] Aristocydes and the women who will be seen about with him.
> Let him not marry another matron or maiden.

This text is a curious combination of the earliest technique (merely listing the victims' names) and a rare example of a simple wish formula.

Just as the athletic curses target the special body parts and skills of the rival wherein his hope of victory was thought to lie, the amatory *defixio* might also list the charms of the rival. See, for example, another curse from Boeotia that, in contrast to the preceding example, seems to have been written by the "other woman" (*DT* 86a, fourth century B.C.):[61] "I deposit (παρατίθεμαι, i.e., for the purposes of binding) Zois the Eretrian, the wife of Kabeirēs, with the Earth and with Hermes, her food, her drink, her sleep, her laughter, her company, her cithara playing, her entrance (πάροδον?), her pleasure, her rear end, her thought, her eyes."

These "separation curses" aim at inhibiting desire and affection, usually in the rival lover but occasionally in the beloved as well. The main purpose is either to restrain any possible erotic attraction or to break any preexisting bond that the two may have developed.

This is in marked contrast to the second kind of *defixio* used in an amatory context, which more properly belongs in the category of aphrodisiac, since its aim is to encourage, rather than inhibit, the sexual desire and activity of the beloved. As these tablets are treated in Winkler's essay on erotic magic (chap. 8), I shall limit my discussion to a portion of one very popular Greco-Egyptian formula, which is adduced purely as a point of comparison:[62]

> Don't ignore [these] names, *nekydaimon,* but arouse yourself and go to every place where Matrona is, whom Tagene bore. You have her οὐσία. Go to her and seize her sleep, her drink, her food, and do not allow Matrona (whom Tagene bore, whose οὐσία you have) to have love or intercourse with any other man, except Theodoros, whom Techosis bore. Drag Matrona by her hair, by her guts, by her soul, by her heart until she comes to Theodoros and make her inseparable from me until death, night and day, for every hour of time. Immediately, immediately, quickly, quickly, now, now.

Many of the details here are characteristic of later *defixiones,* for instance, identification by matronymic[63] and the use of οὐσία (lit., "being" or "stuff," magical slang for bits of the victim's hair, nail trimmings, or threads from the clothing).[64] This curse begins like a traditional κατάδεσμος by restraining Matrona's contact with other men, but it goes on to burn and torture Matrona herself, elements that are often found in the later ἀγωγή spells, designed to "lead" or "drag" (the word derives from the verb ἄγω) the victim into the arms of the performer of the spell.[65] Although the limited number of both types of amatory *defixiones* does not allow us to make any secure generalizations with regard to the patterns of date, provenance, or purpose, I would suggest here that the "separation curses" (from the fourth century B.C. to the third century A.D., primarily from Greece proper) were closer in form and purpose to the original Greek phenomenon of *defixiones* and that the emergence in North Africa and Syria of the aphrodisiac-type of curse in the second century A.D. represents some kind of hybrid flowering of a later, more complex magical tradition.[66]

Judicial *κατάδεσμοι*, which make up our fourth and largest category, were once thought to be posttrial "revenge curses" upon the winning party; subsequent research, however, suggests that they were written, without exception, prior to the final outcome of the trial.[67] They are attempts at binding the opponent's ability to think clearly and speak effectively in court in the hope that a dismal performance will cause him to lose the case. Just as the circus *defixiones* discussed above attempted to bind the parts of a charioteer's body in which his competitive skill lay (i.e., his shoulders, arms, elbows, wrists, and eyes), so judicial curses are primarily concerned with the cognitive and verbal faculties that are so important for success in the law courts (*DTA* 107, Attic, late fifth or early fourth century B.C.):

> *Θερσίλοχος, Οἰνό[φιλος,] Φιλώτιος καὶ εἴ τ[ι]ς ἄλλος Φερενίκωι σύνδικ[ος, πρ]ὸς τὸν Ἑρμῆν τὸγ Χθόν[ι]ον καὶ Ἑκάτην Χθονίαν καταδεδέσθω· Φερενίκο⟨υ⟩ κα[ὶ ψυ]χὴν καὶ νο⟨ῦ⟩ν καὶ γλῶτταν καὶ βο⟨υ⟩λὰς καὶ [τ]ὰ πράττει καὶ τὰ περὶ ἐμο⟨ῦ⟩ βο⟨υ⟩λε[ύ]εται, ἅπαντ᾽ αὐτῶι ἀντία ἔστω καὶ τοῖς μετ᾽ ἐκε⟨ί⟩νο⟨υ⟩ βο⟨υ⟩λεύο⟨υ⟩σιν καὶ πράττο⟨υ⟩σιν.*

> Let Thersilochos, Oino[philos], Philotios, and whoever else is a legal advocate for Pherenikos be bound before Hermes Chthonios and Hecate Chthonia. The soul, the mind, the tongue, the plans of Pherenikos, and whatever else he is doing or plotting with regard to me—let all these things be contrary for him and for those who plot and act with him.

Galen (vol. 12 p. 251 ed. Kühn) scoffs at those who believe in the power of "magic" and specifically mentions the popular belief that the efficacy of a judicial curse lay in its power to inhibit the opposing side's ability to speak persuasively during the trial (*καταδῆσαι τοὺς ἀντιδίκους, ὡς μηδὲν ἐπὶ τοῦ δικανικοῦ δυνηθῆναι φθέγξασθαι*). This was, however, a powerful claim, especially in such communities as Athens, where it was necessary for citizens to speak on their own behalf in court.

Literary evidence suggests that a poor performance in court by a talented orator could often result in the accusation that he had been the victim of binding curses. Aristophanes (*Vesp.* 946–48) alludes to the sudden paralysis of the renowned orator Thucydides, the son of Melesias, during an important trial; the scholiast at that point (preserving some fourth-century-B.C. Attic source) suggests that his tongue had been magically bound.[68] Cicero relates how an opposing attorney suddenly forgot the case he was pleading and subsequently lost the lawsuit; the unfortunate man later claimed that his poor performance was the result of sorceries and incantations (*veneficiis et cantionibus*).[69] In late antiquity, orators and declaimers (both, perhaps, better classified as professional performers) continued to blame witchcraft for sudden lapses of memory and moments of inexplicable stage fright. Libanius tells us in his autobiography (*Orat.* 1. 245–49) how at one point late in his life he became gravely ill and was no longer able to read, write, or speak before his students. After a time the apparatus of a rather bizarre binding ritual was discovered in his lecture room—the twisted and mutilated body of a chameleon. Its severed head had been placed between its hind legs, one of its forefeet was missing and the other positioned in such a way as "to close the mouth for silence." Libanius says that he regained his health after the chameleon was removed.[70]

There are some sixty-seven published examples of Greek judicial binding curses,[71] which with few exceptions date to the classical and hellenistic periods.[72] The oldest examples—those found in Sicily—date to as early as the fifth century B.C. One from the necropolis at Selinus (SGD 95) is inscribed in a boustrophedon pattern and curses the tongues of four, named individuals, as well as the tongues of all their *σύνδικοι*. Three other contemporary Selinuntine curses (SGD 99, 100, and 108) were excavated in or very near the sanctuary of Demeter Malophoros at Gaggara and contain the rather bureaucratic formula *ἐγγράφω τὸν δεῖνα ἐπ' ἀτελείαι* ("I enroll NN for failure"), which, with the exception of one other Sicilian curse (substituting *ἀπογράφω*) is unique. All three date to the first half of the fifth century and similarly curse the tongues of a litigant and his *σύνδικοι*.[73] With the possible (but not probable) exception of eleven judicial curses from Wünsch's *Defixionum Tabellae Atticae*,[74] all of the Attic judicial curses published to date come from the classical period.[75]

Many of the early Attic curses (both those used for judicial purposes and those of unspecified aims) contain names of well-known orators and politicians.[76] This prompted Preisendanz to suggest that all of these curses might be labeled "political curses," but even he admitted that given the blatantly political nature of so many of the lawsuits tried in Athenian courts, such a category would be difficult-if-not-impossible to separate from the category of judicial curses.[77] Although his suggestion is, indeed, of little use in the taxonomy of *defixiones*, his insight into the larger political motivation of some of these curses deserves attention. In a few cases the appearance of female names on ostensibly judicial curses has caused some confusion, especially in the case of *καταδέσμοι* found in Attica, where the barring of women from every facet of courtroom activity was notorious.[78] Such confusion may point to a more generalized meaning of some of the "legal" vocabulary that we find in the binding curses. I suggest that when the terms *σύνδικοι* and *ἀντίδικοι* appear by themselves (i.e., with no other allusion to legal procedure), they might occasionally have looser, political connotations equivalent, perhaps, to the Latin terms *amici* and *inimici*, which often appear on Latin *defixiones*.[79] This would also account for the huge numbers of names that one finds on some allegedly judicial curses, far in excess of any imaginable number of fellow prosecuters or witnesses (e.g., SGD 42 and 107; cf. SGD 48).

Peter Brown describes how in late antiquity, too, the magic that coalesced around smaller arenas of competition often had larger political significance. He points out that the rivalry faced by the charioteer extended beyond his time in the circus and that since he was both the client of local aristocracies and the leader of organized groups of fans, his performance often transcended the realm of mere sport.[80] This connection between athletic and political competition is not limited to later antiquity; in classical Athens intertribal competitions—albeit in a much less organized fashion—often provided arenas for intracity rivalries, where victories in theatrical performances and even athletic events could be interpreted as indicators of the waxing or waning of the political power of the liturgists involved.[81] By inhibiting the performance of actors and athletes, the *defigens* could conceivably restrain the political power of their backers and undermine their popularity with their fans, often their only source of political influence. As was the case in my discussion of the three styles of curse formulas, these categories of different agonistic contexts clearly have a limited heuristic value; indeed, several of the narrowly defined conflicts revealed in the *defixiones* discussed above seem to spill over quite easily into larger arenas, sometimes purely as the result of a common Greek habit of thought, as in the case of the Sicilian binding spell quoted above, which binds rival χορηγοί "both within the contest and without," that is, everywhere. Thus the use of κατάδεσμοι to bind the competitive power of rival businessmen, public performers, or opponents testifying in the courts could often be more than an act of personal rivalry and could fit easily into a larger pattern of political or social competition.

CONCLUSION: MAGIC AND RELIGION

In light of the larger orientation of this volume, we should ask ourselves at this point whether the categories of "magic" and "religion" are of any value in analyzing the phenomenon of the early Greek κατάδεσμοι. It is often a modern assumption that the anonymity and secret burial of the inscribed κατάδεσμοι, like the inaudible whispering or muttering of malevolent verbal prayers, can be attributed to the shame of the agent and that such shame indicates an illicit activity.[82] Often, however, such secrecy is part of the traditional ritual procedure used in approaching the gods for help, while at the same time shielding a person's private affairs from the inquisitive eyes of his neighbors. There are many examples of communication with the gods by means of sealed, written documents (for instance, the oracle questions inscribed on lead tablets at Dodona and elsewhere). As I mentioned above (with regard to the *defixiones* that employ prayer formulas) graves, chthonic sanctuaries, and underground bodies of water are ideal points of contact with the subterranean gods. From a more practical point of view, moreover, keeping potential victims "in the dark" about the existence of the *defixio* prevented them from using specific phylacteries or defensive spells that could ward off the power of the curse. Burying the tablet in areas governed by taboo (e.g., graves and sanctuaries) or sinking it into deep bodies of water would likewise prevent the victim from finding and then loosening the spell (as the Alexandrian Theophilos did by removing the nails that bound his effigy).

The anonymity of the *defigens* and the hidden nature of the curse could also be governed by the relative social or political positions of the man and his victim. Apollo's priest Chrysēs, for example, utters his famous curse along the seashore (*Il.* 1.35–43) only *after* he is out of earshot of the Achaean ships; he is certainly not presented as a shameful personality by the poet. Nor is the hero Pelops, who waits until he is alone on a beach in the dead of night to urge Poseidon to bind Oenomaos (Pindar *Ol.* 1.75–78, quoted above). In Euripides' *Electra* (205ff.) Aegisthus prays out loud to the nymphs for his own continued good fortune and for the destruction of his enemies (a reference, says the messenger, to Orestes and Electra); the undetected Orestes in the meantime prays for the utter reverse but keeps his words "under his breath." His prayer and that of Aegisthus are identical in form and their different degrees of audibility are attributable solely to tactical, not ethical, concerns; Orestes' prayer, like that of Chrysēs and Pelops, is directed against the "powers that be" and therefore requires some degree of caution.

Most of the argument over the comparison of "pious prayers" and the putatively more malevolent "curse tablets" seems inevitably (and unfortunately) to rest on our subjective appraisal of the attitude of the persons performing the acts, an attitude that is to be discovered in the vocabulary they use to express themselves—that is, the "gut feeling" of a modern reader that Chrysēs is a priest operating on a "religious" impulse and using traditional forms of prayer when he directs Apollo to destroy the Greeks but that the average *defigens* is possessed by some inferior "superstitious" impulse and uses "magical" forms of invocation and coercion. The problem of divining the piety or attitude of the *defigens* is, however, enormously complicated by the laconic nature of the early Greek *κατάδεσμοι*. The texts that we have from much later Roman times are more verbose and allow the kind of profitable psychological examination that Versnel offers in his essay (chap. 3). Indeed, by the Roman period we do encounter rather lengthy tablets from Africa, Anatolia, and Asia Minor that entreat, beg, threaten, and command gods and daemons to do what they request. For the preceding centuries, however, any assessment of the psychology or piety of the *defigens* is wholly dependent upon unavoidably subjective inferences drawn from connotations of single words.

A recent debate on some early first-century-B.C. lead tablets from Morgantina underscores the difficulties inherent in establishing criteria for normative piety in the earlier Greek *κατάδεσμοι*. Ten inscribed lead tablets (all but two rolled up) were discovered during the excavation of a sanctuary of some as-yet-unidentified chthonic god or goddess. Nabers argued that of the six legible examples only one was employed as a *defixio* and that the remaining five were "pious prayers offered to the underworld on behalf of persons already dead." The text of the allegedly sole *defixio* (Nabers's no. 6, SGD 120) reads as follows:[83]

> *Γᾶ Ἑρμᾶ θεοὶ*
> *κα[τ]αχθόνιοι*
> *ἀπαγάγετε τὰν* Βενού[*σταν*]
> *τοῦ* Ῥούφο[*υ τὰν*] *δού-*
> [*λαν.*]
>
> Earth, Hermes, gods
> of the underworld
> snatch away Venusta,
> [the slave] of Rufus.

Two of the remaining "pious prayers," however (Nabers's nos. 4 and 5) seem to refer to the same woman. I give the text of number 4 (= SGD 118), as it is undamaged:

> *Γᾶ Ἑρμᾶ*
> *θεοὶ κατα-*
> *χθόνιοι πο-*
> *τιδέξεσθε*

τὰν Βενούσταν τοῦ
Ῥούφου τὰν θεράπαι-
ναν.

Earth, Hermes,
gods of the underworld
receive graciously
Venusta,
the servant of Rufus. (trans. Nabers)

The rest of the tablets have an identical form except that some of them use a third verb παραδέχομαι and name other individuals. Despite the fact that the two tablets quoted here came from the same place and are addressed the same gods in the same form about the same individual, Nabers argued (and convinced a renowned epigraphist)[84] that the first tablet used a verb appropriate to a *defixio* while the second used a verb more fitting to a religious context. But it is clear from the shared placement, form, and function of these two tablets (and from the elsewhere-attested habit of using multiple curses against the same individual) that all of these tablets are in fact *defixiones*.[85] The confusion on the part of these scholars is in itself instructive, for it highlights the fact that a *defixio* employing the prayer formula is exactly that, a prayer to the chthonic deities. Whether the prayer is benevolent or malevolent is immaterial to the pious belief that the gods addressed can and will do what they are asked provided they are approached in a ritually correct manner.

A third-century-B.C. inscribed pillar from Delos preserves a rather lengthy account in epic hexameters of the successful founding of the cult of Serapis on the island.[86] The central miracle in this aretalogy is the god's timely intervention in a lawsuit that threatened the existence of his newly built temple. Because the end result is so strikingly similar to the outcome that is often envisaged by the inscribers of the judicial binding curses discussed above, this miraculous event is of great interest (lines 85–90):

. . . . . φῶτας γὰρ ἀλιτρο⟨νό⟩ους ἐπέδησας
οἵ ῥα δίκην πόρσυνον, ἐνὶ γναθμοῖς ὑπανύσσας
γλῶσσαν ἀναύδητον τῆς οὔτ' ὄπιν ἔκλεεν οὐθείς
οὔτε γ⟨ρ⟩άμμα δίκης ἐπιτάρροθον· ἀλλ' ἄρα θείως
στεῦντο θεοπληγέσσιν ἐοικότας εἰδώλοισιν
ἔμμεναι ἢ λάεσσιν·

For you bound the sinful men who had prepared the lawsuit, secretly making the tongue silent in the mouth, from which (tongue) no one heard a word or an accusation, which is the helpmate in a trial. But as it turned out by divine providence, they confessed themselves to be like god-stricken statues or stones.

The use of the verb ἐπιδέω ("bind up") and the specific mention of the tongue as the target of the paralysis are immediate clues that some sort of judicial binding spell has been employed. There is, however, no explicit mention of any kind of overtly "magical" activity by the priest; in the eyes of the poet Maiistas, who composed this poem, the god's intervention is clearly the result of the frantic prayer of the priest (described in the preceding lines, 43–44), who with tears in his eyes begged the god to protect him from conviction. Both the submissive tone of the prayer and the complete helplessness of the priest fit in perfectly with the genre and purpose of this particular literary product—an aretalogy. The context, however, of the god's intervention is identical to that of the traditional *defixio;* the third-generation Egyptian priest, when faced with the difficult-if-not-impossible task of swaying a native Delian jury, begs Serapis to restrain the attacks of his enemies. Despite the familiar (to modern sensibilities, at least) "religious" mentality of the prayer, both the agonistic setting and the description of the resulting courtroom paralysis point unequivocally to a "binding spell" accomplished by a deity as the result of an urgent prayer. Indeed, even the "defensive" stance of the priest, who considers himself abused by insane and evil men (lines 37–39), can be paralleled from the Syedra inscription or the story of Pelops, in which the *defigentes* seem to envisage themselves striving against unfairly superior opponents.

Jevons saw that it was impossible to distinguish a *defixio* employing a prayer formula from a traditional Greek prayer, and he attempted, instead, to distinguish "magical" *defixiones* from "religious" ones by noting whether or not they invoked deities to perform the curse. In drawing this distinction he applied Nilsson's dictum that a miracle performed without the help of the gods is a "magical" act and one performed with their aid was a "religious" act.[87] Unfortunately great difficulties arise when we recall that *defixiones* not employing a prayer formula are often augmented by prepositional phrases that implicate the gods in the proceedings, for instance, "I bind NN before Hermes the Restrainer," a formulation that is also used on the putatively more pious tombstone curses.[88] The combination, moreover, of different types of formulae on the same tablet or the substitution of one formula for another seems to be completely random, much like the variation of vocabulary in the Morgantina tablets discussed above. Thus, using the different curse formulae to distinguish between "religious" and "magical" *κατάδεσμοι* is a purely artificial exercise that cannot in the end reveal any difference in the social function of the curse or the piety of the *defigens*.

A broadly conceived theoretical dichotomy between "magic" and "religion" is not, therefore, of any great help in analyzing and evaluating the peculiar cultural phenomenon presented by the early Greek *defixiones*. They seem to have evolved from a special form of ritual (a symbolic gesture would have accompanied either incantation or prayer) that was primarily used by individuals involved in often-lopsided agonistic situations, to bind the power of their opponents. As such, they fit easily into the popular competitive strategy of survival and dominance that permeates ancient Greek society, regardless of whether the contests in which they were deployed were international, civic, or personal in scope. The scruple against homicide points quite clearly to the fact that *defixiones* somehow remained within the rules of the game for intramural competition in the Greek city-state. The recurrence of what I have called a "defensive stance" in some of the texts discussed above suggests that the *defigentes* may have perceived such activities as protective in nature and not as aggressive magic at all. Indeed, it is a tempting but, alas, completely unprovable suggestion that the person who would most often employ a binding curse is the one who doubted his or her ability to win without it, that is, that the *defigens* was the perennial "underdog," who, like Chrysēs and Pelops, was protecting himself against what seemed to be insurmountable odds.

Notes

I should like to thank E. Courtney, M. Edwards, M. Gleason, D. Halperin, M. Jameson, D. R. Jordan, R. Kotansky, A. E. Raubitschek, H. S. Versnel, and J. Winkler for their comments and advice on earlier drafts of this essay and to claim as my own all of the deficiencies that may remain. The special debt owed to Jordan's recent published work on *defixiones* is readily apparent in nearly every footnote. It is, however, my special pleasure to acknowledge the inestimable benefits I have received from my private conversations and correspondence with him and from access to his ongoing and as-yet-unpublished work. The excellent article on *defixiones* by B. Bravo ("Une tablette magique d'Olbia pontique, les morts, les héros et les démons," in *Poikilia: Études offertes à Jean-Pierre Vernant,* Recherches d'histoire et de sciences sociales 26 [Paris, 1987]) appeared as the present volume was going to the publisher.

The following oft-cited works will be referred to either by author and date or by the abbreviation given in square brackets.

A. Audollent, *Defixionum Tabellae* (Paris, 1904). [*DT*]

D. R. Jordan, "Two Inscribed Lead Tablets from a Well in the Athenian Kerameikos," *AM* 95 (1980): 225–39.

———, "New Archaeological Evidence for the Practice of Magic in Classical Athens," in *Praktika of the 12th International Congress of Classical Archaeology,* Athens, September 4–10, 1983, vol. 4 (Athens, 1988) 273–77.

———, "Defixiones from a Well near the Southwest Corner of the Athenian Agora," *Hesperia* 54 (1985): 205–55. [Jordan, 1985]

———, "A Survey of Greek *Defixiones* Not Included in the Special Corpora," *GRBS* 26 (1985): 151–97. [SGD]

E. G. Kagarow, *Griechische Fluchtafeln,* Eos Supplementa 4 (Leopoli, 1929).

K. Preisendanz, "Fluchtafel (Defixion)," *RAC* 8 (1972): 1–29.

K. Preisendanz and A. Henrichs, ed., *Papyri Graecae Magicae: Die griechischen Zauberpapyri,* 2d ed., 2 vols. (Stuttgart, 1973–74). [*PGM*]

R. Wünsch, *Defixionum Tabellae Atticae,* Inscr. Gr. vol. 3, pt. 3 (Berlin 1897). [*DTA*]

———, "Neue Fluchtafeln," *RhM* 55 (1900): 62–85, 232–71.

1. L. Robert, *Collection Froehner,* vol. 1 (Paris, 1936), no. 13; M. Guarducci, *Epigrafia greca,* vol. 4 (Rome 1978), 248–49. I give Jordan's text (SGD 64), which is based on some recent, better readings.

2. It is difficult to capture all the connotations of this verb in one English word. Other technical and legal definitions add to its semantic range and may also be important here: "enroll," (*LSJ* ii. 2); "summon by written order," (ii. 3) or "convey [i.e., property by deed]," (ii. 4).

3. The Greek term *κατάδεσμος* is derived from the verb *καταδέω* (which appears in Attic dialect in contracted form *καταδῶ*) "bind down" or "bind fast." The late-Latin term *defixio* (from *defigo* "nail down" or "transfix") seems to be the preferred terminology among scholars today, although its popularity has led to some inconsistencies. Epigraphists and archaeologists often use it as a synonym for "lead curse tablet," i.e., any kind of malevolent prayer inscribed upon lead. I shall use the term to refer to all binding rituals regardless of the medium employed, including, e.g., the different kinds of "voodoo dolls," used in antiquity (see n. 31) or even the bound or twisted bodies of small animals that occasionally accompanied the lead *defixiones* (e.g., the bound rooster mentioned in *DT* 241 and the puppy in *DT* 111–12; cf. the chameleon discovered in the lecture room of Libanius [see p. 16 and n. 70] and

the sewnup mouth of a fish, which aimed at binding the tongues of gossips in Ov. *Fast.* 2.577–78). Wünsch, *DTA* and Audollent, *DT* are the basic collections. Audollent includes Latin *defixiones* in his corpus, but since they are derived directly or indirectly from the Greek practice and develop their own unique characteristics, I shall not deal with them in this essay. See K. Preisendanz, "Die griechischen und lateinischen Zaubertafeln," *APF* 9 (1930): 119–54 and 11 (1933): 153–64 for a full bibliography to that date. His work on the Greek material has now been updated and replaced by Jordan's SGD. Aside from the prolegomena to the above-mentioned corpora and surveys, the best comprehensive discussions of *defixiones* are Kagarow 1929 and Preisendanz 1972.

4. *DTA* has 220 examples (all Attic Greek); *DT* has 166 Greek tablets. Because of the unresolved controversy over their function, I do not include the 436 inscribed lead tablets from the Piraeus (listed together as *DT* 45) and Euboean Styra (listed together as *DT* 80), each containing a single, different name and betraying scant signs of manipulation or nail holes. Audollent included them in his corpus, but other scholars have contended that they were probably used for registration or counting, much like the several hundred lead tablets from the Athenian agora, each of which lists a single cavalryman and a description of his horse (see A. P. Miller, *Studies in Sicilian Epigraphy: An Opisthographic Lead Tablet* [Ph.D.diss., University of North Carolina at Chapel Hill, 1973], 8–9, for discussion). SGD lists another 189 published examples and reports the existence of some 461 others awaiting publication. *DT* has 137 tablets inscribed in languages other than Greek (mostly Latin or other indigenous Italian languages). For a survey of more recently published Latin *defixiones,* M. Besnier ("Récents travaux sur les *defixionum tabellae* latines 1904–14," *Rev. Phil.* 44 [1920]: 5–30) gives a checklist of 61 tablets not in *DT*, and H. Solin surveys an additional 48 in the appendix to his *Eine neue Fluchtafel aus Ostia* Comm. Hum. Litt. vol. 42, pt. 3 (1968): 23–31. For the ongoing discoveries of large numbers of late Latin *defixiones* in Britain, see Versnel's essay (chap. 3).

5. Sometimes several tablets are pierced by the same nail, or several nails are driven through a single tablet (see *DTA*, p. iii). Jordan (1988) provides the best and most recent assessment of the archaeological evidence. Detailed instructions for the manufacture and burial of *defixiones* are preserved in the magical handbooks of the third and fourth centuries A.D. and seem to be in general agreement with the archaeological evidence for the earlier periods, e.g., *PGM* V 304; VII 394, 417; IX; XXXVI 1–35, 231; and LVIII.

6. Tert. *De Anim.* 56.4 and Servius *In Aen.* 4.382. There is no direct testimony about such beliefs in the classical period, but the scanty archaeological evidence, where available, seems to corroborate some such belief. The idea seems to be that the ghost of the ἄωροι ("those who are untimely dead") would remain in or near the grave until they have completed allotted time on earth (*DT*, pp. cxii–xv). See A. D. Nock, "Tertullian and the *Ahoroi,*" in *Essays on Religion in the Ancient World,* ed. J. Stewart, vol. 2 (Cambridge, Mass., 1972), 712–20. Originally *Vig. Christ.* 4 (1950): 129–41. Jordan (1988, 273) points out that "in every period of antiquity when we have been able to estimate the age of the dead who have curse tablets in their graves, that age proved to be young." For similar beliefs about the special status of the βιαιοθάνατοι ("violently killed people") see J. H. Waszink, *RAC* 2 (1954) 391–94.

7. Jordan (1980, 231 n. 23) gives the following list of published *defixiones* found in chthonic (usually Demeter) sanctuaries: ten from the fifth century B.C. found in or near the sanctuary of Demeter Malophoros in Selinous (SGD 99–108); thirteen second-century-B.C.(?) examples discovered in a Demeter sanctuary on Cnidus (*DT* 1–13—see Versnel's essay [chap. 3] for a discussion and bibliography on these much-debated tablets); ten first-century-B.C. tablets from the shrine of an as-yet-unidentified chthonic deity in Morgantina (SGD 115–20, discussed in detail at the end of this essay); and fourteen examples of Roman date from the sanctuary of Demeter and Kore at Corinth. He also discusses some unpublished

examples that date to the hellenistic period: one text from a small, early hellenistic(?) Demeter shrine on Rhodes (see SGD, p. 168); and seventeen lead tablets of hellenistic date excavated from a well in the Athenian agora, which probably—like the votive offerings found amongst them—came from a nearby rectangular shrine of a chthonic female deity (Demeter?). To this list I would add the four lead voodoo dolls excavated from the supporting wall of the first-century-B.C. sanctuary of Zeus Hypsistos on Delos (see A. Plassart, *Les sanctuaires et les cultes des Mont Cynthe,* Expl. Arch. de Délos 11 [Paris, 1928], 292–93), although I should perhaps point out that the deity here could hardly be called "chthonic."

8. W. S. Fox, "Submerged *Tabellae Defixionum,*" *AJP* 33 (1912): 301–10; Jordan 1980, 225–39; idem 1985, 205–55.

9. See n. 56 for a description of the charioteer curses found buried in hippodromes in Syria (SGD 149) and Apamea (unpublished). A curse against bronze workers (SGD 20) was found in the mud-brick wall of a house in the industrial district of ancient Athens (see R. S. Young, *Hesperia* 20 [1951]: 222–23), and Dugas has published four bronze voodoo dolls from Delos that were discovered amongst the ruins of a house dating to the hellenistic period (see n. 31). Cf. also the alleged discovery of inscribed lead tablets (*nomen Germanici plumbeis tabulis insculptum*) in the walls and floors of Germanicus' house after his mysterious death (Tac. *Ann.* 2.69 and Dio Cass. 58.18) and the placement of a more grisly binding spell in the lecture room of Libanius in order to inhibit his ability to teach and declaim (p. 16 and note 70).

10. Wünsch (*DTA*, pp. ii–iii) discusses the plentitude and cheapness of lead in Attica. See Jordan 1980, 226–29 and Miller (see see n. 4), 1–30, for more recent discussions. I am unconvinced by the arguments of Kagarow (1929, 24–25) and Guarducci—see n. 1—(pp. 240–41) that the great majority of Athenians were illiterate and had to depend on professional magicians. It is really a question of emphasis, for just as there are examples of ostraca mass-produced for "the lazy and the illiterate," we can imagine that some equally small percentage of the *defixiones* were similarly manufactured. R. Meiggs and D. Lewis (*Greek Historical Inscriptions* [Oxford, 1969], 40–45) discuss the cache of 191 ostraca inscribed by fourteen hands and all bearing the name Themistocles, and give a balanced assessment of the questions these ostraca have raised about Athenian literacy.

11. The fifteen *defixiones* reportedly found in the same well at Kourion on Cyprus seem to be written by the same person who used the same elaborate formula over and over again (*DT* 22–35 and 37). P. Aupert and D. Jordan (*AJA* 85 [1981]: 184) identify the provenance of these tablets as Amathous, not Kourion, and report the existence of more than one hundred more tablets probably from the same deposit, of which random samples have been found to contain texts similar to those published by MacDonald. R. Wünsch (*Sethianische Verfluchungstafeln aus Rom* [Leipzig, 1898]) published a cache of forty-eight lead tablets (= *DT* 140–87) found in a columbarium on the Appian Way, all of which have similar formulae and drawings on them, suggesting that they were the product of one individual or group of individuals working from a model. Jordan (1985) has identified the hand of an anonymous professional scribe of the third century A.D. who carefully inscribed more than twenty lead curse tablets discovered in four wells in different parts of the Athenian agora. The existence of two other tablets in a very inferior script of the same text suggests the existence of a handbook from which the "master magician" was training "apprentices."

12. Jordan 1988.

13. Jordan (1980, 226, n.6) gives a detailed list of the numerous examples from Attica, the Black Sea, and elsewhere, all dating to the classical period.

14. Jordan (1988, 273–74) points to two instances (SGD 1 and 2) in which the rolled-up *defixio* was placed in the right hand of the deceased (as if they were scrolls that he or she were meant to read?).

15. This might be the idea behind the "gift giving" mentioned in another Attic curse (SGD 54): "I send NN as a gift to the Earth and the underworld gods."

16. *DTA* 107, 109; SGD 62. Cf. *DT* 96 *inimicorum nomina ad . . . inferas* and see Wünsch, *DTA*, p. iii for a good discussion.

17. The rite of "cursing the name" is a commonly observed phenomenon in traditional societies throughout the world today. For a good—albeit much outdated—discussion of the cross-cultural parallels, see F. B. Jevons, "Graeco-Italian Magic," in *Anthropology and the Classics* (Oxford, 1908), 93–120. The only analogous ancient ritual with which I am familiar is the ancient Egyptian practice of painting the name of an enemy on a simple earthenware bowl and then shattering it. K. Sethe, *Die Ächtung feindlichen Fürsten, Völker und Dinge auf Tongefässsherben des Mittleren Reiches,* Abhandlungen der preussischen Akademie der Wissenschaften (1926), no. 5 provides the classic discussion. The aim of this ritual, however, was to destroy the victim(s) completely. The particular features of the Greek practice (e.g., folding and piercing with a nail) seem to be a unique Hellenic invention.

18. E.g., Wünsch, *DTA*, pp. ii–iii and Audollent, *DT*, p. xlii. Kagarow (1929, 5–6) believes—wrongly, I think—that the sympathetic action (i.e., piercing the tablet) was the original ritual and that the verbal aspect was a later addition that reinforced and eventually replaced the action as people began to forget its original meaning. See S. Tambiah, "The Magical Power of Words," *Man* 3 (1968): 175–208, for a critique of the tendency of modern scholars (in anthropology, but his point is applicable to classical scholars as well) to underrate the antiquity and importance of verbal magic as "performative utterance." For a similar combination (at the dawn of literacy) of verbal charm and magical object into an inscribed amulet, see Kotansky's essay (chap. 4).

19. Listed together as *DT* 109. See Wünsch 1900, 268–69 and Preisendanz 1972, 5 for discussion. Jordan informs me, however, that these tablets are now among the missing and that he is suspicious about their description in Audollent and their inclusion in the corpus. Another more easily verifiable example is the use of both inscribed and uninscribed "voodoo dolls" in classical Greece (see n. 31).

20. The early Sicilian *κατάδεσμοι* show a unique propensity for using compounds of the Greek verb *γράφω* (write), and on the face of it they may offer evidence for the importance of the act of writing. But with the possible exception of SGD 88 (which alone uses an uncompounded form of *γράφω*), the compound forms of *γράφω* used in the early Sicilian *defixiones* (*ἐγγράφω* and *ἀπογράφω*) and occasionally in those found in Attica (*καταγράφω*) all seem to have legal or technical meanings without any explicit emphasis on the basic meaning of the stem, e.g. "register," "summon" or "accuse." See also p. 1 and n. 2.

21. C. A. Faraone, "Aeschylus' *ὕμνος δέσμιος* (*Eum.* 306) and Attic Judicial Curses," *JHS* 105 (1985): 150–54. B. M. W. Knox has arrived independently at the same conclusion in a forthcoming essay entitled "Black Magic in Aeschylus' *Oresteia*."

22. Kagarow 1929, 44–49, with graph of formula frequencies, p. 45.

23. This is a simplification of Kagarow 1929, 28–34, which sets up five groups with numerous subgroups. His fifth category (*Kontaminationsformeln*) is too widely conceived to be of any help in analyzing the formulae.

24. Audollent (*DT*, pp. vii–viii) gives a list of more than twenty alternative verbs, of which the most frequent are *καταδίδωμι*, *καταγράφω* and *παραδίδωμι*. Wünsch (*DTA*, p. iii) and Jevons—see n. 17—(p. 109) both suggest that *καταδῶ* (shorthand for *καταδῶ ἥλοις* "I [trans]fix with nails"; cf. Pind. *Pyth.* 4.71) is the Greek equivalent for the Latin *defigo* and alludes directly to the practice of "nailing" the lead tablet. Kagarow (1929, 25–28), gives a sophisticated discussion of the two semantic fields into which these verbs fall: (1) literal binding (verbs compounded with *δέω*) and (2) verbs with technical or legal connotations that either "register" the victims before an imagined underworld tribunal (i.e., compounds of

γράφω) or those that simply "consign" the victims to the control of the chthonic deities (i.e., compounds of τίθημι and δίδωμι). For more discussion of the compound forms of γράφω see n. 20.

25. See, e.g., fifth-century binding curses such as SGD 91, which identifies itself as "the εὐχά of Apelles" and SGD 1 and 107, which summon the victim "before the holy goddess (Persephone)."

26. Kagarow (1929, 41–44) gives a survey of formulaic expressions shared by traditional Greek prayers and the *defixiones* that employ the "prayer" formula. He also points out (p. 34, n. 1) that the *anadiplosis* of the names of gods in the *defixiones* occurs almost exclusively in those employing the prayer formula. For this feature of many traditional Greek prayers, see E. Norden, *Agnostos Theos* (Leipzig, 1923), 2–9 and n. 1. For an excellent discussion of the traditional forms of prayer in *PGM,* see Graf's article (chap. 7). These close affinities are at the root of the controversy over the Morgantina lead tablets (see pp. 18–19 and notes).

27. In the Attic curse tablets the verb κατέχω (hold fast) is easily the most frequently used verb in the prayer formulas and (as Jevons, pp. 112–13, suggests; see n. 17) must be connected with the epithets of Hermes Katochos and Earth Katochē, the two most frequently invoked deities on the early Attic curse tablets. Both verb and epithet are virtually unknown on *defixiones* found outside of Attica or areas deeply influenced by Athenian culture, such as Euboea. See R. Ganschinietz, "Katochos," *RE* 10, 2 (1919): 2526–34.

28. Another very rare approach is the use of a verb of request and an infinitive, e.g., *DTA* 100 (Attic, fourth century B.C.): "I beg (ἱκετεύω) that you oversee these affairs." Versnel, in fact, points out that such a formula belongs to the group of atypical *defixiones* that he labels "borderline" cases (see chap. 3). By the second century A.D., however, this form of curse grows into one of the most popular, especially when compounds of the verb ὁρκίζω are employed and there is an overt emphasis on compelling daemons or gods to bind a victim. The use of the verb ὁρκίζω, however, cannot be used per se as a criterion for calling a ritual a "magical act." In chap. 2 Strubbe notes the popularity of this verb on tombstone curses of the traditional "religious" type.

29. On the use of lead for daily correspondence, lists of cavalry, oracle requests, and other purposes, see n. 10. Kotansky discusses two fourth-century B.C. lead amulets in his contribution to this volume (chap. 4).

30. The recent discovery in a villa outside of Pompeii of wooden writing tablets coated with reddish "gum lac" instead of the wax probably accounts for Ovid's designation of the wax as *poenicea,* as well as the ms. variant *sanguinea* (cf. J. Reynolds, *JRS* 61 [1971]: 148).

31. Ch. Dugas, "Figurines d'envoûtement trouvées à Délos," *BCH* 39 (1915): 413–23; Preisendanz, "Die Zaubertafeln," (see n. 3), 163–64; and Jordan 1983 all provide detailed surveys. See note 33 below for late antique dolls of wax or clay found in Egypt and used in erotic magical spells. I use the term "voodoo doll" simply as the most familiar modern equivalent in English to *Rachepuppe* or *figurine d'envoûtement,* without implying any connection whatsoever to the Afro-Carribean religious practices of the island of Haiti.

32. A fourth-century-B.C. Cyrenean inscription (Meiggs-Lewis [see n. 10], no. 5, lines 44–49) describes and paraphrases the oath of the seventh-century Theran colonists of Cyrene, which contains the usual conditional self-imprecation. This imprecation, however, involves the hitherto-unknown use of wax voodoo dolls in the oath ceremony: "They made wax figurines and burned them saying: 'Whosoever does not abide by these oaths, but transgresses them, may he waste away (lit., melt or drip down) and run to ruin just as these dolls do, the man himself, his family, and his possessions.'" For a careful and thorough study of the inscription, its relation to Hdt. 4.145–59 and the probably direct or indirect archaic source for the oath and curse, see A. J. Graham, "The Authenticity of the ΟΡΚΙΟΝ ΤΩΝ ΟΙΚΙ-ΣΤΗΡΩΝ of Cyrene," *JHS* 80 (1960): 95–111. For the significance of this rather early

attestation of voodoo dolls, see A. D. Nock, "A Curse from Cyrene," *ARW* 24 (1926): 172–73.

33. For the Greco-Egyptian combination of a lead tablet with wax or clay dolls (bound and/or "nailed"), see SGD 152–53 and 155 (and fig. 7), which all seem to have been manufactured and inscribed according to the recipe found at *PGM* IV.335–408. For some terra-cotta examples from Italy, see *DT* 200–207 and Solin (see n. 4), no. 33.

34. The early habit of writing in retrograde is usually attributed to the Phoenician origin of the alphabet; see L. H. Jeffery, *Local Scripts of Archaic Greece* (Oxford, 1961), 43–50.

35. The use of stone or bronze implements in rituals is often thought to indicate an origin in the Stone or Bronze Age, e.g., the use of a flint knife in the ritual circumcision of the Jews or the use of brazen instruments to collect magical herbs (Soph. Frag. 534 with Pearson's comments). J. Z. Smith (*Imagining Religion* [Chicago, 1982], 53–56) gives a wonderfully clear exposition of the process by which originally nonsignificant or even accidental elements of the rite are continually repeated due to the conservative nature of rituals in general and how aetiological rationalizations (similar to those in the *similia similibus* formulas discussed above) are invented later on to explain their new-found significance.

36. S. J. Tambiah, "Form and Meaning of Magical Acts: A Point of View," in *Modes of Thought*, ed. R. Horton and R. Finnegan (London, 1973), 199–229. Cf. G. E. R. Lloyd, *Magic, Reason, and Experience* (Cambridge, 1979), 2–3 and 7.

37. E. E. Evans-Pritchard, *Witchcraft, Oracles, and Magic among the Azande* (Oxford, 1937), 450, quoted by Tambiah on p. 204.

38. The goal of death or destruction is very rarely mentioned in the texts of the early Greek *κατάδεσμοι*. The verb *ὄλλυμι* ("destroy") and its compounds, which are so characteristic of the other forms of Greek cursing, only appear five times in the published *defixiones*, and in three out of these five instances it is a tentative restoration to a damaged tablet (*DTA* 75a [*bis*] and SGD 89). A third-century B.C.-tablet from the Chersonese contains a wish that is very much like a traditional curse: "May they be destroyed with their families." L. H. Jeffery ("Further Comments on Archaic Greek Inscriptions," *BSA* 50 [1955]: 73, no. 2) provides the best reading of SGD 104 (Selinuntine, fifth century B.C.), apparently the only other secure example of the traditional cursing formula ("May they be utterly destroyed, they and their kin"), which is then followed by a list of names. Occasionally idiosyncratic phrases do occur that seem to imply the destruction of the victim: "I bind these men *in tombs" (DTA* 55 and 87); "I do away with him (*ἀφανίζω*) and bury him under (*κατορύττω*)" (*DT* 49; cf. SGD 48 and 49); "Restrain him until he comes down to Hades" (*DT* 50, Attic, late?). The last-mentioned must, however, refer to the intended length of the curse (i.e., "May he be restrained for the rest of his life"); for the repeated misinterpretation of similar locutions in Greek poetry, see D. Young, *Pindar Isthmian 7: Myth and Exempla*, Mnem. Supp. 15 (Leiden, 1971), 12–14 and 40–42. The formula *καταδῶ καὶ οὐκ ἀπολύσω τὸν δεῖνα* ("I bind NN and I will not release him," cf. *DTA* 158 and SGD 18) seems to imply that a binding curse could have a limited duration or be loosened at a later date. In such cases it is difficult to imagine that the curse resulted in death. See Strubbe's discussion (chap. 2) about the "loosening" of the so-called scepter curses on tombstones and Versnel's report (chap. 3) on the use of *λύω* and its derivatives in the prayers for justice.

39. Paus. 9.38.5.

40. L. Robert (*Documents de l'Asie Mineure Méridionale* [Geneva, 1966], 91–100) and J. Wiseman ("Gods, War, and Plague in the Times of the Antonines," in *Studies in the Antiquities of Stobi*, vol. 1, edd. D. Mano-Zissi and J. Wiseman [Beograd, 1973], 174–79) attempt unsuccessfully to redate the inscription to the reign of Lucius Verus using numismatic evidence. See E. Maroti, "A Recently Found Versified Oracle against the Pirates," *Acta Ant.*

16 (1968): 233–38 for a good refutation of Robert, an article of which Wiseman was apparently unaware. With regard to this inscription see K. Meuli and [R. Merkelbach] ("Die Gefesselten Götter," in *K. Meuli: Gesammelte Schriften,* vol. 2 [Stuttgart, 1975], 1077–78), who discuss the myths concerned with the binding of Ares and give two parallels for the binding of his images: Paus. 3.15.7 (the bound statue of Enyalios at Sparta) and Pliny *HN* 35.27 and 93 (the famous picture painted by Apelles that was set up in the Forum of Augustus at Rome: *Belli imaginem restrictis ad terga manibus;* cf. Verg. *Aen.* 1.294–96 and Servius, ad loc.).

41. Olympiodorus of Thebes in *FHG* 4.63.27 (= Frag. 27 Blockley). Valerius, the governor of Thrace at the time and Olympiodorus' eyewitness source, was told by the local people that the statues were consecrated by ancient rites to prevent the depredations of the barbarians. The statues were excavated and carried off, and soon afterward the area was invaded by three successive waves of barbarian tribes: the Goths, the Huns, and the Sarmatians. On the chronological problems of the dating of these events, see J. F. Matthews, "Olympiodorus of Thebes and the History of the West (A.D. 407–25)," *JRS* 60 (1970): 96 and R. C. Blockley, *The Fragmentary Classicising Historians of the Late Roman Empire,* vol. 1 (Liverpool, 1981), 164, n. 20.

42. *Narratio Miraculorum Sanctorum Cyri et Joannis* (= *PG* 87.3, col. 3625); the relevant portions of the Greek text appear in *DT*, pp. cxxii–iii.

43. E.g., Versnel, in chap. 3, suggests that the long-standing debate over the presentation of the Cnidian tablets is a red herring of sorts. He points out that we now have enough examples of both publically displayed and secretly buried curses against unknown thieves and criminals to show that either method was acceptable; i.e., the most important function of these *πινάκια* is that they be delivered to or nearby the abode of the deity, where it serves as a "legal" cession to the god of either the guilty party or the stolen property. Any use as a publicly displayed warning is probably a secondary function.

44. *I. Cret.* 2 (17) 28: *παραδίδωμι τοῖς καταχθονίοις θεοῖς τοῦτ' τὸ ἡρῶιον φυλάσσειν*. See Jordan 1980, 228, n. 16 for discussion of this text and the two Attic parallels.

45. In addition to amatory, circus and judicial curses, Audollent (*DT*, p. lxxxix) included public proclamations against unknown thieves in his four types; in chap. 3 Versnel rightly reclassifies the proclamations as "judicial prayers," which have a social context different from that of the *defixiones*. Kagarow (1929, 50–55) described five types, adding phylacteries written on lead. Phylacteries are defensive rather than offensive magical operations and as such obviously fall outside the definition of a binding curse. Preisendanz (1972, 9) is impressed by the number of well-known politicians whose names appear on the tablets and suggests still another category, "political *defixiones,*" but this is too broad and cannot be adequately distinguished (as he himself admits) from judicial curses. I discuss this phenomenon in the section that follows my survey of judicial curses.

46. E.g., *DTA* 12 (shield maker) and 30 (innkeepers); SGD 3 (silversmith), 11 (innkeepers), 20 (bronze workers), 48 (painter, flour[?]seller and scribe), 72 (seamstress), 129 (doctor), and 170 (ship's pilot). There are however, some examples of people described by profession who appear as victims in judicial curses, a fact suggesting that designation by trade may have merely been another way of identifying people, like a demotic or a patronymic.

47. They are all from Greece or Sicily and date to the classical or hellenistic periods: *DTA* 68–75, 84–87; *DT* 47, 52, 70–73, 92; and SGD 20, 52, 73, 75, 88, and 124.

48. L. H. Jeffery—see n. 38—(pp. 67–84, no. 18) prefers to read ΚΕΡΔΟΝ as a proper name in the accusative. For *kerdos* as the object of a verb of binding, see *DTA* 86: *καταδῶ . . . ἐργασίας κέρδη*.

49. Although one might argue that *διδάσκαλος* could merely mean "teacher," Wünsch

points out (ad loc.) that the term ὑποδιδάσκαλος only occurs in the context of dramatic competition, e.g., Pl. *Ion* 536a. Cf. also *SIG* 692 A 31 (Delphi, second century B.C.) and Cic. *Fam.* 9.18.4.

50. The hereditary nature of the curse and the significant name of Eunikos ("Good at Winning") may suggest a professional actor; see Miller (n. 4), 80–83 for a detailed discussion.

51. This tablet seems to be closely modeled on some Greek prototype; see H. S. Versnel, "May he not be able to sacrifice . . . : Concerning a Curious Formula in Greek and Latin Curses," *ZPE* 58 (1985): 247–48 and 269. Jordan (SGD, p. 167) mentions an unpublished Corinthian *defixio* (inv. MF 69–118) from an underground bath complex cursing a "retired(?) mimic actress."

52. SGD 24–26. See Jordan 1985, 214–18, nos. 1–3 for these translations and an excellent commentary on the curses against Eutychian. No. 4 (= SGD 27) is a curse against the wrestler Attalos, the ephebe, son of Attalos ("Let him grow cold and not wrestle"); and no. 5 (= SGD 28) is against a Macedonian wrestler called Petres, the pupil of Dionysios (". . . if he does wrestle, in order that he fall down and disgrace himself").

53. SGD 157. See D. Wortmann, "Neue magische Texte," *BJ* 168 (1968): 108–9, no. 12 for text and discussion.

54. SGD 29. See Jordan 1985, 221–23, no. 6 for this translation and commentary. Jordan mentions (p. 214) two other unpublished examples of racing curses that have been unearthed recently at Corinth and Isthmia (both of late Roman date).

55. *DT* 145–87. For a detailed discussion see Wünsch, *Sethianische Verfluchungstafeln* (see n. 11). *DT* 141–44 and 153 are written in Latin and are not included in my reckoning. Here and in the case of the binding curses found at Hadrumetum against charioteers and those found at Carthage against *venatores* (discussed below) my separation of Greek and Latin texts found at the same place and dating to the same period is admittedly artificial, since they are all the product of the same social environment.

56. There are thirteen Greek examples from Carthage (*DT* 234–44 and SGD 138–39); two from Hadrumetum (*DT* 285, SGD 144); one from Lepcis Magna in Libya found buried in the starting gates of the circus (SGD 149); one from Damascus (SGD 166); and one from Beirut (SGD 167), which curses the horses of the blue faction. Jordan (SGD, pp. 192–93) also mentions the discovery of charioteer curses at Apamea that bind the limbs of the drivers "so that they cannot drink or eat or sleep" and mentions the excavation of others (presumably still unread) from the *spina* of the hippodrome at Antioch-on-the-Orontes. See also Jordan, *SGD*, pp. 166–67 for a description of an unpublished lead tablet from Corinth that appears to bind the performance of someone (an athlete ?) "in the circus." There are a number of extant Latin *defixiones* of similar date from North Africa that bind charioteers, most notably a group of twenty-two from Hadrumetum (*DT* 272–84 and 286–95).

57. Wünsch (1900, 248–59) gives an excellent discussion of this third-century-A.D. Carthaginian circus curse (*DT* 242).

58. *DT* 246–47, 249–50, and 252–53 (Carthage, second or third century). Three Latin examples were also found there (*DT* 248, 251, and 254), as well as an earlier example (first century?) from Caerleon in Wales; see R. Egger, *Ö Jh* 35 (1943): 108–10.

59. *DTA* 78, 89, 93(?); *DT* 68, 69, 85, 86, and 198; SGD 30–32, 57, and 154. The discovery of two additional late-hellenistic Corinthian divorce curses have been announced by S. G. Miller, *Hesperia* 50 (1981): 64–65.

60. The syntax of this tablet is sometimes difficult. See the detailed commentaries of E. Ziebarth, "Neue attische Fluchtafeln," *GöttNachr.* (1899), p. 132, no. 1; Wünsch 1900, p. 70, no. 1; and Audollent ad *DT* 85.

61. See Kagarow 1929, 51 for a good discussion.

62. SGD 156, lines 48–62. For the identical formula in SGD 152 and 153 and the close

correspondences to the recipe for an aphrodisiac *defixio* in *PGM* IV. 367ff., see D. Wortmann, "Neue Magische Texte," *BJ* 169 (1969), no. 1; S. Kambitsis, "Une nouvelle tablette magique d'Égypte," *BIFAO* 76 (1976): 213–30; D. Martinez, *P. Mich. 6926: A New Magical Love Charm* (Ph.D. diss. Univ. of Mich., 1985); and D. R. Jordan, "A Love Charm with Verses," *ZPE* 72 (1988): 289.

63. D. R. Jordan, "*CIL* VIII 19525 (B).2: *QPVVLVA* = q(uem) p(eperit) vulva," *Philologus* 120 (1976): 127–32.

64. Jordan (1985, 251), discusses the five extant examples of hair or other kinds of *ousia* rolled up inside *defixiones*. Four of them come from late antique Egypt, and a fifth (of uncertain purpose) dates to the third century A.D. and was found in the Athenian agora.

65. Another important difference between the traditional *defixiones* discussed in this paper and the later aphrodisiac curses is the inclusion of the name of the operator of the spell, which rarely occurs in the former. For a discussion of this genre of aphrodisiac spell, see Winkler's contribution (chap. 8).

66. There are twenty-three published aphrodisiac curses on lead all dating between the second and fourth centuries A.D., from Carthage (*DT* 227, 230–31); Hadrumetum (*DT* 264–71, 304) and Egypt (*DT* 38; SGD 151–53, 155–56, 158–61). Jordan (SGD) mentions unpublished and partially read tablets from Egypt (p. 191), Carthage (pp. 186–87) and Tyre (p. 192). For examples in other media, see, e.g., R. W. Daniel ("Two Love Charms," *ZPE* 19 [1975]: 249–64), who publishes one papyrus and one linen example (both dating to the third or fourth century A.D. and of unknown [Egyptian?] provenance). See also *PGM* O[stracon] 2 (Egyptian, second century A.D.).

67. E. Ziebarth ("Neue attische Fluchtafeln" [see n. 60], 122) asserted that judicial curses were enacted by the losers of a lawsuit, after a decision had been rendered. He was refuted by Wünsch (1900, 68), who argued that the formulas of judicial binding curses all seemed to point to a future event and that they were therefore employed beforehand or while cases were still pending. Audollent (*DT*, pp. lxxxviii–ix, n. 2) supported this view. Years later, Ziebarth, (*Neue Verfluchungstafeln aus Attika, Boiotien und Euboia,* SBAW 33 [Munich, 1934]: 1028–32) adopted the compromise view that a judicial curse was enacted while the trial was going on but only after its author had come to the conclusion that he was about to lose his case. P. Moraux (*Une défixion judicaire au Musée d'Istanbul,* Mém. Acad. Roy. Belg. 54.2 [Brussels, 1960]: 42) reviews the debate and concludes that although none of the curses seem to have been enacted after the final outcome of the trial, it is impossible to know at what point before or during the trial the litigants wrote the curses. There seems to be a trade-off between the practical desire to inhibit damaging evidence as early as possible and the litigants' sense of urgency later on. Kagarow 1929, 53–54 gives a chart listing the sixteen different terms that point to a confrontation in the courts, e.g., *σύνδικοι, ἀντίδικοι, δικαστήριον*, etc.

68. C. A. Faraone, "An Accusation of Magic in Classical Athens (Ar. *Wasps* 946–48)" *TAPA* 119 (1989): 149–61.

69. *Brut*. 217 and *Orat*. 128–29.

70. C. Bonner ("Witchcraft in the Lecture Room of Libanius," *TAPA* 66 [1932]: 34–44) interprets this as a form of *envoûtement* directed against Libanius' oratorical abilities; the cutting off of the one forefoot was directed against the hand with which the orator gesticulated and the position of the other attempted to silence him, as Libanius himself seemed to realize. For the placement of the front foot over the mouth, see Plassart's (see n. 7) discussion of four lead voodoo dolls from Delos, of which the two male dolls had nails driven into their eyes, ears, and mouth and the right hand twisted up to cover the mouth entirely. Bonner ("Witchcraft") and Peter Brown ("Sorcery, Demons, and the Rise of Christianity" in *Witchcraft Confessions and Accusations,* ed. Mary Douglas [London 1970]) discuss the popularity of magic and accusations of magic among the declaimers of the late empire.

71. *DTA* 25, 38, 39, 63, 65–68, 81, 88, 94, 95, 103, 105–7, and 129; *DT* 18, 22–35, 37, 39, 43, 44, 49, 60, 62, 63, 77, and 87–90; and SGD 6, 9, 19, 42, 49, 51, 61, 68, 71, 89, 95, 99, 100, 108, 133, 162–64, 168, 173, 176, and 179.

72. The exceptions are SGD 162 (fifth century A.D., Egypt), 163 and 164 (third century A.D. Palestine), 168 (second century A.D., Upper Maeandros Valley), and 179 (third or fourth century A.D., provenance unknown). To these six late examples we must add the sixteen tablets found on Cyprus (*DT* 18, 22–35, and 37), which have little statistical importance because they were written by the same individual against many of the same people (see n. 11). All of the remaining examples date to classical or hellenistic times and were discovered in Sicily (four tablets), Attica (twenty-eight), Olbia (five), Megara (two), and one example each from Melos, Corcyra, Eretria, and Emporion in Spain. Two additional tablets are from mainland Greece, but their exact provenance is a mystery.

73. W. M. Calder, "The Great Defixio from Selinus," *Philologus* 107 (1963): 163–72, has plausibly suggested that a fourth tablet found in the same area (SGD 107) is also judicial in nature; it curses seventeen men, who can be grouped together into seven interrelated families, a relationship that suggests testamentary litigation similar to that attested in Attic law, e.g., the lawsuits involving the Dikaiogenes family (Isae. 5) or the Hagnon family (Isae. 11).

74. *DTA* provides the most extensive collection of Attic curses, including seventeen judicial curses (*DTA* 25, 38–39, 63, 65–68, 81, 88, 94–95, 103, 105–7, and 129). Wünsch, however, was cautious, almost agnostic, in his dating of the tablets, assigning all to the third century B.C.—and then only tentatively—unless some overwhelming evidence pointed to an earlier or later date (see his introduction, p. i); accordingly, of the Attic judicial curses enumerated above, he assigned only *DTA* 38 and 107 with confidence to the fourth century B.C. A. Wilhelm ("Uber die Zeit einiger attischer Fluchtafeln," *Ö Jh* 7 [1907]: 105–26) argued that Wünsch greatly underestimated the antiquity of the *DTA* curses and by way of example he redated a number of them to the fourth century B.C. (including four of the judicial curses, *DTA* 65, 66, 95, and 103) and a few (including *DTA* 38) to the fifth century B.C., using a combination of paleographic and prosopographic evidence. The tablets themselves have since disappeared, and as a result most of them have never been properly redated.

75. Audollent gives five examples of Attic judicial binding curses: four from the fourth century B.C. (*DT* 49, 60, 62–63); and one that he was unable to date (*DT* 77). Attic *defixiones* published subsequent to these two major collections include nine judicial binding curses: four have been assigned to the fourth century B.C. (SGD 19, 42, 49, and 51); and two to the late fifth–early fourth century B.C. (SGD 6 and 9). Three of the unpublished inscribed lead dolls described by Jordan (1988) date to the end of the fifth century. Jordan (SGD, p. 162) reports, however, the discovery of seventeen tablets from a well in the agora (inv. IL 1695, 1704–19), which were found in a late-fourth-to-early-third-century-B.C. context and seem at first glance to be judicial curses.

76. See Wünsch's commentary on *DTA* 28, 47–51, 87, 89, and 167 and idem 1900, 63, where he argues that the Demosthenes and Lycurgus mentioned on *DT* 60 are the famous Athenian orators. A. Wilhelm—see n. 74—(pp. 105–26) gives prosopographical notes on *DTA* 11, 24, 30, 42, 65, 84 and SGD 18 identifying several prominent Athenians, including the famous orator Callistratus of Aphidna. E. Ziebarth—see n. 67—(pp. 1028–32) traced many of the individuals mentioned on SGD 48 to the political circle of Demades. L. Robert—see n. 1—(pp. 12–13, no. 11) published a lengthy judicial curse tablet (SGD 42) listing several politicians from the early fourth century, most notably Aristophon from Azenia. Jordan (1988) shows that three rather rare names inscribed on lead voodoo dolls excavated in the Kerameikos (SGD 9 and two of the unpublished dolls) are probably those of active politicians accused in speeches written by Lysias: Mnesimachos in Lysias Frag. 182 (Sauppe);

Mikines in Lysias frags. 170–78 (Sauppe); and Theozotides, the father of one of Socrates' students (Nikostratos mentioned at *Pl. Ap.* 33c), who was also accused in a speech by Lysias (*P. Hibeh* 14). Jordan (1980, 229–39) also discusses a *defixio* that curses members of the Macedonian ruling circle during the occupation of Athens: Kassander (the king), Pleistarchos (his brother), Eupolemos (his general in Greece), and Demetrios the Phalerian (the governor of occupied Athens).

77. Preisendanz 1972, 9.

78. E.g., *DTA* 106 (Attic, third century B.C.); *DT* 49 (Attic, fourth century B.C.); and *DT* 87 (Corcyra, third century B.C.); cf. *DT* 61.

79. This suggestion presupposes the existence of some formal or at least organized system of patronage and political alliance similar perhaps to that at Rome, for which see M. Finley, *Politics in the Ancient World* (Cambridge, 1983), 76–84 and P. J. Rhodes, "Political Activity in Classical Athens," *JHS* 106 (1986): 132–44. For the Latin curses against *inimici* and their *amici,* see the index to *DT.*

80. P. Brown (see n. 70), 25. A curse tablet from Beirut (SGD 167, third century A.D.) points, perhaps, to the larger political ramifications of competition in the hippodrome when it curses thirty-four different drivers and/or horses (it is often not clear which is which), identifying them all as members of the "blue" faction.

81. Rhodes (see n. 79), 136. One might see some kind of political competition in the curses against χορηγοί (discussed above). The fifth-century Sicilian curse (SGD 91) curses the rival χορηγοί of Eunikos, and it is written by a certain Apelles "on account of his φιλία for Eunikos." For the strong political connotations of the term φιλία (= *amicitia*), see W. R. Conner, *The New Politicians of Fifth Century Athens* (Princeton, 1971), 30–66 passim.

82. See Versnel (chap. 3, pp. 62–63) for the widely held opinion that *defixiones* are self-admittedly shameful because they are hidden and anonymous; and idem, "Religious Mentality in Ancient Prayer," in *Faith, Hope and Worship,* ed. H. S. Versnel and F. T. van Straten (Leiden, 1981), 26–28 for a similar discussion of silent and malevolent prayer.

83. N. Nabers, "Lead *Tabellae* from Morgantina," *AJA* 70 (1966): 67–68; idem, "Ten Lead *Tabellae* from Morgantina," *AJA* 83 (1979): 463–64. The restoration at the end of the text was suggested to me privately by L. Koenen; Nabers prints the following: τοῦ Ῥ[ούφο[υ] . . . ιου|. . .

84. L. Robert, *BE* (1966): 518.

85. Guarducci (see n. 1), 250–51; Jordan 1980, 236–38. Nabers provided several parallels from Greek tragedy (e.g., Soph. *Phil.* 819 and Eur. *Alc.* 744) and later sources for the "pious" use of δέχομαι *uncompounded,* but Jordan points out that the imperative παραδέξασθε would be the proper Greek equivalent (for that period) for παραδίδωμι ὑμῖν, a fairly common expression in the corpus of *defixiones*. It is probable that ποτιδέξασθε, like most of the compound verbs used in the early *defixiones,* is modeled on a legal or technical usage, i.e., it does not mean "receive graciously" (so Nabers, using *LSJ* i), but rather something more technical or bureaucratic modeled along the lines of "admit into citizenship" or "undertake a liability upon oneself, guarantee" (*LSJ* ii. 2 and 6).

86. *IG* XI, pt. 4, no. 1299; I. U. Powell, *Collectanea Alexandrina* (Oxford, 1925), 68–71. For direct parallels between this description and the texts of contemporary *defixiones,* see H. Engelmann, *The Delian Aretalogy of Serapis,* EPRO 44 (Leiden, 1975), 53–54.

87. Jevons, "Graeco- Italian Magic" (see n. 17), 109–15. He believes that the "magical" form (i.e., the performative utterance) historically preceded the "religious" (i.e., the "prayer formula") and that after both were in use side by side, combination and contamination resulted in the "hybrid" state of the majority of the extant examples. E. G. Kagarow ("Form und Stil der Texte der Fluchtafeln," *ARW* 21 [1922]: 496–97) argues instead for the historical priority

of the third type, the *similia similibus,* the others having developed from it. Both scholars were, perhaps, unduly influenced by the overly simplistic evolutionary models that had been put forth by the anthropologists of their days.

88. For a similar usage in the funerary imprecations, see Strubbe's essay (chap. 2, pp. 34–35) on formulae such as *ἔσται αὐτῶι πρὸς* plus name of god ("He will have to reckon with the god NN") or *ἕξει πρὸς* plus name of god ("He will give answer to the god NN").

2

"Cursed be he that moves my bones"

J. H. M. Strubbe

The last line of the epitaph of Shakespeare, engraved on the stone slab that covers his grave in Holy Trinity Church at Stratford-upon-Avon, warns, "Curst be he yt moves my bones." The malediction was designed to frighten off the sexton of the church and his successors, who sometimes had to dig up an old grave in order to make room for the newly deceased.[1] Almost the same prohibition and malediction occur in the closing lines of an epitaph from Synnada in Asia Minor: *τὶς οὖν π̣[ο]τε τὰ ὀστέα σ[αλεύσε?]ι, κατάρᾳ αὐ[τῷ γένοι]το* ("and whoever thus will move[?] these bones, may he have a curse").[2] Both maledictions had the same purpose: they assured the undisturbed rest of the deceased. But while imprecations written on a gravestone were rather exceptional at the time of Shakespeare's death, they were very common in antiquity, especially in Asia Minor (Anatolia). My concern here will not be with Shakespeare's malediction but with the ancient funerary imprecations that are found in the Greek epitaphs of Asia Minor. By funerary imprecations (I adopt the term used by P. Moraux to denote this kind of curse) I mean curses that are clearly and publicly written on the gravestone by the owner of the tomb (who does not conceal his identity) to warn any potential wrongdoer that evil will befall him in case he should violate the grave in defiance of the legitimate prohibitions to do so.[3] Although these imprecations were used by pagans, Christians, and Jews alike, I will restrict this study to the pagan formulae.[4] A further restriction in this study has to do with the different groups of funerary imprecations that can be discerned. I distinguish two main categories. The first contains all the imprecations that do not specify the punishments awaiting the wrongdoer. I will call this group the "nonspecific" group. The second category includes the imprecations in which the punishments wished for are more or less clearly specified. I will refer to this group as the "specific" group. The nonspecific group contains many different types of imprecations.

The violator of the tomb may be declared to be *ἀσεβής* (impious) or, less frequently, *ἱερόσυλος* (sacrilegious) or said to be guilty of *ἀσεβεία* (impiety) or, rarely, of *ἱεροσυλία* (sacrilege). The name of a god or several gods is often added. The following texts may give an illustration. At Telmessus in Lycia an epitaph says that the man who will bury a strange corpse in the grave *ἀσεβὴς ἔστω θεοῖς καταχθονίοις καὶ ἐκτὸς ὀφειλέτω τῷ* Τελμησσέων δήμῳ *, ε′ ("will be impious towards the gods of the underworld and besides he will pay 5,000 denarii to the people of Telmessus").[5] At Aperlae, also situated in Lycia, the person who will place another into the tomb *ὑπεύθυνος ἔσται ἀσεβείᾳ καταχθονίοις θεοῖς καὶ ὑποκείσεται τοῖς διατεταγμένοις κ[αὶ ἔ]ξωθεν Ἀπερλειτῶν τῷ δήμῳ* * *μ(ύρια) β'* ("will be liable of impiety towards the gods of the underworld and he will fall under the regulations and besides he will pay 2,000 denarii to the people of Aperlae").[6] In a number of related texts it is said that the offender will be *ἔνοχος* (or *ὕποχος* or other variants) to the god or gods or that he *ἐνέχεται* (or *ὑποκεῖται* or other variants) to the god or gods. These formulae indicate that the wrongdoer will fall under the power of the god(s). They may be considered as shortened expressions of the above-mentioned formulae. The following illustration comes from the region of the Söğüt Gölü near Tyriaeum in Pisidia: *[ἔ]νοχος ἔστω πᾶσι θεοῖς καὶ Σελήνῃ καὶ Λητῷ* ("He will be liable to all the gods and Selene and Leto").[7] Another example is found at Olympus in Lycia: *ἐκτίσει ὁ θάψας τῇ πόλει *φ' καὶ ἔσται ὑπόδικος τοῖς κατακτονίοις θεοῖς* ("The man who buries [a strange corpse] will pay 500 denarii to the city and he will be liable to the gods of the underworld").[8] The implication of all the formulae of the first type is not only that the violator of the grave will be penalized by the human law[9] but also that he will be punished by the god or the gods for his impious or sacrilegious deed. It is interesting to note that many of the legal terms referring to punishment by the human law (such as *ἔνοχος* or *ὑποκεῖσθαι*, etc.) recur in the imprecations discussed here.

In Lycia, especially along the south coast, the violator of the tomb is often called *ἁμαρτωλός* (a wrongdoer). Here, too, the name of a god or several gods is frequently mentioned. An example from Antiphellus gives the following text: *ὁ δὲ παρὰ ταῦτα ποιήσας ἁμαρτωλὸς ἔστω θεοῖς καταχθονίοις καὶ εἰσοίσει προστείμου τῷ ἱερωτάτῳ ταμείῳ *, αφ'* ("The man who acts against these [prohibitions] will be a wrongdoer towards the gods of the underworld, and he will pay a fine of 1,500 denarii to the most sacred [i.e., imperial] treasury").[10] Another example from Rhodiapolis warns, *εἰ δὲ μή, ἁμαρτωλὸς ἔστω θεοῖς πᾶσι καὶ πάσαις* ("If not, he will be a wrongdoer towards all the gods and goddesses").[11] As to the exact meaning of *ἁμαρτωλός* there is some dispute. According to some the word denotes a sinner, but according to others, among whom is K. H. Rengstorf (whom I follow), the word refers to "ganz allgemein den Gedanken des Übergriffs," and there are no good reasons for interpreting *ἁμαρτωλός* as "Sünder im Sinne einer qualitativen Aussage."[12] In any case one can assume that the term *ἁμαρτωλός* contains a threat of divine punishment by the offended god(s).

The formulae *ἔσται αὐτῷ πρὸς* ("he shall have to reckon with") and *ἕξει πρὸς* ("he shall give an answer to") followed by the name of a divinity contain only a vague threat that the violator of the grave will be liable to the god(s). This seems to imply that the god(s) will bring an undefined punishment. The following texts illustrate these formulae. Near the Söğüt Gölü in Pisidia an epitaph warns the offender that *ἔστι αὐτῷ πρὸς Ἥλλιων κὲ Σελήνην* ("he will have to reckon with Helios and Selene").[13] At Termessus in Pisidia the man who tries to violate the grave *ἐκτείσει προστείμου Διὶ Σολυμεῖ *, α' καὶ ἕξει πρὸς τοὺς κα-*

τοιχομένους ("will pay a fine of 1,000 denarii to Zeus Solymeus, and he shall give an answer to the departed").[14] In the greater part of the texts the divinity is nameless, for example, at Eumeneia in Phrygia: *ἔσται αὐτῷ πρὸς τὸν θεόν* ("He shall give an answer to the god")[15] or at Laodiceia Combusta in Phrygia: [*ἔξ*]*ι πρὸς τὸν θεόν* ("He will have to reckon with the god").[16] Such formulae were adopted by Christians and Jews but were never exclusively used by them.[17]

In some parts of Asia Minor a god or several gods are adjured not to permit any violation of the grave. An illustration of this usage is given by a tomb inscription from the neighborhood of Sidamaria in Lycaonia: *ἐνορκίζω δὲ Μῆνας τόν τε οὐράνιον καὶ τοὺς καταχθονίους μὴ ἐξ*[*εῖ*]*ναί τινι πωλῆσαι τὸ περίβολον τοῦ τά*[*φ*]*ου, μήτε ἀγοράζειν ἐκτὸ*[*ς*] *τ*[*οῦ ἀδελφοῦ*?] ("I adjure the Mens, the one in heaven and those in the underworld, not to allow anyone to sell or to buy the precinct of this tomb, except to my brother [?]").[18] It also happens that the potential violator is adjured by the god(s) not to desecrate the tomb. This is attested, for example, at Elaeussa-Sebaste in Cilicia, where an epitaph says. *ἐξορκίζομεν ὑμᾶς τὸν ἐπουράνιον θεὸν καὶ Ἥλιον καὶ Σελήνην καὶ τοὺς παραλαβόντας ἡμᾶς καταχθονίους θεοὺς μηδένα κτλ. ἐπενβαλεῖν τοῖς ὀστοῖς ἡμῶν ἕτερον πτῶμα* ("We adjure you by the heavenly god [Zeus] and Helios and Selene and the gods of the underworld, who receive us, that no one [. . .] will throw another corpse upon our bones").[19] Very often the adjuration is abbreviated and the deity is nameless; this is the origin of the well-known formula *τὸν θεόν σοι*, for example at Cotyaeum in Phrygia: *τὸν θεόν σοι, μὴ ἀδικήσεις* ("[I adjure] you by the god, do not harm [this tomb]").[20] This shortened formula was adopted by Christians, but it was never an exclusive Christian use.[21] The verb most frequently used in the funerary adjurations is *ὁρκίζειν*, often with prepositions such as *ἐν-*, *ἐπ-*, *κατ-*, or *ἐξ-*.[22] It is most interesting to note that this same verb also occurs in *defixiones*, in which daemons are frequently adjured.[23] What the gods invoked in the funerary adjurations were expected to do when someone violated the grave is not expressed in the texts but it is beyond doubt that the gods were thought to inflict some kind of punishment on the offender.

The four types of texts discussed so far correspond to the definition of funerary imprecations given above. Sometimes the erectors of the gravestones referred to the inscriptions as *ἀραί* (curses). In an epitaph from Canytelis near Elaeussa-Sebaste in Cilicia the violator of the grave is first warned that he will be *ἠσεβηκὼς εἴς τε τὸν Δία καὶ τὴν Σελήνην* ("impious towards Zeus and Selene") and *ἀσεβὴς εἴς τε τοὺς προγεγραμμένους θεοὺς καὶ τὸν Ἥλιον* ("impious towards the above-mentioned gods and Helios") and this is summarized at the end of the text as *ἐνεχέσθω ταῖς ἀραῖς* ("May he be submitted to these curses").[24] In an inscription from the territory of Olba in Cilicia the usurper is adjured by the gods of the underworld and Helios Patrios (*ὁρκίζω τού⟨ς⟩ χθονίους θεοὺς* [*καὶ*] *τὸν πάτριον Ἥλιον*) and at the end of the text is added, "These are the curses" (*αἵδε ἀραί*).[25] In the imprecations discussed so far the evil wished for is never specified. The nature and the degree of the punishment are left up to the god (or gods) to be decided as if by a judge. For this reason I have called this group the nonspecific group. In the second group of funerary imprecations, which I call the specific group, the evil wished for is more or less clearly specified. The punishment is fixed by the person who has set up the epitaph. As soon as the prohibition against the grave violation is transgressed, the punishment will automatically occur. This evil does not depend on the judgment of a god: it operates directly through the force of the (written) word itself.

The division of the funerary imprecations into a nonspecific group in which divine agents play an important role and a specific group in which a divine agent is not necessarily involved brings to mind the distinction made by K. Latte between two groups of curses.[26] On the one hand Latte discerned a group of *Greek* curses bringing evil for the wrongdoer by the "Zauberkraft des ausgesprochenen Wortes;" if the offender thereby became ἐναγής (unclean) in the sight of the gods, this was only a result and not the essence of the curse. On the other hand Latte saw a group of *Anatolian* curses rooted in a totally different, oriental religiosity. In Asia Minor the offense was regarded as ἁμαρτία (sin); the curse cut off the way to the gods and made the wrongdoer unclean, who subsequently became the focus of the wrath of the gods. As examples of the Anatolian curses Latte cited some formulas of our nonspecific group. The imprecations of the specific group in Asia Minor were explained by him as "übernommene echthellenische Wendungen."[27] Latte's division of the imprecations into two groups with different ethnic origin is not convincing. First of all, it is not certain that in Anatolia offences (including violation of the grave) were regarded as a sin, because the meaning of ἁμαρτία, ἁμαρτωλός as "a sin, a sinner" is questionable, as I have indicated above.[28] Also a division between a more "magical" practice (which would be Greek) and a more "religious" practice (which would be Anatolian) in regard to curses is problematic. It is nowadays generally agreed that such a theoretical distinction cannot be made.[29] I therefore think that there is no fundamental distinction between Greek and Anatolian curses nor between nonspecific and specific imprecations. The distinction I make is only a heuristic one and somewhat artificial. It is, for example, difficult to say whether the wish for the wrath of the gods belongs to the nonspecific or to the specific group. One could argue on the one hand that the intended punishments are not specified but on the other that some evils (such as cruel death, blindness, or natural disasters) were definitely regarded as the result of the anger of the gods and that these well-specified punishments were intended.[30] Likewise the term ἀσεβής (impious), which I have ranged in the nonspecific group, could have well-defined implications in the mind of the ancients, namely the exclusion from taking part in the sacrifices.[31]

I will discuss only the group of specific funerary imprecations because these contain much more detailed and varied information than the standard formulas of the nonspecific group.[32] The number of these specific funerary imprecations is at this moment (as far as I have been able to collect the scattered evidence) somewhat higher than 350.[33] The texts come from all parts of Asia Minor. I propose to study here, after setting the material in its historical and psychological context, the information that the texts give on the curses themselves and on their relation with the "orthodox" religion involving priests, cults, and gods.

THE ANATOLIAN AND GREEK TRADITIONS

The Greek funerary imprecations of Asia Minor are rooted in two traditions: Greek and oriental. All over the Greek world, including the Greek cities of Asia Minor, it was customary to protect material and immaterial objects from potential wrongdoers by means of imprecations.[34] I propose to call these the "nonfunerary" imprecations. In Anatolia the Greek custom coincided with the indigenous oriental tradition: in this country and in the Near East imprecations were commonly employed to protect such things as treaties, statues, and contracts from the time of the Sumerians and the Akkadians onwards.[35] The objects that were safeguarded by nonfunerary imprecations in the Greek world belonged to the public, the religious, and the private spheres, for example property and property rights of individuals and temples, the constitution of a city-state, laws, treaties between cities, *asylia* of temples, private foundations. Some imprecations were directed against enemies of the city or against religious offenders. Many conditional imprecations were imbedded in the self-cursing oath. The number of the nonfunerary imprecations is very large; I will give some examples taken from the oldest attestations (i.e., from the seventh century B.C. onwards).[36]

At Cymae in Italy a Protocorinthian aryballos, dated in the seventh century B.C., is protected from theft by the following imprecation: hòs δ' ἄν με κλέφσει, θυφλὸς ἔσται ("The one who will steal me, will become blind").[37] Near Camirus on the isle of Rhodes a σᾶμα, which is probably a votive monument, is protected from damage: Ζεὺ δέ νιν ὅστις πημαίνοι λειόλη θείη ("May Zeus completely destroy him who injures this"). This text presumably dates from the first quarter of the sixth century B.C.[38] A well-known and very extensive imprecation also dating from the early sixth century B.C. is the Amphictyonic oath. It concerns the plain of Cirrha, which was dedicated to the Delphic gods. The Amphictyons swore an oath not to till the sacred plain nor to let another till it. The text cited by Aeschines is as follows:

> ἐναγὴς ἔστω τοῦ Ἀπόλλωνος καὶ τῆς Ἀρτέμιδος καὶ τῆς Λητοῦς καὶ Ἀθηνᾶς Προναίας. καὶ ἐπεύχεται αὐτοῖς μήτε γῆν καρποὺς φέρειν, μήτε γυναῖκας τέκνα τίκτειν γονεῦσιν ἐοικότα, ἀλλὰ τέρατα, μήτε βοσκήματα κατὰ φύσιν γονὰς ποιεῖσθαι, ἧτταν δὲ αὐτοῖς εἶναι πολέμου καὶ δικῶν καὶ ἀγορᾶς, καὶ ἐξώλεις εἶναι καὶ αὐτοὺς καὶ οἰκίας καὶ γένος ἐκείνων. καὶ μήποτε ὁσίως θύσειαν τῷ Ἀπόλλωνι μηδὲ τῇ Ἀρτέμιδι μηδὲ τῇ Λητοῖ μηδ' Ἀθηνᾷ Προναίᾳ, μηδὲ δέξαιντο αὐτοῖς τὰ ἱερά.

> Let them be under the curse of Apollo and Artemis and Leto and Athena Pronaea. The curse goes on: that their land bear no fruit; that their wives bear children not like those who begat them, but monsters; that their flocks yield not their natural increase; that defeat await them in camp and court and market-place; and that they perish utterly, themselves, their houses, their whole race. And never may they offer pure sacrifice unto Apollo, nor to Artemis, nor to Leto, nor to Athena Pronaea, and may the gods refuse to accept their offerings.[39]

Equally famous are the so-called *dirae Teiae,* imprecations probably making part of a *Bürgereid* at Teus in Ionia (c. 480–450 B.C.). They were directed against those who endangered the city life of Teus, for example, by obstructing the grain supply or by treacherous intrigues or disobedience to the magistrates. Even the steles and the text of the inscription were protected by an imprecation. The intended evil is nearly identical for every kind of offense: *κε̂νον ἀπόλλυσθαι καὶ αὐτὸν καὶ γένος τὸ κένο* ("He will expire, himself and his race").[40]

From the examples given above one important point becomes clear, namely, that many of the funerary imprecations used by the Greeks of Asia Minor to protect their graves correspond very closely to the nonfunerary imprecations of the Greek world: they are rooted in the same tradition. In fact some imprecations such as the wish for death, blindness, infertility of the earth, destruction of the race (and this list could easily be extended) are found in nonfunerary and funerary imprecations alike, sometimes even in identical words. For example the imprecation *μή(τε) γῆν καρποὺς φέρειν*, which occurs in the Amphictyonic oath, is also found in many epitaphs in all regions of Asia Minor.[41] The same formula is also known in *defixiones*.[42] A second important point that must be emphasized is the fact that imprecations were very rarely used to protect graves from violation in the Greek world outside Asia Minor. I know only some twenty cases, of which only two can be confidently dated before the imperial period.[43] The protection of graves by means of imprecations seems to be alien to the Greeks of the Greek homeland. In Asia Minor the practice was very common. Here the second great influence on the funerary imprecations, the oriental tradition, becomes manifest.

In the Near East and in Anatolia there existed a long tradition of protecting the tomb with imprecations.[44] The oldest example is found in the epitaph of the Phoenician king Ahiram, which dates probably to the latter part of the eleventh century B.C. The text of this imprecation, which is directed against the king or governor who might violate the grave, is as follows: "May the sceptre of his rule be torn away, may the throne of his kingdom be overturned, and may peace flee from Byblos; and as for him, may his inscription be effaced (. . .)!".[45] Another Near Eastern example is found in the grave inscription of Sin-zer-ibni, priest of Sahar, who died at Nerab near Aleppo in the early seventh century B.C.: "May Sahar and Shamash and Nikkal and Nusk pluck your name and your place out of life, and an evil death make you die; and may they cause your seed to perish!"[46] The indigenous oriental tradition led to the emergence of funerary imprecations in Asia Minor written in the Lycian language from the sixth to the fourth century B.C. and in the Lydian language in the fourth century B.C. in the period of the Persian supremacy.[47] I cite an example from each of the two epichoric languages. At Antiphellus in Lycia the potential violator is warned; "The assembled (or confederate?) gods and the Lycian treasurer(?) shall punish(?) him!"[48] At Sardis in Lydia the offender is threatened, "Artemus will bring destruction(?) for him, (his) property, land(?)!"[49]

From the moment Asia Minor was liberated from Persian rule, funerary imprecations began to appear in Greek. The oldest instances are found in Lycia at the end of the fourth century B.C. The first comes from Telmessus, the second from Antiphellus. Both are bilingual texts, and the Greek imprecation is generally thought to be inspired by the contemporary Lycian examples. The imprecation written in the Lycian language in the text of Telmessus is difficult to understand; it means something like, "He will punish(?)!"[50] Its Greek counterpart (clearly not a translation of the Lycian text) is: ἐξώλεα καὶ πανώλεα εἴη ἀστῶι πάντων ("May there be for him complete ruin and destruction of all [or everything?]"). The Lycian imprecation in the text of Antiphellus warns, "Let the mother of this precinct(?) here (i.e., Leto) and the municipality(?) of Wehñta (i.e., Phellus) judge(?) him!"[51] The Greek imprecation gives only a partial translation of this text and is much harsher: ἡ Λητὼ αὐτὸν ἐπιτ⟨ρί⟩ψ⟨ε⟩ι ("Leto will destroy him"). Only a few decades later, probably in the first quarter of the third century B.C., comes a third Greek funerary imprecation, which is now in the J. P. Getty Museum (California).[52] Its provenance is not known but on the basis of the names of the two goddesses who will bring the punishment (Artemis Medeia and Ephesia) I have a strong suspicion that it comes from Lydia.[53] If this is right, then the imprecation may be inspired by the Lydian examples of the preceding century. The text of the Getty imprecation is as follows: ἡ Ἄρτεμις ἡ Μηδεία καὶ ἡ Ἐφεσία καὶ οἱ θεοὶ ἅπαντες αὐτὸν καὶ τοὺς ἐγγόνους ("The Median and the Ephesian Artemis and all the gods [no verb expressed] him and his descendants"). During the remainder of the hellenistic period the number of imprecations against grave-violators remained very small.[54] One example from Pinara in Lycia may date approximately between the middle of the second century and the middle of the first century B.C.[55] Another text from the neighborhood of Olba in Cilicia also seems to be hellenistic.[56] It is only in the first century B.C. that some more texts emerge. One example comes from Mytilene on Lesbos,[57] another from Philomelium, a city in Phrygia.[58] There are in addition about fifteen texts that show some Roman influence and that probably should be dated some time after the beginning of the Roman occupation of Asia Minor; but they could equally well belong to the Roman imperial period.[59]

A very large number of funerary imprecations can be assigned with certainty or with good probability to imperial times.[60] The attribution of the texts to different centuries, however, is problematic, for often the only criteria that are available, such as the letter forms or the personal names, are not very reliable. As far as I have been able to date the texts, the following results appear. Fifteen texts may date in the first century A.D., while twenty-three date in that century or later. Fifty-seven texts may date in the second century A.D., while thirty-two may date in that century or later. Another forty-five belong to the second or third century. Ninety-one texts seem to date in the third century or the early fourth century A.D. Only two or three texts certainly date in the (early) fourth century A.D. It is not at all certain that the growing number of imprecations is a sign of an increase in their popularity or in the belief in their efficacy, and it is dubious whether the growth reflects an increasing need to protect the graves from violation. As R. MacMullen has recently warned, the frequency of epigraphic attestation of behavior or activities does not permit us to draw conclusions about their actual prominence, decline, or the like; apart from economic and demographic factors, the number of inscriptions was influenced by "epigraphic habit," which was controlled by many forces, such as urbanization and hellenization, literacy and culture, fashion and psychological attitude.[61] These factors fluctuated over the centuries and probably varied from one region of Asia Minor to the next. This could explain the fact that in northeastern Lydia the greatest number of funerary imprecations (eleven of eighteen attestations, all exactly dated by the Sullan or Actian era) date to the last quarter of the first century A.D. and the first quarter of the second century A.D. It is interesting to note, however, that the period of greatest frequency of funerary imprecations in this region does not coincide with the period of greatest frequency of epitaphs.[62] Apparently the habit of having *imprecations* and that of having *epitaphs* inscribed on the grave were influenced by different factors.[63]

An important point I have so far neglected, though it is fundamental for the understanding of the funerary imprecations of Asia Minor, is the question why the Greeks of Anatolia so frequently protected their graves from violation with imprecations while the Greeks of the Greek homeland did it very rarely. I would suggest that a difference in ideas about the dead and the afterlife may be responsible.[64] From earliest times the Anatolians sometimes built the tombs of the deceased in the shape of the houses of the living. The custom to bury the dead in a "grave house" is already attested in the second part of the third millennium B.C. in the graves of the dynasts of Alacahüyük and Gedikli.[65] In Hittite texts of the fourteenth to thirteenth centuries B.C. one reads that the ashes of the dead king were placed in a house of stone (a mausoleum or a rock-cut chamber).[66] The concept of the grave house was widely spread in the Phrygian and Lydian periods and continued on in hellenistic times, during which it manifested itself for example in sarcophagi with architectural ornaments. In the Roman imperial period the concept gave rise to the representation of a door on gravestones in some parts of Phrygia.[67] The fact that the tomb was built in the form of a house implies a certain idea about the afterlife, namely, that the dead body continues living—that it still has feelings, needs, and desires.[68] The dead body needs a house to live in; this grave house has to stand and to remain undisturbed forever. Therefore the Anatolians protected it from violation.

The ideas of the Greeks in the Greek homeland were different. As far as it is possible to learn anything about the original ideas of the Greeks,[69] it looks as if they did not attach the same importance as the Anatolians to material aspects of the afterlife. They were more concerned about the burial, the funerary rites and the remembrance of the name. The Greeks seem to have believed that the psyche (soul) left the body at the moment of death and went down to the underworld. There it lived a life that was only a weak reflection of the existence on earth.[70] As long as the name of the deceased person as an individual continued to be remembered, the psyche had an individual life and the deceased enjoyed a kind of immortality.[71] As to the dead body, the Greeks imagined that it stayed in two places alike, in the grave and in the underworld. In the underworld the dead body did not cease to live, but it did not know any more the needs and desires of the living. Only great offenders were thought to have corporal feelings while being punished in Tartarus.[72] The body, as far as it was thought to live on in the grave, was equally believed to be insensible—free from corporal feelings and needs.[73]

The ideas about the afterlife may account for the almost fanatical concern of the Anatolians for the fate of the grave. They protected their tombs by means of legal measures (such as fines), imprecations, or both at the same time. Why some persons preferred imprecations is of course unknown. Perhaps it had something to do with the belief in the inefficacy of civil justice.[74] Indeed, the violator of the grave did his criminal work in the cemetery outside the town and therefore had a good chance to escape unnoticed and unpunished.[75] It has been noted that in circumstances or places or periods in which human law is vitiated by its powerlessness, unsteadiness, partiality, or even absence, people who suffer injustice often resort to curses as a means of *Selbsthilfe*.[76] By the power of the curse, which operates independently from the human law, the culprit nevertheless gets the due penalty.

It is beyond doubt that the authors of the funerary imprecations were moved by noble sentiments and that they aimed at entirely justified goals. It is however remarkable that they asked only for the punishment of the offender, that is, revenge; they never requested that the harm done to the tomb or to the corpse should be repaired. In this respect the funerary imprecations resemble the prayers for vengeance discussed by H. S. Versnel (chap. 3). It is not surprising, therefore, that some of the terms used in the latter are also found in the former. For example the verb *μετέρχομαι* (punish) occurs in a funerary imprecation at Kalos Agros between Chalcedon and Nicomedeia in Bithynia: *μετελθῇ αὐτὸν ὁ θεός* ("May the god punish him").[77] A term derived from the verb *κολάζω* (inflict a punishment) is attested at Assus in Mysia: *αὐτοῖς πᾶ[σαν ἀκ]ο̣λ̣ο̣υ̣θῆσαι κόλασιν* ("[I ask] to pursue them with all punishment").[78] And the wish that the offender not be concealed from the god Helios is found in an imprecation at Parium in Mysia: *μὴ λάθυ τὸν Ἥλιον ἀλλὰ πάθυ ἃ καὶ αὐτή* ("May he not stay hidden from Helios, but may he suffer what she [has suffered]").[79]

THE POWER OF WORDS

The force of a curse is based on a more general belief in the efficacy of the word. This power is increased if the word is spoken by a person of higher status, such as a king, a priest, parents, the dying, or the dead.[80] Funerary imprecations against violators of the grave must have been regarded as very powerful, since they were the wish of the dying or the dead. In some inscriptions it is explicitly stated that the prohibitions and the imprecations are recorded in the testament, as is the case in an epitaph from Halicarnassus on the Doric coast of Caria, where one reads, *κατὰ τὰς ἐν ταῖς διαθήκαις ἀράς* ("according to the curses in the will").[81] In one case in Nacrason in Mysia the text of the inscription with the imprecations is a copy of the will itself.[82]

The force of the cursing word could be increased by a variety of rhetorical devices, such as repetition, rhythm, and the use of triplets. These phenomena are common to both "magical" and "religious" liturgy.[83] In the funerary imprecations a word is often repeated with only slight variations, such as *ὤλης ἐξώλης ἀπόλοιτο* ("May he die, dead and gone")[84] or *κακὸς [κα]κῶς ἐξώλης γένοιτο* ("May he, an evil man, be evilly destroyed").[85] Many funerary imprecations are metrical. This indicates nothing special when the epitaph itself is metrical, but often a prose epitaph is followed by a metrical (interdiction and) imprecation. Such is the case with the so-called North Phrygian curse formula, *τὶς ἂν προσοίσει χεῖρα τὴν βαρύφθονον, οὕτως ἀώροις περιπέσοιτο συμφοραῖς* ("Whoever will lay a hand heavy of envy against [this tomb], may he fall foul in the same way of untimely fates [i.e., of the fate of untimely dead children]"), in iambic trimeters,[86] with the East Phrygian curse formula, *ὀρφανὰ τέκνα λίποιτο χῆρον βίον οἶκον ἔρημον*; ("May he leave orphaned children, an empty [i.e., childless] life, a desolate house behind him"), a (bad) dactylic hexameter;[87] and with some other imprecations.[88] In many imprecations three elements of the malediction are put very closely together, such as *οἴκῳ βίῳ τῷ σώματι αὐτοῦ* ("for his house, life, body")[89] or *ἄτεκνος ἄτυμβος ἀναυχίστευτος ὀλῖται* ("He will die without children, tomb, relatives").[90] The triple repetition of a funerary imprecation is not attested, but such a repetition occurs in adjurations.[91] In Lycaonia, near the city of Perta, the god who will bring the punishment—namely, Men—is once augmented to nine Mens: *ἕ[ξει] κεχολωμέν[ους]* Μῆνας *αἰνέα* ("He will incur the anger of the nine Mens").[92]

The power of a curse could also be enhanced by accompanying gestures, such as the touching of the earth or of the accursed person, the performance of a sympathetic action, or the raising of the hands.[93] This last act is perhaps once attested in the text of a funerary imprecation. In an epitaph near Hadrianutherae in Mysia the owner of the tomb says, *χεῖρας ἀεί[ρω]* ("I raise my hands").[94] Raised hands are depicted on three gravestones that have an imprecation. It is well known that this gesture is frequent on the tombs of children and young persons, who, it seemed, could not have died a natural death but must have been killed in a criminal way (if not taken away by a god). The raising of the hands is the symbol of the invocation to Helios for divine vengeance.[95] This explanation of the raised hands, however, seems excluded in at least one of the three cases, a tomb near Laodiceia Combusta in Phrygia erected by a daughter for her parents; there is no sign that they died a premature or a violent death.[96]

A curse could become so powerful that it became a bad daemon, Ἀρά ("Curse").[97] Such a personification of the curse is only once attested in a tomb imprecation, namely at Neocaesareia in Pontic Cappadocia. This epitaph, which contains very extensive imprecations, was set up by an intellectual who had studied in Athens under Herodes Atticus. He was largely inspired by the maledictions that Herodes had engraved on herms in Attica for the protection of the statues of his dearest departed.[98] The author of the epitaph made several additions to his example. One of these is that Ἀρὰ *ἡ πρεσβυτάτη δαιμόνων* ("Curse, the oldest of daemons"), together with other gods, will penalize and hurry on the violator of the grave.[99] Ara as a daemon was sometimes identified with Erinys; she apparently had her home in Hades.[100]

The power of a curse is always two-sided: the word can bring harm but also profit.[101] There is a very narrow relation between cursing and blessing and both frequently occur together in many cultures, even in a funerary context.[102] In the funerary imprecations of Asia Minor the aspect of the blessing is almost completely absent. In a text from the neighborhood of Pissia in Phrygia a blessing follows the imprecation: *ὅσα εὖ ἐμοί, διπλᾶ σοι θεός* ("The good [you do] to me, god [will give back] to you in double").[103] In the above-mentioned inscription from Neocaesareia the imprecations are followed by blessings for those who preserve the prescriptions without alterations and observe them: *πολλὰ καὶ ἀγαθὰ εἶναι τούτωι καὶ αὐτῶι καὶ πατρίδι καὶ οἴκωι καὶ τῆι ἔπειτα μνήμηι καὶ ἐκγόνοις* ("And much good will come for him, for himself and for his fatherland and for his house and for his remembrance later and for his posterity").[104]

An essential characteristic of the curse is that it is irrevocable: once it has been spoken, it usually cannot be stopped. The only ones who had the power to revoke the curse and to finish the punishment were the persons who spoke the curse and the god(s) who were invoked.[105] Apparently the gods could be placated by the offender by some kind of atonement, for example a sacrifice or, in northeastern Lydia, a confession, but even this possibility is expressly denied in a small number of funerary imprecations.[106] In a fragmentary epitaph from the neighborhood of Eumeneia in Phrygia it is wished that the violator may become the irreconcilable (*ἀδιάλυ[τος]*) enemy(?) of the gods(?).[107] In Nacrason in Mysia one of the elements of an extensive imprecation is that the gods and the heroes will be enraged and implacable (*ἀνεξειλάστους*).[108] Near Saittae in Lydia an imprecation contains the wish that the transgressor will find (Men) Axiottenos implacable through the generations (*[ἀ]νεξείλαστον τέκνα τέκνων*).[109] And in Tabala in Lydia it is said that the violator will find enraged the insoluble (*ἄλυτα*) scepters (of the gods) in Tabala.[110]

Another characteristic of a curse is that it often strikes not only the wrongdoer himself but also his *οἶκος* (house, household) or his *γένος* (posterity), even when they are totally innocent or yet unborn.[111] In a very large number of funerary imprecations it is wished that the relatives of the offender will perish or that they will have a curse or will suffer from the wrath of the gods together with the violator, as in the following text from Aphrodisias in Caria: *ἐξώλη ἀπόλοιτο σὺν τέκνοις καὶ παντὶ τῷ γένει* ("May he die, dead with his children and with all his posterity").[112] But it also often happens that the wish for evil affects only the children or the relatives of the violator, so that the latter will be a witness of their untimely death or misery. An example of such a wish is *ἄωρα προθοῖτο τέκνα* ("May he place upon a bier his untimely dead children") in an epitaph from the neighborhood of Appia or Alia in Phrygia.[113] For the extension of the imprecation to the descendants special abbreviated formulas were in use, such as *τέκνα τέκνων* (or *τέκνοις*) (to his children's children).[114] The material property of the violator is sometimes equally affected by the imprecation, as it is part of the *οἶκος*. Most often the evil wish is for the destruction of the possessions, so that the violator stays alive but in utter misery. This may be illustrated by an example from Nacrason in Mysia: *πρόσρειζα δὲ καὶ πανόλεθρα ἀρθείη καὶ ἀφανισθείη πάντα* ("May everything from its roots and with total ruin perish and disappear").[115]

Very rarely funerary imprecations are directed against the whole society in which the violator lives. It may occur in the epitaph from Neocaesareia, where one of the imprecations is, *μὴ γυναῖκες τίκτοιεν κατὰ φύσιν* ("May the wives not bear children according to nature").[116] This imprecation asks that the children that are born may not resemble their parents but will be deformed, monstrous. The birth of such children was regarded as a sign of the wrath of the gods against the community as a whole.[117] But it is not at all certain that the author of this text really intended to strike the whole society. In fact this malediction is placed between other imprecations in which the violator alone is the object. The author of the text may thus have intended to strike only the wife of the offender. The use of the plural *γυναῖκες* (wives) may be caused by the fact that the author has taken over this imprecation from very old oaths, such as the Amphictyonic oath cited above, without making the necessary adaptations.[118] In another text from the territory of Nicaea in Bithynia it is asked that the violator *λυμῷ γε . . . τ . . δὲ ἐξαπόλοιτο* ("May die from an epidemic disease[?]").[119] If the meaning of the wish is properly understood, a large number of people from the society must have been the object of the attack, for it is characteristic of a plague to infect numerous people.[120]

As a result of the dangerous contagion, a cursed person had to be banished from society, usually to a place far away from human habitation.[121] There may be an allusion to such an expulsion in the inscription from Neocaesareia, in which is said, *τούτῳ μὴ πατρὶς οἰκοῖτο* ("May his fatherland not be inhabited by him").[122]

In northeastern Lydia the cursing of the potential violator of the grave sometimes went together with the erection of a scepter or several scepters. This is attested in a steadily growing number of texts.[123] The most striking example, in which the procedure is most fully described, comes from the territory of Silandus: *ἐπηράσαντο μή τις αὐτοῦ τῷ μνημείῳ προσαμάρτῃ, διὰ τὸ ἐπεστάσθαι σκῆπτρα* ("They have established an imprecation that no one should do wrong to his grave monument, through the erection of scepters").[124] In a second text from the neighborhood of Saittae the scepter itself speaks the imprecation: *ὃς ἂν τοῦτο ἄρῃ ἢ κατεάξῃ, τῶν θεῶν κεχολωμένων τύχοιτο· περὶ τούτου σκῆπτρον ἐπηράσ⟨α⟩το* ("Who will displace or break this [gravestone], may he find the gods enraged; concerning this the scepter has established an imprecation").[125] A third text, again from the territory of Silandus, does not mention the imprecation: *ἵνα μή τις προσαμάρτῃ τῇ στήλῃ ἢ τῷ μνημείῳ, σκῆπτρα ἐπέστησαν τοῦ Ἀξ[ι]οττηνοῦ καὶ Ἀναείτιδος* ("In order that no one should do wrong to this stele or to this grave monument, they have set up scepters of [Men] Axiottenos and Anaeitis").[126] Apart from these texts, in which the erection of a scepter is explicitly mentioned, there are several funerary imprecations in which it is wished that the violator will find the scepters enraged.[127] It is not certain that we have here the same procedure; possibly the wrath of the scepters is only a variation of the wrath of the gods. According to a common belief the gods are really embodied in their representations.[128] The erection of scepters is also mentioned in confession inscriptions from northeastern Lydia.[129] In one case it is done in order to prevent a crime (theft of clothes) from being committed; in three cases the crime (theft, poisoning) has been committed but the culprit is unknown or the suspect denies guilt. All the cases, funerary and expiatory, have in common the fact that the wrongdoer is not known.[130] This again illustrates the above-mentioned function of the imprecations in relation to the execution of justice.

The scepter was undoubtedly erected as the symbol or the incarnation of the judicial power of the god. This is an old and widespread image.[131] By this action the crime was transferred to the juridical authority of the god in order that the offender might be unmasked and punished. The erection of the scepter presumably was the work of the priest: two reliefs on expiatory steles show the priest with a long stick, which must be the scepter of the god.[132] The spot where the scepter was placed is not known. It does not seem likely that the holy object was set up near the place of the (future) crime (in funerary context, the grave). It is more probable that it was erected inside the temple area.[133] In the first and second of the above-cited texts there is a close connection between scepter and imprecation. This is also the case in one confession inscription: a scepter is erected and apparently simultaneously imprecations are placed in the temple.[134] These were perhaps written on a πιττάκιον (tablet).[135] One can suppose that the tablet, which is generally assumed to contain an exposition of the issue, its transference to the god, and the imprecations for the culprit, was fixed in an easily visible place in the temple area so that everyone could read it.[136] It is not known if the same procedure was followed in funerary cases, but it does not seem improbable. After the wrongdoer had been punished by the god and had confessed his fault or paid an atonement, the divine involvement and the imprecations could be dissolved. This was done by the removal of the scepter (λυθῆναι τὸ σκῆπτρον) and the imprecations, as is attested in an expiatory text.[137] The mention of ἄλυτα σκῆπτρα (insoluble scepters) in a funerary imprecation seems to refer to a similar possibility of stopping the punishment, a possibility that is here, however, expressly denied.[138]

Apart from the erection of scepters in northeastern Lydia there is no sign that the priests played a role with regard to funerary imprecations. As to the influence of the imprecations upon the cult of the gods, for example, upon the sacrifices, there is no such explicit mention in the texts.[139] The failure of the sacrifice of the wrongdoer may be implied in the very common wish that the gods will be angry (κεχολωμένοι or the like.)[140] and in the less frequent formula that the gods will not be well disposed (ἵλεως).[141]

FUNERARY IMPRECATIONS AND THE GODS

In almost one-third of the funerary imprecations a god or several gods are named and are expected to inflict the punishment on the wrongdoer. Formal prayers to the gods to take action against the violator of the tomb are rare. A person from Eirenopolis in Cilicia prays, καταρῶμαι τοὺς καταχθονίους ὠλίαν ἐξωλίαν ("I pray to the gods of the underworld for utter destruction").[142] A citizen of Nicaea in Bithynia, who died and was buried in Philippopolis in Thrace, prays, [ἐν]εύχομαι τῷ Κενδρεισῳ Ἀπόλλωνι—πανσπερμεὶ ἐ[ξολέσθαι] ("I pray to Apollo Cendreisos to destroy [the violator] with all his seed").[143] And a woman from the neighborhood of Nacoleia in Phrygia addresses the following order to Helios Teitan: τὴν αὐτὴν [χ]άριν ἀντάποδος ("Do him the same 'favor' in return").[144] In none of the texts is there any sign of submissiveness on the part of the authors to the mighty gods, which is a characteristic aspect of prayers for justice or vengeance.[145]

The gods named in the tomb imprecations may be anonymous (θεός, θεά, θεοί) or specified by name, for example, Apollo, Hecate, Helios, Leto, Men Axiottenos, Nemesis, Pluto, Selene, or Zeus Olympios. About thirty different gods are named in the texts. Some gods appear only once or twice. This is occasionally due to the fact that the imprecation was set up by an intellectual who diverged from popular belief. Two examples are the imprecation engraved by the man from Neocaesareia, who studied in Athens under Herodes Atticus, and an imprecation set up by a certain P. Varius Aquila, a Roman citizen from Assus in Mysia who was obviously influenced by the Second Sophistic. The first text contains the unique mention of

Hermes Chthonios and Zeus Olympios.[146] The second text mentions Ge, Kore, and Pluto,[147] the first of whom does not occur in any other funerary imprecation, while the two other gods are found again in the text of Neocaesareia alone. In some cases a god is mentioned only once because he or she was a foreign god.[148] Thus there is only one attestation of Atarknateis (= Atargatis), a North Syrian goddess, who is obviously to be connected with a family or group of Syrian immigrants (in the region of northeastern Lydia).[149] Another unique mention is that of the *θυοὶ Περσῶν* (gods of the Persians) at Acipayam in Pisidia.[150] The Persian gods were introduced in the valley of Acipayam by Persian colonists.[151] A third example is the goddess Daeira, mentioned in the inscription of Neocaesareia; she is an Athenian divinity belonging to the Eleusinian cult. Her name must have been picked up by the author of the text during his stay in Athens.[152] In some cases a god or a group of gods is rarely mentioned because the god was only a local god, worshipped by a small group of devotees, such as the gods in Iaza, the scepters (= gods) in Tabala, or the gods in Tamasis (all in northeastern Lydia).[153]

Some gods are mentioned more or less frequently in the funerary imprecations. The most "popular" gods were the *καταχθόνιοι θεοί* (gods of the underworld). In a number of texts they are named together with the *οὐράνιοι θεοί* (heavenly gods). The gods of the underworld and heaven joined together formed the group of all gods,[154] the *πάντες θεοί* (or, briefly, *θεοί*), who are also frequently mentioned in the funerary imprecations. I think it is fair to say that the gods as a whole and the gods of the underworld were the most important agents mobilized to act for the punishment of the violator. In second place come the lunar gods, Men and Selene, and the related goddess Hecate with her Erinyes. The sun god Helios ranks third, often in the company of a lunar god. Less popular are Zeus and Meter, the Anatolian mother goddess who occurs under different names and forms in the texts, such as Leto, the Pisidian goddess, and perhaps also Anaeitis.

The reason why a specific god was chosen by an individual to act as agent is almost never mentioned in the texts.[155] The gods may have been chosen because the Greeks were convinced that the gods punished all crimes, especially the crimes against themselves. The gods of the underworld may have been chosen because the dead, having departed from the world of the living, belonged to the realm of the chthonic gods.[156] These were not only gods who had their home in Hades, as Pluto, Hecate, or Men Katachthonios, but also gods who were in some way related to the underworld, such as the lunar god, the sun god, and all fertility gods.[157] A number of these gods, like the *καταχθόνιοι θεοί*, Hecate, and the Erinyes also play a role in *defixiones*.[158] A third group of gods (Helios, Zeus Olympios, and Nemesis) may have been preferred because they were all-seeing gods: they saw everything that happened on earth, even the hidden crimes. Moreover these gods were truth-loving gods and executors of revenge. In prayers for vengeance they are frequently invoked, especially Helios.[159] The gods of this third group defended justice; their role in the funerary imprecations once more illustrates the connection of these texts with the execution of justice.[160] The choice of a particular god by the person who set up an imprecation depended on many factors that we cannot uncover now. Personal religiosity may have played a role, but the personal preference seems to have been strongly influenced by the local custom. There are regions in Asia Minor in which a marked preference existed for one or another god, for example, for Men Katachthonios in eastern Phrygia and Lycaonia, for Selene in western Cilicia and for the Pisidian gods (who are presumably Hecate and Helios) in the valley of Acipayam in West Pisidia.

Notes

The following oft-cited studies will be referred to by author's last name and date of publication:

A. E. Crawley, "Cursing and Blessing," in *Encyclopaedia of Religion and Ethics* (New York, 1911), 4:367–74
K. Latte, *Heiliges Recht: Untersuchungen zur Geschichte der sakralen Rechtsformen in Griechenland* (Tübingen, 1920).
P. Moraux, *Une imprécation funéraire récemment découverte à Néocésarée* (Paris, 1959).
A. Parrot, *Malédictions et violations de tombes* (Paris, 1939).
L. Robert, "Malédictions funéraires grecques," *CRAI* (1978): 241–89.
W. Speyer, "Fluch," *RAC* 7 (1969): 1160–1288.
J. Strubbe, "Vervloekingen tegen grafschenners," *Lampas* 16 (1983): 248–74.
R. Vallois, "APAI," *BCH* 38 (1914): 250–71.
H. S. Versnel, 'May he not be able to sacrifice . . . ': Concerning a Curious Formula in Greek and Latin Curses," *ZPE* 58 (1985): 247–69.

1. For the full text of the epitaph, see G. E. Bentley, *Shakespeare: A Biographical Handbook* (New Haven, 1961), 67. For its purpose, see S. Schoenbaum, *Shakespeare's Lives* (Oxford, 1970), 4.

2. W. H. Buckler, W. M. Calder and W. K. C. Guthrie, *Monumenta Asiae Minoris Antiqua* [*MAMA*], vol. IV (Manchester, 1933), no. 84, lines 4–6. The editors restore in line 5, σ[κυβλίσε]ι (will desecrate); one could also think of σ[κυλεύσε]ι or σ[κυλήσε]ι (will rob).

3. P. Moraux, *Une défixion judiciaire au Musée d'Istanbul,* Mém. Ac. Roy. Belg. vol. 54, pt. 2 (Brussels, 1960), 3–5. It is clear that funerary imprecations differ in many ways from *defixiones;* see the essays of C. Faraone and H. S. Versnel (chaps. 1, 3). But in the course of this article some points of contact will appear, for example in terminology.

4. Under this rubric, however, I include not only the imprecations written down by pagans, but also the imprecations engraved by Christian and Jewish people in so far as they took over formulas and ideas from the pagan world. For an example of the use of a pagan formula by a Christian see W. M. Ramsay, *The Cities and Bishoprics of Phrygia,* vol. I, pt.2 (Oxford, 1897), no. 658, frag. a, line 24 (from the Phrygian Pentapolis): ἄ⟨ω⟩[ρ]α τέκνα ⟨ἔ⟩χωσι (or ⟨τύ⟩χωσι) ("May they have untimely dead children"). The use by a Jew is attested, for example, in W. H. Buckler and W. M. Calder, *MAMA*, vol. VI (Manchester, 1939), no. 316, frag. b, lines 2–5 (from the territory of Acmonia in Phrygia): λάβοιτ[ο ἀπρ]οσδόκητον ὅ[ποῖ]ον καὶ ὁ ἀδελφὸς αὐτῶν ("May he receive a sudden death, just as their brother").

5. E. Kalinka, *Tituli Asiae Minoris* [*TAM*] vol. II, pt.1 (Vienna, 1920), no. 51, lines 15–20. The variant ἠσεβηκώς is found, for example, at Canytellis near Elaeussa-Sebaste in

Cilicia; see R. Heberdey and A. Wilhelm, *Reisen in Kilikien, ausgeführt 1891 und 1892,* DenkschrWien, vol. 44, no. 6, Philos.-hist. Kl. (Vienna, 1896): pts. 58–59, no. 133, lines 8–10: *ἔστα[ι] ἠσεβηκὼς εἴς τε τὸν Δία καὶ τὸν Ἥλιον καὶ τὴν Σελήνην καὶ εἰς τὴν Ἀθηνᾶν* ("He will have committed an act of impiety towards Zeus and Helios and Selene and Athena").

6. Ph. Le Bas and W. H. Waddington, *Voyage archéologique en Grèce et en Asie Mineure: Explication des inscriptions,* vol. III. pt. 5, *Asie Mineure* (Paris, 1870), no. 1299, lines 4–6. For *ἱερόσυλος* see, for example, E. Kalinka, *TAM* II.2 (Wien, 1930), no. 521, lines 16–25 at Pinara in Lycia. For *ἱεροσυλία* see, for example, W. Judeich, in *Altertümer von Hierapolis,* ed. C. Humann, C. Cichorius, et. al. (Berlin, 1898), no. 270, line 3 at Hierapolis in Phrygia: *ἔνοχος ἔσται ἱεροσυλίᾳ* ("He will be liable of sacrilege"). The combination of *ἀσεβής* and *ἱερόσυλος* is also attested, for example, *ἀσεβὴ[ς] ἔσται κα[ὶ ἱ]ερόσυλος* ("He will be impious and sacrilegious"), at Sidyma in Lycia; cf. Kalinka (see, n. 5), no. 221, lines 13–14.

7. C. Naour, *Tyriaion en Cabalide: Epigraphie et géographie historique* (Zutphen, 1980), 74, no. 31, lines 8–9. One finds *ὑποκεῖσθαι* in the same region: R. Heberdey and E. Kalinka, *Bericht über zwei Reisen im südwestlichen Kleinasien,* DenkschrWien vol. 45, no. 1, Philos.-hist. Kl. (Vienna, 1897): 7, no. 20, lines 6–7: *ὑ[π]οκείσθω Ἡλίῳ καὶ Σελήνῃ* ("He will fall under Helios and Selene").

8. E. Kalinka, *TAM,* vol. II, pt. 3 (Wien, 1944), no. 1081, lines 3–5.

9. For violation of the grave as a judicial offense and its punishment, see E. Gerner, *ZSav* 61 (1941): 237–43, 247–48; J. S. Creaghan, *Violatio Sepulcri: An Epigraphical Study* (Ph.D. diss., Princeton University, 1951, microfilm), 116. The potential wrongdoer is often threatened with a charge for *τυμβωρυχία* (violation of the grave) or *ἀσεβεία* (impiety) or *ἱεροσυλία* (sacrilege), for example, at Termessus in Pisidia (R. Heberdey, *TAM* vol. III, pt. 1 [Wien, 1941], no. 224, lines 11–12): the offender will pay a fine *καὶ ἔνοχος ἔσται ἐνκλήματι τυμβωρυχίας* ("and he will fall under the charge of violation of the grave"). We know that in some cities there existed a *νόμος τυμβωρυχίας* or *ἀσεβείας* (law concerning violation of the grave or impiety), for example, at Olympus in Lycia; cf. Kalinka (see n. 8), no. 953A, lines 7–8: the offender will pay a fine *καὶ ὑποκεῖσθαι αὐτὸν τῷ τῆς τυμβωρυχίας νόμῳ* ("and he will fall under the law of grave robbery"). At Sidyma in Lycia the wrongdoer will pay a fine *καὶ ὑ[ποκ]ειμενος [ἔστω] τῷ τῆς ἀ[σε]βείας νόμῳ* ("and he will fall under the law of impiety"); cf. Kalinka (see n. 5), no. 246, lines 23–25.

10. Le Bas and Waddington (see n. 6), no. 1275, lines 9–11.

11. Kalinka (see n. 8), no. 923, line 4.

12. K. H. Rengstorf, "*ἁμαρτωλός*," *TWNT* 1 (1933): 321–22. The scholars who interpret *ἁμαρτωλός* as "sinner" consider the violation of the grave as a sin, for example, W. Arkwright (*JHS* 31 [1911]: 271 and 277) and Parrot (1939, 107). *Ἁμαρτωλός* is also attested in *defixiones*; see chap. 3, p. 64, where the term seems to denote "the culprit, the criminal." See also n. 28.

13. Heberdey and Kalinka (see n. 7), 8, no. 23, lines 2–3. In Termessus in Pisidia the offender will have to reckon with the deceased, for example, Heberdey (see n. 9), no. 365, lines 4–6: *ἢ ἔσται αὐτῷ πρὸς τοὺς κατυχομένους* (and he will pay a fine).

14. Heberdey (see n. 9), no. 509, lines 8–11.

15. Buckler, Calder, and Guthrie (see n. 2), no. 360, lines 11–13.

16. *SEG* VI, no. 300, lines 11–13.

17. For these formulae, commonly called the "Eumeneian" and the "Laodiceian" formulae, see M. Waelkens, in *Actes du VIIe congrès international d'épigraphie grecque et latine, Constantza 1977,* ed. D. M. Pippidi (Paris, 1979), 126–28; A. Strobel, *Das heilige Land der Montanisten* (Berlin, 1980), 74–83. The formula *δώσει λόγον τῷ Θεῷ* ("He will give an

account to God"), however, was exclusively Christian; see L. Robert, *Hellenica,* vols. XI–XII (Paris, 1960), p. 407.

18. E. N. Lane, *Corpus Monumentorum Religionis Dei Menis,* I (Leiden, 1971), 98–99, no. 156, lines 10–18.

19. J. Keil and A. Wilhelm, *JOAI Beib.* 18 (1915): 45–48, lines 3–9.

20. E. Pfuhl and H. Möbius, *Die ostgriechischen Grabreliefs,* vol. II (Mainz, 1979), 517–18, no. 2161C, lines 1–4. For the formula and its interpretation, see L. Robert, *Hellenica,* vol. XIII (Paris, 1965), 100–3.

21. Cf. Waelkens (see n. 17), 127.

22. Also attested is (ἐν)ὁρκῶ. Other verbs such as ἐπαρῶμαι, are rarely used. (Of the latter there is an attestation at Germanicopolis in Cilicia: *SEG* VI, no. 784.)

23. A. Audollent, *Defixionum Tabellae* (Paris, 1904), lviii and index 474–76; Th. Drew-Bear, *BASP* 9 (1972): 93–94.

24. E. L. Hicks, *JHS* 12 (1891): 231, no. 11 (revised by Heberdey and Wilhelm [see n. 5], 58, n. 2), lines 5–6, 9–10, 10–11.

25. J. Keil and A. Wilhelm, *MAMA*, vol. III (Manchester, 1931), no. 56, lines 1–2, 4. There is no need to interpret the last words as "These are the Arai [i.e., the Erinyes]" as does Parrot (1939, 121); cf. Creaghan (see n. 9), 50.

26. Latte 1920, 77–78.

27. Ibid., 78 and n. 48.

28. The curses cited by Latte do not at all prove that the offence was a sin; there is no indication in the texts that someone called ἀσεβής (impious) was really considered a sinner. Concepts of sin may have existed in some regions of Asia Minor, for example in northeastern Lydia (cf. n. 128), but I do not think that it was common to the whole of Asia Minor. For a critique of Latte, see also G. Björck, *Der Fluch des Christen Sabinus*, Papyrus Upsaliensis 8 (Uppsala, 1938), 109.

29. H. S. Versnel, *Mnemosyne* 29 (1976): 395–96 and 389, n. 65 with bibliography; Speyer 1969, 1163–1164. For a summary of recent discussion on the concepts of "magic" and "religion," see H. S. Versnel, *Lampas* 19 (1986): 68–71.

30. Strubbe 1983, 266 with bibliography in n. 102; R. Parker, *Miasma: Pollution and Purification in Early Greek Religion* (Oxford, 1983), 235–36; Speyer 1969, 1187, 1189.

31. On this aspect see Versnel, 1985, 247–69.

32. In the specific group I have included all the funerary imprecations except the above-mentioned formulas of the nonspecific group.

33. I have edited and commented on all these texts in my doctoral dissertation *Arai Epitymbioi: Een uitgave en studie van de heidense vervloekingen tegen eventuele grafschenners in de Griekse funeraire inscripties van Klein-Azië* (Gent, 1983). Here I cite only the most important earlier editions that are readily accessible. The main results of my investigations are published in Strubbe 1983, including a map of Asia Minor with the ethnological boundaries of the different peoples (p. 252) and a survey of the geographical distribution of the imprecations over the different regions (pp. 252–53).

34. Much material is collected in E. Ziebarth, *Hermes* 30 (1895): 57–70; Latte 1920, 61–80; Speyer 1969, 1203–9; M. Guarducci, *Epigrafia Greca*, vol. IV, *Epigrafi sacre pagane e cristiane* (Roma, 1978), 222–36.

35. Collections of material can be found in the doctoral dissertation of S. Gevirtz, *Curse Motifs in the Old Testament and in the Ancient Near East* (University of Chicago, 1959) and in an article by the same author in *Vetus Testamentum* 11 (1961): 137–58; see also Speyer 1969, 1170–74. For the relation between funerary and public imprecations, cf. Björck (see n. 28), 107–11.

36. Some "nonfunerary" imprecations may be even older, for example, the Attic Bouzy-

gean curses, which are said to have been instituted by the mythical ancestor Bouzyges; see Speyer 1969, 1204.

37. P. Friedländer and H. B. Hoffleit, *Epigrammata: Greek Inscriptions in Verse, from the Beginnings to the Persian Wars* (Berkeley, 1948), no. 177C.

38. P. A. Hansen, *Carmina Epigraphica Graeca Saeculorum VIII–V a. Chr. n.* (Berlin, 1983), no. 459.

39. Aeschin. *Or.* 3. 111; translation by Ch.D. Adams, *The Speeches of Aeschines* (London, 1948), 393–95. Many elements of the Amphictyonic oath occur in hellenistic oaths; see Moraux 1959, 20–22. A new parallel can be added: M. Wörrle, *Chiron* 8 (1978): 201–46 (*SEG* XXVIII, no. 1224), from Telmessus in Lycia.

40. W. Dittenberger, *SIG,* 3d ed. (Leipzig, 1915), no. 37–38; P. Herrmann, *Chiron* 11 (1981): 1–30 (*SEG* XXXI, no. 984–85).

41. For example, at Seleuceia on the Calycadnus in Cilicia: Heberdey and Wilhelm (see n. 5), 105, no. 185, lines 7–8: *τύχοι τῶν καταχθονίων θεῶν πάντων κεχολωμένων καὶ μήτε θάλασσα αὐτῷ πλωτὴ εἴη μήτε γῆ καρπὸν ἐνένκαι* ("May he find all the gods of the underworld enraged, and may the sea not be navigable for him, and may the earth not bear fruit"). Another example near Thyateira in Lydia (G. Radet, *BCH* 11 [1887] 453–54, no. 15, lines 5–8): *μήτε οἱ θεοὶ ἵλεως αὐτῶι γένοι[ν]το, μήτε τέκνων [ὄνη]σις μήτε γῆ καρπο[φόρος* ("May the gods not be well disposed to him, may there be no pleasure of children, may the earth not bear fruit"). Radet restored *[ποίη]σις* in lines 7–8; at the end of line 8 other possibilities are *καρπο[φορήσοιτο* or *καρπὸ[ν]/καρπο[ὺς φέροι* or *δοίη*. Two elements of the preceding imprecation run parallel with the Amphictyonic oath. In the following funerary imprecation from Halicarnassus on the Doric coast of Caria even three elements run parallel: *μηδὲ γῆ καρποφορήσοιτο αὐτῷ μηδὲ θάλασσα πλωτὴ μηδὲ τέκνων ὄνησ⟨ι⟩ς μηδὲ βίου κράτησις ἀλλὰ ὤλη πανώλη* ("May the earth not bear fruit for him, and may the sea not be navigable, may there be no pleasure of children, no control of life but [may he become] dead and gone") (G. Hirschfeld, in *The Collection of Ancient Greek Inscriptions in the British Museum*, vol. IV, pt. 1, ed. G. Hirschfeld and F. M. Marshall [Oxford, 1893], no. 918, lines 3–5).

42. See Versnel 1985, 254, n. 21.

43. They are, at Aegina in the fourth century B.C., the somewhat vague threat *αὐταυτὸν αἰτιάσῃ* ("He will accuse himself") (Dittenberger [see n. 40], no. 1236) and at Calydon in Aetolia in the third century B.C., *[ἐ]πικατάρατον* (cursed) (G. Klaffenbach, *IG*, vol. IX, pt. 11, 2nd ed. (Berlin, 1957), no. 148.

44. Much material is collected in the works cited in n. 35 and in Parrot 1939, 9–106.

45. Translation by J. C. L. Gibson, *Textbook of Syrian Semitic Inscriptions*, vol. III, (Oxford, 1982), 12–16, no. 4.

46. Translation by J. C. L. Gibson, *Textbook of Syrian Semitic Inscriptions*, vol. II (Oxford, 1975), 93–97, no. 18.

47. The emergence occurred simultaneously with the economic growth of Lycia and Lydia under Persian rule, for which see C. G. Starr, *Iranica Antiqua* 11 (1975): 84–87, 94–99.

48. E. Kalinka, *TAM* vol. I (Wien, 1901), no. 59. Translation by Ph. Houwink ten Cate, *The Luwian Population Groups of Lycia and Cilicia Aspera during the Hellenistic Period* (Leiden, 1961), 93. The exact meaning of the verb of the imprecation (*tubeiti*) is doubtful. It is translated as "shall strike/hit him" by E. Laroche (in *Fouilles de Xanthos*, vol. VI, *La stèle trilingue du Létôon*, ed. P. Demargne, H. Metzger, et. al. [Paris, 1979], 107) and by G. Neumann (*Neufunde lykischer Inschriften seit* 1901, DenkschrWien, vol. 135, Philos.-hist. Kl. (Vienna, 1979), ad N 314b.

49. R. Gusmani, *Lydisches Wörterbuch, mit grammatischer Skizze und Inschriftensammlung* (Heidelberg, 1964), no. 5. The last words of the imprecation are problematic; they should perhaps be translated as "(and) whatever he possesses."

50. Kalinka (see n. 48), no. 6. Translation by Houwink ten Cate (see n. 48), 88. For the meaning of the verb of the imprecation, *tubeiti,* see n. 48. The second word of the imprecation, *punamaθθi,* is enigmatic.

51. Kalinka (see n. 48), no. 56. For translations, see Houwink ten Cate (see n. 48), 93 and T. R. Bryce, *AnatSt* 31 (1981): 86. The meaning of the verb (*qasttu*) is not certain; see Bryce, ibid., 88 (to strike/punish?).

52. A. N. Oikonomides, *ZPE* 45 (1982): 115–18 (*SEG* XXXII, no. 1612).

53. J. and L. Robert (*BE* [1982]: no. 280) seem to suggest a provenance from Lycia. Artemis Ephesia had a temple at Ephesus and possessed territory in the Caystrus valley in Lydia; see R. Meriç et al., *Die Inschriften von Ephesos* VII, pt. 2 (Bonn, 1981), 296. She also occurs as the divinity who punishes the violator of the tomb (together with Artemis of Coloe) in a bilingual Lydian-Aramaic inscription that comes from Sardis in Lydia (Gusmani [see n. 49], no. 1 from the early? fourth century B.C.). Artemis Medeia is—correctly I believe—identified by S. M. Sherwin-White (*ZPE* 49 [1982]: 30) with the Persian Artemis. This goddess is also well known in Lydia; see J. and L. Robert, *Fouilles d'Amyzon en Carie*, vol. I, *Exploration, histoire, monnaies, et inscriptions* (Paris, 1983), 117. Her presence there is explained by the settlement of Persian colonists in the region, for example, in Sardis, Hypaepa, Hierocaesareia; see L. Robert, *BCH* 107 (1983): 508. I think that a provenance from the regions north or south of Mt. Tmolus in Lydia would not be unlikely.

54. The economic situation of Asia Minor during the hellenistic period was miserable. The country was distressed by many wars, and after the coming of Rome the land suffered heavily from exploitation, see T. R. S. Broughton, in *An Economic Survey of Ancient Rome*, vol. IV, ed. T. Frank (Baltimore, 1938), 503–98. The number of imprecations of the nonspecific group was also extremely small in the hellenistic period. There are, for example, only one or two instances of the *ἁμαρτωλός* formula (cf. supra) in the third century B.C.: Kalinka (see n. 8), no. 923 and possibly no. 520. There are four attestations in the following centuries B.C.: Kalinka, ibid., nos. 797–98; G. E. Bean, *AnzWien* 99 (1962): 4, no. 1; E. Petersen and F. von Luschan, *Reisen in Lykien, Milyas, und Kibyratien* (Vienna, 1889), no. 58. All examples come from Lycia.

55. Kalinka (see n. 6), no. 524 lines 4–5: [*μήτε γῆ μήτε θάλασσ*]*α καρπὸν φέροι, ἀ*[*λ*]*λ' ἐξώλεις* [*καὶ*] *πανώ*[*λ*]*ε*[*ις εἶεν*] ("May neither the earth nor the sea yield fruit, but may they become dead and utterly destroyed"). The letterforms of the inscription and the archaeological data of the monument indicate the date.

56. Keil and Wilhelm (see n. 25), no. 111 (Pfuhl and Möbius [see n. 20], 496, ad nos. 2069–70), lines 4–8: *βάλοι αὐτό*[*ν*] *ὁ Ἥλιος καὶ ἡ Σελήν*[*η*] ("May Helios and Selene throw him down"). According to the first editor, E. L. Hicks (*JHS* 12 [1891]: 260, ad no. 36), the letters are not later than the Christian era. This date is confirmed by the type of the grave, a *naiskos* hewn into the rocks with the representation of a soldier.

57. F. Hiller von Gaertringen, *IG*, vol. XII, *Supplementum* (Berlin, 1939), no. 83 (Pfuhl and Möbius [see n. 20] 535–36, no. 2232, lines 4–6: *ἐξξώλης καὶ πρωώλης γένοιτο αὐτὸς καὶ γένος τὸ κήνω* ("May he become a dead and deceased man, himself and his race"). Mytilene was a flourishing harbor town and trading place and the residence of many Romans; see D. Magie, *Roman Rule in Asia Minor to the End of the Third Century after Christ* (Princeton, 1950), 415.

58. W. M. Calder, *MAMA*, vol. VII (Manchester, 1956), no. 201, line 20: *οὗτος τὰν αὐτὰν μοῖραν ἐμοὶ λαχέτωι* ("May he obtain the same fate as I"). Philomelium was a

well-situated trading center, where many Italian business men had settled down; see W. Ruge, "Philomelion," *RE* XIX, pt. 2 (1938): 2522. The text is dated by W. Peek, *Griechische Vers-Inschriften*, vol. I (Berlin, 1955), no. 1870.

59. The recovery from the economic disasters of the hellenistic period started under the Julio-Claudii. From that time Asia Minor became very wealthy but its prosperity began to decline from the dynasty of the Severi onwards; cf. Magie (see n. 57), 688–723; Broughton (see n. 54), 733–34, 794–97, 903–13.

60. The number of nonspecific imprecations is also very large during the imperial period. Some of the formulas, such as *ἔσται αὐτῷ πρὸς τὸν θεόν* ("He shall have to reckon with the god"), only emerged at the beginning of the third century A.D.; cf. Waelkens (see n. 17), 127.

61. R. MacMullen, *AJP* 103 (1982): 233–46, especially 244. I now doubt if the view that I expressed earlier on this point (Strubbe 1983, 250–51) is correct. Further investigation is necessary.

62. Of the dated epitaphs published by P. Herrmann (*TAM* vol. V, pt. 1 [Wien 1981]), 24 texts date in the first half of the first century A.D., c. 40 in the second half; c. 86 in the first half of the second century A.D. and c. 122 in the second half; c. 95 in the first half of the third century A.D. and c. 22 in the second half; only 3 date after 300 A.D.; see, now, R. MacMullen, *ZPE* 65 (1986): 237–38.

63. For a good understanding of the role and the importance of funerary imprecations in a given region one should also take into account the number of epitaphs with regulations for the protection of the grave other than imprecations, such as fines or the threat of legal action. As to northeastern Lydia, for example, it is most striking that the number of fines is extremely low. In *TAM* vol. V, pt. 1 one finds only a dozen examples, and these come without exception from the western part of the region, i.e., west of the rivers Hyllus(?)-Hermus.

64. Cf. R. Lattimore (*Themes in Greek and Latin Epitaphs* [Urbana Ill., 1942], 108–12 and 117), who noted in a general way a special attitude toward death in Asia Minor in contrast to the rest of the Greek world.

65. M. Waelkens, *Antike Welt* 11 (1980): 3.

66. O. R. Gurney, *Some Aspects of Hittite Religion* (Oxford, 1977), 59–63; H. Otten, *Hethitische Totenrituale* (Berlin, 1958).

67. Waelkens (see n. 65), 4–11; idem., *Die kleinasiatischen Türsteine: Typologie und epigraphische Untersuchungen der kleinasiatischen Grabreliefs mit Scheintür* (Mainz, 1986).

68. E. Vermeule, *Aspects of Death in Early Greek Art and Poetry* (Berkeley, 1979), 48.

69. The original ideas of the preclassical times cannot easily be detected because from early times the Greeks have been influenced by other peoples, for example, the Egyptians; cf. Vermeule (see n. 68), 69–82. Of course there are many uncertainties and contradictions, as ideas about the afterlife are neither logical nor uniform.

70. J. Bremmer, *The Early Greek Concept of the Soul* (Ph. D. diss., Free University of Amsterdam, 1979), 69–70, 84–87; Vermeule (see n. 68), 23–24.

71. Vermeule (see n. 68), 8, 27; S. Humphreys, *The Family, Women, and Death: Comparative Studies* (London, 1983), 152–53, 170.

72. Vermeule (see n. 68), 48, 55.

73. Vermeule (see n. 68), 54, 74.

74. One inscription from Termessus in Pisidia, dating after 212 A.D. because of the *Aureliusnomen*, mentions the possibility that the violator will not be deterred by the fine; in that case a curse will come into action (R. Heberdey [see n. 9], no. 742, lines 5–9): *ἐπεὶ ὁ πειράσας ἐκτείσει τῷ ἱερωτάτῳ ταμείῳ* *, *αφ΄· εἰ δέ τις κὲ τούτου καταφρονήσει, σχήσει ἀτεκνία* ("because the man who tries [sc. to open the grave] will pay to the very

sacred [i.e., imperial] treasury 1,500 denarii; and if someone will despise this [sc. penalty], he will be childless").

75. For the location of cemeteries beyond the city gates, see B. Kötting, "Grab," *RAC* XII (1983): 375–76 and D. Kurtz and J. Boardman, *Greek Burial Customs* (London, 1971) 188–89. There is no evidence that cemeteries were guarded. Only in Lycia did there exist a special institution that had functions with regard to the grave (including, possibly, its protection), namely the *miñti;* see T. R. Bryce, *AnatSt* 26 (1976): 183–84. It is well attested in Lycian inscriptions but occurs only three times in the Greek inscriptions of Lycia, as the *μίνδις-μενδῖται*: Kalinka (see n. 5), no. 40 and no. 62 (at Telmessus); Petersen and von Luschan (see n. 54), no. 27 (at Cyaneae). The texts date from the fifth and fourth centuries B.C. The institution apparently did not appeal to the Greeks.

76. This is attested in Greece and elsewhere, for example, in Western Europe from the eighth century to the end of the eleventh century A.D.; see L. K. Little, *Annales ESC* 34 (1979): 43–60, esp. 47, 57–58. Little has shown that the *anathema* was frequently used in this period, during which the Carolingian rulers were weak and unable to give military or judicial help to the injured. For the same phenomenon in Babylonia, Israel, and elsewhere, see W. Wiefel, *Numen* 16 (1969): 218; Gevirtz (see n. 35), 258–59; J. Hempel, *ZDMG*, n.s. 4 (1925): 39 and n. 2; 41, n. 1. No allusions to the failure of human justice are found in the funerary imprecations, cf. Versnel's essay (chap. 3, n. 34).

77. F. W. Hasluck, *JHS* 25 (1905): 63, line 15 (the text may be Christian). The term is also attested at Neocaesareia in Pontic Cappadocia; see Moraux 1959, 11, line 9. The verb *ἐκδικέω*, however, is not attested in funerary imprecations.

78. R. Merkelbach, *Die Inschriften von Assos* (Bonn, 1976), no. 71, lines 11–12. The imprecation of Assus comes very close to the text of a tablet from Cnidus; see Versnel's essay (chap. 3, pp. 72–73).

79. P. Frisch, *Die Inschriften von Parion* (Bonn, 1983), no. 29, lines 6–8. (I do not follow the interpretation of Frisch, who takes *αὐτή* as the damaged *εἰκών*; I take it to refer to the woman who may have died an untimely death.) For Helios, common in both types of texts, see below.

80. Speyer 1969, 1165–67, 1194; Vallois 1914, 254–55 and n. 7; Crawley 1911, 370.

81. A. Maiuri, *ASAtene* 4–5 (1921–22): 470–71, no. 13 (*SEG* vol. IV, no. 196), lines 1–2; see also Heberdey and Wilhelm (see n. 5), 54–55, no. 123, lines 4–6 with Add. p. 164 (from the neighborhood of Elaeussa-Sebaste in Cilicia): *κατ' ἐντολὴν καὶ διαθήκην Ἀρίου τοῦ ἀνδρὸς ἐντέλλομαι καὶ κελεύω καὶ διατάσσομαι* ("According to the order and the will of my husband Arius I order and command and stipulate"). The interdictions and the imprecation follow. The juridical term *διατάσσομαι*, which frequently occurs in wills, is not rare in funerary imprecations, especially at Aphrodisias in Caria; see, for example, J. M. R. Cormack, in *MAMA*, vol. VIII, ed. W. M. Calder and J. M. R. Cormack (Manchester, 1962), nos. 544, 550, 566, 577.

82. P. Herrmann and K. Z. Polatkan, *Das Testament des Epikrates und andere neue Inschriften aus dem Museum von Manisa,* SBWien vol. 265, no. 1, Philos.-Hist. Kl. (Vienna 1969), 8–17, no. 1.

83. See the contribution by F. Graf (chap. 7).

84. For example, E. Schwertheim, *Die Inschriften von Kyzikos und Umgebung*, vol. I (Bonn, 1980), no. 500 (*SEG* XXVIII, no. 943), lines 13–14 (from the neighborhood of Cyzicus in Mysia). Another variation is *ὤλης πανώλης* (deceased and totally dead). For other variants and their attestations, see J. and L. Robert, *Hellenica,* vol. VI (Paris, 1948), 14–15. These expressions became fixed formulas that were often used without any regard to the grammatical context, such as *ὤλη πανώλη γένοισαν* ("May they become deceased and

totally dead") in Th. Ihnken, *Die Inschriften von Magnesia am Sipylos* (Bonn, 1978), no. 28, lines 10–11 (from Magnesia in Lydia) or in an abbreviated form such as ἀλλὰ ὤλη πανώλη (but dead and totally deceased) in Hirschfeld and Marshall (see n. 41), no. 918, line 5 (at Halicarnassus on the Doric coast of Caria). For other examples see J. and L. Robert, *Hellenica* vol. VI (Paris, 1948), 14–15 and Robert (see n. 20) 132–33.

85. A. Geissen, *ZPE* 56 (1984): 300, no. 3, lines 9–10 (unknown provenance, perh. Smyrna?); the same expression in Moraux 1959, 11, line 8 and in literary texts, for example, Men. *Dysc.* 442. For repetitions of the same word, see A. Henrichs, *ZPE* 39 (1980): 12 and n. 9.

86. As to the meaning of the imprecation I follow Robert (1978, 259–62), who gives many examples of the formula; see also E. Gibson, *ZPE* 28 (1978): 17–18.

87. For a variety of reasons I doubt if L. Robert's interpretation of χῆρον βίον (an empty life) as the life of the widow (empty by the dead of her husband) is right; see Strubbe 1983, 255–56, where one finds many examples of the formula. Λίποιτο is awkward in the meter.

88. Calder (see n. 58), no. 210, lines 8–10 and no. 246, lines 6–7 (both from Claneus in Phrygia), written in iambic senarii. Other interesting cases are the following texts: (1) E. Haspels, *The Highlands of Phrygia* (Princeton, 1971), 314, no. 41, frag. c, lines 1–3 (near Metropolis in Phrygia), with an imprecation written in perfect iambic trimeters and the epitaph composed in poor dactylic hexameters; and (2) Calder (see n. 58), no. 201, lines 19–20 (from Philomelium in Phrygia), with metrical epitaph in Ionic dialect followed by a metrical prohibition and imprecation in Doric dialect. Such metrical formulas, if not orally transmitted, were perhaps collected in books that were in the hands of stonecutters and masons; see Th. Drew-Bear, in *Arktouros: Hellenic Studies Presented to B. M. W. Knox on the Occasion of His 65th Birthday,* ed. G. Bowersock (Berlin, 1979), 316; E. Gibson, *The "Christians for Christians" Inscriptions of Phrygia* (Missoula, Mont., 1978), 94; for similar collections of *defixio* formulae, see H. S. Versnel, *Hermeneus* 55 (1983): 204 and his contribution to the present volume (chap. 3, p. 91 and n. 143).

89. Calder (see n. 58), no. 199, lines 9–11 (from Philomelium in Phrygia); the verb indicating the nature of the penalty is suspended, cf. Robert (see n. 20), 97–98. Compare W. M. Calder, *MAMA*, vol. I (Manchester, 1928), no. 437, lines 4–7 (near Amorium in Phrygia): ὁ θεὸς αὐτῷ προσκόψαιτο ὁράσει, τέκνοις, βίῳ ("May the god cut him off from sight, children, life"); and R. Merkelbach and J. Nollé, *Die Inschriften von Ephesos,* vol. VI (Bonn, 1980), no. 2304, lines 4–5 (from Ephesus on the west coast of Asia Minor): μὴ ἐνπλήσθοιτο μήτε βίου μήτε τέκνων μήτε σώματος ("May he not have full measure of life, children, health[?]").

90. Peek (see n. 58), no. 819, frag. f, lines 6–14 (from the neighborhood of Appia in Phrygia). Compare Judeich (see n. 6), no. 339, lines 10–11 (from Hierapolis in Phrygia): ἀλλὰ ἄτεκνος καὶ ἄβιος καὶ πηρὸς π[αθή]ματι παντὶ ἀποθάνοι ("But may he die with every suffering, childless and without life and cripple").

91. Calder (see n. 81), no. 234, line 3 and no. 234A, line 7 (Lane [see n. 18], 97–98, no. 155), both near Savatra in Lycaonia. Oaths, too, were often repeated three times; see R. Hirzel, *Der Eid: Ein Beitrag zu seiner Geschichte* (Leipzig, 1902), 82, n. 4.

92. Calder (see n. 81), no. 234B, lines 2–4. Nine Mens also occur in adjurations; see the texts cited in n. 91. The reading of W. M. Calder in no. 234A, lines 7–8 ἐνορκῶ τρὶς θ' Μῆνας ("I adjure thrice the nine Mens"); is preferable to Lane's reading of it as ἐνορκῶ τρῖς θ(εοὺς) Μῆνας ("I adjure the three gods Men"); see Versnel 1985, 262, n. 55.

93. Speyer 1969, 1167, 1201–03; Vallois 1914, 267–69; Crawley 1911, 369.

94. L. Robert, *Villes d'Asie Mineure* (Paris, 1935), 387, n. 2. The text of this inscription is not yet fully published. L. Robert indicates the contents only vaguely as follows: "L'im-

précation protégeant la tombe." There is thus no certainty that we have to do with an imprecation belonging to the group of funerary imprecations to which our study has been restricted.

95. G. Pfohl, "Grabinschrift I," *RAC* XII (1983): 480–81; S. Mitchell, *Regional Epigraphic Catalogues of Asia Minor*, vol. II (Oxford, 1982), 104, ad no. 110; P. Lambrechts and R. Bogaert, in *Hommages à M. Renard,* vol. II (Brussels, 1969), 413, n. 1 with further bibliography. See also Versnel's essay (chap. 3, p. 70).

96. W.M. Calder, *MAMA*, vol. I (Manchester, 1928), no. 294. Two forearms with outstretched hands are depicted on the back of the stele. The first of the two doubtful cases is Calder, ibid., no. 399 (from the region southeast of Nacoleia in Phrygia). The imprecation, which wishes that the violator may receive the same fate (*χάριν*, lit. "favor") as the deceased may suggest that the departed died an abnormal or violent(?) death. The second case is Heberdey and Kalinka (see n. 7), 53, no. 74 (from Oenoanda in Pisidia). The grave is a family tomb, designed for husband, wife, and children according to the main inscription, which is engraved in a *tabula*. There is a second inscription in a different lettering under the border of the lid of the sarcophagus. It is very fragmentary and may record the death of a wife. Two hands are depicted besides this second inscription. They could refer to the abnormal death of the wife, who perhaps was not the same person as the wife of the first inscription; it is conceivable that the tomb was sold or usurped after the death of the wife.

97. Speyer 1969, 1196; Vallois 1914, 256–58. The same occurs with non-Greek peoples, for example, in Babylonia; see S. Mercer, *JAOS* 34 (1915): 284, 305.

98. Moraux 1959, 46–50. The text of the inscription is to be found on p. 11 (*SEG* XVIII, no. 561). For Herodes' curses, see Moraux 1959, 13–14.

99. Moraux 1959, 11, line 12. Why Ara is called the oldest of daemons is not clear; see Moraux, ibid., 39–40. Ara was not foreign to popular belief. She is also named in an imprecation on a grave monument at Mopsuestia in Cilicia, which is directed against a man who had done wrong to his brother (V. W. Yorke, *JHS* 18 [1898]: 307, no. 3; cf. J. Zingerle, *JOAI Beibl.* 23 [1926]: 59).

100. Moraux 1959, 39; cf. Speyer 1969, 1196; E. Wüst, "Erinys," *RE Suppl.* VIII (1956): 86–87. The Erinyes are attested a few times as guardians of the grave. The three Erinyes are depicted on a monument with an imprecation at Anazarba in Cilicia, and according to the inscription they protect the deceased, a eunuch (*ἄγονον εὐνοῦχον φυλάσσομεν*): Heberdey and Wilhelm (see n. 5), 38, no. 94 frag. c, line 1 (E. Pfuhl and H. Möbius, *Die ostgriechischen Grabreliefs,* vol. I [Mainz, 1977], 498–99, no. 2084 with plate 299); for the relief, see also L. Deubner, *AM* 27 (1902): 262–63. The Erinyes are perhaps mentioned in another inscription from the territory of Cyzicus in Mysia: Schwertheim (see n. 84), no. 83, lines 7–8: *φυλάσσου]σιν δὲ οἱ δαίμονες ọἱ τεταγμένοι ἀπὸ ἀναπ̣[αύσεως* ("The daemons, who have been ordered for the rest[?], protect [sc. the grave monument]"). The parallel with the inscription from Anazarba suggests, I believe, that the identification of the daemons with the Erinyes is the correct one, not that with the Manes as J. and L. Robert (*BE* [1980], no. 401) suggest. The fact that the gravestone belongs to a Christian family is no problem, because the Erinyes were not considered real pagan deities; see Robert, 1978, 148. I am convinced that the restoration of J. Zingerle (*Philologus* 53 [1894]: 347–48) to the text of G. Doublet and G. Deschamps, *BCH* 14 (1890): 630, no. 35, lines 1–2 (from Neapolis in Caria)—*ἐπάρατος ἔσṭα[ι Ἐριν]|ῦσ[ι - -]* ("He will be cursed to the Erinyes")—is not correct. In Caria *ἐπάρατος* is never followed by the name of a divinity, in contrast to the usage in Lycia. I suggest restoring *ἐπάρατος ἔσṭα[ι τέκνα τέκν]|υς [- -]* ("He will be cursed to his children's children").

101. Speyer 1969, 1161, 1164, 1166; Crawley 1911, 367, 369.

102. Speyer 1969, 1172; Lattimore (see n. 64), 121. Blessings are also found in Shake-

speare's epitaph; they precede the imprecation (line 3) "Bleste be ye man yt spares thes stones"; see n. 1.

103. J. G. C. Anderson, *JHS* 19 (1899): 306–7, no. 246, line 19.

104. Moraux 1959, 11, lines 16–18 with commentary on pp. 42–43.

105. Speyer 1969, 1169–70, 1189–91; Wiefel (see n. 76), 219–20; Vallois 1914, 264. Compare *defixiones* that sometimes say that the only one who can loose the spell is the *defigens;* see Versnel 1985, 262, n. 59.

106. Versnel 1985, 261–62 has collected all the evidence not only in Greek funerary imprecations but also in adjurations, "nonfunerary" imprecations, Jewish and Christian imprecations, and Latin texts. The wish that the violator of the grave may never get something perhaps also refers to the irrevocability of the imprecation, for example, C. Naour, *ZPE* 44 (1981): 18–21, no. 1 (*SEG* XXXI, no. 1003), lines 7–9 (near Saittae in northeastern Lydia): μηδέποτε τοῦ 'Αξιοττηνοῦ Μηνὸς ἵλεως τύχοιτο ("May he never find Men Axiottenos well disposed"). Versnel, ibid., 263 argues that the state of ἀσέβεια (impiety) to which the violator is often condemned (see above) is also irrevocable, for the impious is not allowed to sacrifice to the enraged gods, who could only be placated by atoning sacrifices.

107. Th. Drew-Bear, *Nouvelles inscriptions de Phrygie* (Zutphen, 1978), 102–3, no. 40, lines 10–11.

108. Herrmann and Polatkan (see n. 82), 8–17, no. 1, line 99.

109. S. Bakir-Barthel and H. Müller, *ZPE* 36 (1979): 182–83, no. 36 (*SEG* XXIX, no. 1179), line 8.

110. The inscription is not yet published. It is mentioned by G. Petzl, *ZPE* 30 (1978): 260 (Herrmann [see n. 62], 62, testimonium B3). For a discussion of the "scepters of the gods" see below.

111. The same phenomenon is visible in a wide variety of regulations, for example, in oaths (cf. Parker [see n. 30], 186) and in the confession inscriptions of northeastern Lydia (see E. Varinlioğlu, *EA* 1 (1983): 83 with n. 40). It is also attested outside the Greek world; see, for example, J. Scharbert, *Solidarität in Segen und Fluch im alten Testament und in seiner Umwelt* (Bonn, 1958).

112. Cormack (see n. 81), no. 570, lines 9–10.

113. J. and L. Robert (see n. 100), no. 493 (*SEG* XXX, no. 1501), lines 4–5. For further examples of the formula and similar wishes, see Robert 1978, 263–64.

114. Τέκνα τέκνων occurs, for example, in Heberdey and Kalinka (see n. 7), 53, no. 74, line 13 (from Oenoanda in Pisidia); τέκνα τέκνοις, for example, ibid., 8, no. 22, lines 20–21 (from the neighborhood of the Söğüt Gölü near Tyriaeum in Pisidia). For the abbreviated formula, cf. Robert (see n. 20), 96–97 and idem 1978, 282–83; but I do not believe that it was used as an independent curse nor that it was typical of Jews or those imitating Jewish culture.

115. Herrmann and Polatkan (see n. 82), 8–17, no. 1, lines 102–3; also Ramsay (see n. 4), no. 498 bis (from the neighborhood of Sebaste in Phrygia): μήτε τῶν ἰδίω[ν τι] ἀκμά[ζη] ("May none of his property flourish").

116. Moraux 1959, 11, line 7.

117. M. Delcourt, *Stérilités mystérieuses et naissances maléfiques dans l'antiquité classique* (Liège, 1938), 9–28; cf. also Moraux 1959, 23.

118. The wish for the birth of monstrous children does not occur in Herodes' imprecations; it is a personal addition made by the author from Neocaesareia; see Moraux 1959, 14.

119. S. Şahin. *Katalog der antiken Inschriften des Museums von Iznik (Nikaia)* vol. II, pt. 2 (Bonn, 1982), no. 1251, line 4. The curse of a plague may also be present in the wish that the air will not be pure and healthy (Herrmann and Polatkan [see n. 82], 7–18, no. 1, lines

99–100 [from Nacrason in Mysia]: μήτε - - ἀὴρ καθαρὸς ἢ ὑγιεινός) and in the wish that the violator will never use a clear fountain (J. Franz, *CIG* vol. III [Berlin, 1853], no. 4190, lines 5–6 [from Nazianzus in Cappadocia]: μήπ[οτε καθαρᾷ χρήσεται] πηγῇ). Impure air and water were thought to cause diseases, especially epidemic diseases; see Strubbe, 1983, 263–64.

120. L. Robert, *REA* 42 (1940): 309, n. 2. For λοιμός, see A. Patrick, in *Diseases in Antiquity,* ed. R. Brothwell and A. T. Sandison (Springfield Ill., 1967), 245.

121. This *apopompe* is frequently attested in antiquity; see Versnel 1985, 254 with bibliography; Speyer 1969, 1167, 1184–86, 1197–98. It also occurs in later periods (see F. Pradel, *Griechische und süditalienische Gebete, Beschwörungen und Rezepte des Mittelalters* [Giessen, 1907], 356–360) and in other cultures than the Mediterranean, (see H.-P. Hasenfratz, *Die toten Lebenden: Eine religions-phänomenologische Studie zum sozialen Tod in archaischen Gesellschaften* [Leiden, 1982], esp. 14–24, 33–34, 38–41.

122. Moraux 1959, 11, line 6, with commentary on the exact meaning of the wish on p. 22.

123. See the recent collections by C. Naour, in *Travaux et recherches en Turquie,* vol. II (Paris, 1984), 47–48; idem, *EA* 2 (1983): 119–21; L. Robert, *BCH* 107 (1983): 519–20.

124. Herrmann (see n. 62), no. 160, lines 5–8.

125. Naour (see n. 123), 45–46, no. 11 (*SEG* XXXIV, no. 1231), lines 7–9.

126. Herrmann (see n. 62), no. 172, lines 6–9.

127. Near Silandus: C. Naour, *EA* 2 (1983): 118–19, no. 2, lines 8–10 and 121, no. 9, lines 3–5. Near Saittae: Herrmann (see n. 63), no. 167A (*SEG* XXVIII, no. 917), lines 13–15; H. Malay, *ZPE* 47 (1982): 112–13, no. 1 (*SEG* XXXII, no. 1222), lines 4–6. At Tabala: see n. 110. It is not improbable that scepters were erected in these cases, because the wrath of the scepters(= the gods) is the same penalty that is exacted in the imprecation spoken by the scepter (in the second of the three texts, see n. 125). If so, one may possibly go even further and suppose that scepters were also erected in the cases in which the wrath of the gods was wished for (especially of Men Axiottenos and/or Anaeitis, see the third of the three texts [cf. n. 126]) even without any mention of the scepters in the text, as for example, in Herrmann, ibid., no. 173 (compare the confession inscription published in L. Robert, *BCH* 107 [1983]: 520, in which no scepter is mentioned but the relief shows the priest with the scepter of the god!). If that supposition is right, implying that the erection of scepters was the normal usage in this context in that region, why is the erection of the scepters mentioned in only three cases? If the hypothesis is wrong, why did three families have recourse to the extraordinary procedure of the scepters? As far as I see, there is nothing in the three epitaphs that gives any clue. This may speak against the second hypothesis.

128. L. Robert, *BCH* 107 (1983): 520; J. Zingerle, *JOAI Beibl.* 23 (1926): 13–14. The region of northeastern Lydia was characterized by a special religiosity; see H. W. Pleket, in *Faith, Hope, and Worship: Aspects of Religious Mentality in the Ancient World*, ed. H. S. Versnel (Leiden, 1981), 177–78; F. S. Steinleitner, *Die Beicht im Zusammenhange mit der sakralen Rechtspflege der Antike* (Munich, 1913), 76–82.

129. Herrmann (see n. 62), nos. 159, 231, 317, 318 from the territories of Silandus and Collyda. All texts, funerary and expiatory, come from a small region in northeastern Lydia, north and south of the middle Hermus. For the confession inscriptions of northeastern Lydia in general, see Versnel's essay (chap. 3, pp. 75–79, with a collection of texts in n. 77).

130. Cf. O. Eger, in *Festschrift P. Koschaker,* vol. III (Weimar, 1939), 290. But the scepters apparently could not be erected in all judicial cases: the gods had to become involved in the case—for example, by perjury; see P. Herrmann and E. Varinlioğlu, *EA* 3 (1984): 4 and 6, ad no. 3. The theory of E. N. Lane (*Corpus Monumentorum Religionis Dei Menis*, vol. III [Leiden 1976], 27–29) that the gods were involved through the oath does not seem very likely.

The involvement of the gods in cases of violation of the grave is evident, for this crime was also an offence against the gods as an act of impiety.

131. Steinleitner (see n. 128), 101, n. 1; Eger (see n. 130), 291, n. 33; Versnel's essay (chap. 3, p. 76). It has been remarked that the number of scepters does not necessarily correspond to the number of gods involved, as in the third text from the territory of Silandus; see C. Naour, *EA* 2 (1983): 120.

132. L. Robert, *BCH* 107 (1983): 518–22. As to the problem of the further role and the judicial power of the priests and of the *Tempelgerichtsbarkeit,* see E. Varinlioğlu, *EA* 1 (1983): 84–85; P. Herrmann, in *Studien zur Religion und Kultur Kleinasiens: Festschrift für F. K. Doerner zum 65. Geburtstag,* vol. II, ed. S. Şahin, E. Schwertheim et al., (Leiden, 1978), 421 with n. 25; see also Versnel's essay (chap. 3, pp. 75–79).

133. The scepter was not placed on an altar, as was supposed in earlier publications; see L. Robert, *BCH* 107 (1983): 522; Herrmann (see n. 62), 77, ad no. 231. On one of the two expiatory texts with relief, published by L. Robert, ibid., the scepter stands on a base.

134. Herrmann (see n. 62), no. 318, lines 9–11: *ἐπέστησεν σκῆπτρον καὶ ἀρὰς ἔθηκεν ἐν τῷ ναῷ* ("She set up a scepter and laid down curses in the temple"); cf. lines 24–27: *ἐπεζήτησαν λ̣υθῆναι τὸ σκῆπτρον καὶ τὰς ἀρὰς τὰς γενομένας ἐν τῷ ναῷ* ("[The gods in Azitta] requested that the scepters should be dissolved and the curses that were made in the temple"). Cf. also Steinleitner (see n. 128), 100–104 and Versnel's essay (chap. 3, p. 76).

135. In the expiatory inscription published in Herrmann (see n. 62), no. 251, lines 6–7, it is said that the injured has given a tablet (*πιττάκιον ἔδωκεν*). In another expiatory inscription, ibid., no. 362, lines 3–6, it is told that someone has overthrown and removed the tablet (*τὸν βεβληκότα τὸ π[ι]νακίδιον κ⟨α⟩ὶ ἠρκό⟨τ⟩α*).

136. See the bibliography given by I. Diakonoff, *BABesch* 54 (1979): 163, n.121. It is logical to suppose that the tablet was placed near the scepter; perhaps it was attached to the base on which the scepter stood (see n. 133). For the analogous practice of placing prayers for justice or vengeance publicly in the temple area, see Versnel's essay (chap. 3, pp. 72–74).

137. Cf. n. 134; see also G. Petzl, *ZPE* 30 (1978): 260, n. 48.

138. *Ἄλυτα σκῆπτρα* can hardly refer to the removal of the scepters themselves, for the tomb had to be guarded from violation forever.

139. It is mentioned, though very rarely, in "nonfunerary" imprecations and in Greek and Latin *defixiones;* see the collection of texts by Versnel 1985, 247–55.

140. See Versnel 1985, 250, 259. For the term *ἀσεβής* and its implications, see above.

141. For the term, see Versnel 1985, 255 with n. 23, 260–61, with literature and a collection of examples (including tablets of Cnidus and the prayer for justice of Artemisia; cf. also Versnel's contribution to this volume (chap. 3, pp. 69, 73). For the wrath of the gods, see Versnel 1985, 259, n. 44. For *μὴ ἵλεως* in funerary imprecations, see the text from Saittae in Lydia cited in n. 106; Herrmann (see n. 62), no. 101, also from the territory of Saittae; and the text cited in n. 41 near Thyateira in Lydia.

142. G. E. Bean and T. B. Mitford, *Journeys in Rough Cilicia 1964–1968,* DenkschrWien vol. 102 (Vienna, 1970): 207–8, no. 234, lines 6–8.

143. G. Mihailov, *Inscriptiones Graecae in Bulgaria repertae* vol. III, pt. 1 (Sofia, 1961), no. 998, lines 6–8.

144. Calder (see n. 58), no. 399, line 3.

145. See Versnel's essay (chap. 3. p. 70).

146. Moraux 1959, 11, lines 10, 12. For commentary on the gods, see ibid., 38–39. For the author, see ibid. n. 98.

147. Merkelbach (see n. 78), no. 71, lines 6, 10–11. For the author, see ibid. 96.

148. Some foreign gods are mentioned more frequently. For example, Anaeitis, the Per-

sian Anahita who was introduced by Persian colonists (see n. 53), occurs not infrequently in funerary imprecations of northeastern Lydia. Nor are the *καταχθόνιοι δαίμονες* (daemons of the underworld), who are the Latin *Di Manes,* any rarer. They are mostly attested in and around the bigger cities that played an important role in commerce and industry and where native Romans had settled down as businessmen or governors, like Cyzicus, Acmonia, Cibyra, Smyrna. Many of these inscriptions show marked Roman influence; a good example is G. Petzl, *Die Inschriften von Smyrna,* vol. I (Bonn, 1982), no. 210 (from Smyrna on the west coast of Asia Minor).

149. A. M. Fontrier, *Mouseion* (1886): 77, no. 565, line 7 with restorations by K. Buresch, *Aus Lydien: Epigraphisch-geographische Reisefrüchte* (Leipzig, 1898), 117 n. and 118 (from an unknown place in Maeonia).

150. Robert 1978, 280 (*SEG* XXVIII, no. 1079, with a new restoration of line 5 by H. W. Pleket), lines 1–2.

151. Robert 1978, 283–86.

152. Moraux 1959, 11, line 12. For a commentary on Daeira, see ibid., 30–38.

153. For the gods in Iaza, cf. Herrmann (see n. 62), no. 468A, line 3; for the scepters in Tabala, see n. 110; for the gods in Tamasis, see Herrmann, ibid., 50, ad no. 156.

154. For the sake of completeness the gods on the land and the gods in the sea are added in the inscription of Nacrason, published by Herrmann and Polatkan (see n. 82), 8–17, no. 1, lines 97–98: *θεοὺς - - ἐπουρανίους τε καὶ ἐπιγείους καὶ ἐναλίους καὶ καταχθονίους* ("the gods in the heaven and on the land and in the sea and in the underworld"). The gods of the land and the sea are not mentioned in other funerary imprecations.

155. A unique case is *Ἡλίου τε τοῦ πάντα ἐφορῶντος* ("and Helios who sees everything") in Moraux 1959, 11, line 10. I do not think this is a meaningless *epitheton ornans* in imitation of Homer, for example, *Od.* 11.109 (*ὃς πάντ' ἐφορᾷ*, "who sees everything").

156. It is not important here whether the spirits of the dead were thought to descend to the underworld or to reside in the grave.

157. For the relation of the moon with the underworld, see for example W. Drexler, "Men," *Roscher,* vol. II, pt. 1 (Leipzig, 1890–97) 2768–2769; F. Schwenn, "Selene no. 1," *RE* vol. II, pt. A, sec. 1 (1921): 1137. For this reason Selene plays an important role in magic too; see Schwenn, ibid., 1139–40. For the sun and the underworld, see A. Dieterich, *Nekyia: Beiträge zur Erklärung der neuentdeckten Petrusapokalypse,* 2nd. ed. (Leipzig, 1913), 21–23. For fertility gods see, for example, Th. Schreiber, "Artemis," *Roscher* vol I, pt. 1 (Leipzig, 1884–90), 570–73.

158. See Versnel's essay (chap. 3, p. 64); for the Erinyes see especially his n. 17; Audollent (see n. 23), index, pp. 461–64.

159. Moraux 1959, 27; Versnel's essay (chap. 3, p. 70 with nn. 45–46).

160. It is not surprising, therefore, to find the god Hosios Dikaios, who in some way has to do with justice, in a funerary imprecation: at Hadrianutherae in Mysia the owner of the tomb says [*θεῷ* ʽO]*σίῳ τε Δικαίῳ χεῖρας ἀεί*[*ρω*] ("I raise my hands to the god Hosios and Dikaios"); see the reference given in n. 94.

3

Beyond Cursing: The Appeal to Justice in Judicial Prayers

H. S. Versnel

In 1972 a lead tablet was found in Italica (Spain) with the following, partially mutilated text (second century A.D.):[1]

> Domna Fons Fọyi [. . .]
> ut tu persequaris ṭuas
> res demando quiscun-
> que caligas meas tel-
> luit et solias tibi
> illa demando {ut} ut
> illas abọitor si quis
> puela si mulier siue
> [ho]mo inuolauit
> [. . .] illos persequaris.

Except for a few sections, the translation is not difficult:

> O Mistress (*domina*) Spring Foyi . . . , I ask that you track down (or claim) your possessions. Whoever has stolen my shoes and sandals (*telluit,* perfect of *tollo*) I ask that you. . . . (?) Whether it is a girl, a woman or a man who stole them . . . pursue them.

The text is written on lead and had obviously been deposited in the spring that is addressed in the opening line; both characteristics naturally reminded scholars of the *defixiones* discussed at the start of this volume. The text, however, has remained puzzling, and no satisfactory interpretation has been offered to date. How can we explain that the stolen possessions of a person are at the same time called the property (lit., "the affairs") of a goddess? The text is also peculiar, because of, among other things, the double use of *persequaris* with two different objects and hence two different meanings.

I want to show that the text of the Italica tablet, as well as a number of related texts, cannot be understood if they are considered as *defixiones* in the traditional meaning of that term. Secondly I shall demonstrate that in this respect the isolation of Greek and Latin texts has been a hindrance, since it is only in comparison with the Greek material that many Latin "curse" texts can be correctly interpreted; but even then we have to look outside the limited domain of the *defixiones*. These new interpretations are offered within the larger framework of an overview of the texts that, although reminiscent of the *defixio,* in reality form another category, which I shall refer to as "judicial prayers" or "prayers for legal help." I have attempted to give a complete survey of these prayers insofar as they are epigraphical or papyrological in character—with the exception of a few categories that have already been described at length or form a separate group, such as, for example, the public prayer for revenge or the confession inscriptions. Rather than cite all texts completely here, I have restricted myself, as far as the argument allows it, to translation.[2] I am not the first to study the judicial prayer: Ziebarth, Steinleitner, Zingerle, Björck, and Latte have already come to important conclusions.[3] But the quantity of new material, particularly the Latin, has increased to such an extent that substantial interpretative progress has become possible. In addition it seems that the research into these types of texts has practically been at a standstill for several decades—so much so that very recently a prayer for revenge on papyrus was published[4] without the (very able) editor even alluding to the genre to which it belonged or the relevant literature dealing with it.

"*Defixiones,* more commonly known as curse tablets, are inscribed pieces of lead, usually in the form of small, thin sheets, intended to influence, by supernatural means, the actions or the welfare of persons or animals against their will" is the definition of D. Jordan.[5] For a complete discussion of these texts, I can refer to the contribution by Faraone (chap. 1). I shall make only a few introductory remarks indispensable to my argument. If most *defixiones* are lead tablets (there are a few exceptions), this purely formal characteristic must not seduce us into claiming that the reverse is also true, namely, that all short texts written on thin sheets of lead are *defixiones*. Their implicit or explicit purpose is another and far more important element; the victim must be "bound" (according to Greek terminology) or "nailed down" (as the Latin puts it)[6]—which may include a wide variety of different meanings from "making powerless and unable to take action" to "making ill" (often with a detailed enumeration of the bodily parts to be afflicted) to "killing" (only rarely). In addition, most *defixiones* were either buried in the grave of an *ἄωρος*, that is, "a person untimely dead" or in chthonic sanctuaries or placed in wells for reasons explained by Faraone. The older examples normally do not give more than the name(s) of the victim(s), a practice maintained throughout classical antiquity, or the simple formula "I bind NN," which betrays the "mechanical" and more or less "automatic" procedure usually associated with magic. The involvement of the gods or the daemons in the action seems to be a result of an evolution that (though the first attestations already date from the fifth century B.C.) reaches perfection only in the imperial period. Even then, these supernatural helpers are instructed or compelled to go about their destructive task. One may, then, for reasons of systematic classification, speak of "the prayer formula," but we should realize that we are employing the minimum criteria of a prayer. This is apparent from the overt imperative tone in which the gods or daemons are ordered to do whatever the author of the tablet wishes them to do.[7]

Most noticeable is the fact that notions of supplication or vow are absent or extremely rare in the *defixiones*.[8] If and when they occur, they should arouse our suspicions about the nature of the text; as we shall see, these aberrations are indeed often accompanied by other elements that are atypical for the taxonomy of the *defixiones*. The nature of the gods addressed also changes over time. In the earlier period they belong exclusively to the sphere of death and the underworld and appear to be liable to forms of manipulation that the Olympian deities would not easily tolerate. As for the social context, purposes, and functions of the curses, Audollent and others[9] made a classification of four categories: (1) rivalry in the theater, amphitheater, and circus; (2) competition in the world of love; (3) rivalry connected with litigation; and (4), damages of any kind (especially theft and slander) caused to the author by someone else. Faraone has quite rightly added, at Jordan's suggestion, a category for commercial curses, which he substitutes for category number 4 above. Just as Winkler argues in chapter 8 for a functional interpretation of the amatory curses in the context of the Mediterranean struggle for life and honor in a thoroughly competitive society, so Faraone interprets the curses concerning the circus, lawsuits, and commerce as just another means of fighting with envious and often hostile rivals on a day-to-day basis. Both interpretations seem to me to be particularly convincing and revealing.

I must, however, draw special attention to a supposition that although conceivable, may be mistaken. The intended victims in all four of Faraone's categories are not being cursed because they are guilty of any crime or misdeed against the *defigens* but rather because they are his rivals with regard to social prestige or economic position, and any attack against their social position will result in an increase of his own honor. Accordingly, we find in the *defixiones* neither justification for the cursing action nor the names of their authors. The absence of the author's name (to which there are only a few exceptions) may be partly explained by the fear of being accidently cursed by the "nailing" of one's own name (which would be inscribed next to the victim's) or by fear of countermagic. There are, however, strong social pressures that also encourage such anonymity. As one anthropologist put it, the basic rule of the Mediterranean "amoral familist" (his designation for the "able protector" of other ethnologists) is, "Maximize the material, short-term advantage of the nuclear family; assume that all others will do likewise."[10] And although in this competition practically any means or method (including "black magic") is employed, a sharp and consistent distinction is maintained between exploits of which one may publicly boast (e.g., to have outmaneuvered competitors by cunning or even wily tricks; cf. the ambiguous "Odyssean" meaning of πονηρός in modern Greek) and actions that are *never* confessed either publicly or privately.

This is indeed a classic instance of a double standard of morality, a fact that may vitiate the argument in the essays of Winkler and Faraone that all methods used to further the aims of the family were also morally approved or socially tolerated. Individuals may indeed try to spoil the milk, the crops, or the procreation of a neighbor by magical means and even feel that this is inevitable if they wish to protect their own families. Nevertheless, they know full well that the act is strongly condemned by all other members of the society, just as they themselves would publicly condemn similar attempts made by others. There is at this point a tension between public morality and private enterprise in the modern Mediterranean, as there was in the ancient. Acts of black magic (real or suspected) were denounced as threats to social stability and cohesion on account of their harmful intentions (*maleficium*) and their relationship to the uncanny (*superstitio*). Nor is this negative attitude toward malevolent magic restricted to philosophers such as Plato[11] or to official legislation such as the Laws of the Twelve Tables at Rome, which explicitly forbade incantation for harmful purposes (followed by many laws in the imperial period that outlawed magic). We have evidence that similar attitudes existed in more or less private circles; the terror and disapproval that followed the fortuitous find in the nineteen sixties of a magical curse in a grave in modern Greece and the official execration of its author by the village priest[12] have a splendid parallel in an event narrated by a Latin inscription from Tuder (*Corpus Inscriptionum Latinarum* 11.3639), in which Iuppiter O. M. is thanked for bringing to light a buried *defixio* that cursed a number of *decuriones* and praised for having liberated the city from fear. A public slave is charged with the nefarious act. Surely his name was *not* on the tablet, just as there was no hint of the name of the person who inscribed the *defixiones* that were discovered in the walls and floors of the house of Germanicus after his supposedly unnatural death (Tac. *Ann.* 2.69), a discovery that similarly caused terror in the masses and subsequently led to the execution of a scapegoat of senatorial rank. There can be little doubt that this social abhorrence is the main reason why people did not add their signatures to what must be unconditionally labeled as an instrument of black magic. Accordingly these tablets are often and correctly described as *Schadenzauber,* "magic tablets," "magical curses," and so on.

Now although some conception of the *defixio* certainly existed in antiquity, we should always remember that the definition of this term is in some sense a modern creation. In this connection, a remark made by Björck has never been sufficiently appreciated: "Man möchte sagen dass der Begriff der Tabella Defixionis nicht so sehr in der Wirklichkeit verankert ist wie vielmehr in Audollent's Sammlung."[13] He alludes here to the existence of texts on lead tablets that satisfy some but not all the characteristics discussed above and that nevertheless are generally referred to as *defixiones* because they were included in Audollent's corpus. As we might expect, this pertains above all to Audollent's fourth category, which, eliminated from the discussion of traditional early Greek *defixiones* by Faraone, becomes the central focus of my essay. This category is the least extensive of the four, and it will quickly become evident that all the examples Audollent gives (summed up on his p. xc) belong to a "borderline" group and that some even fall completely outside the boundaries of the *defixio*. The texts in Audollent's collection that mention the name of the sender are few and mostly belong to his fourth category (p. xlv).[14] A group of judicial *defixiones,* particularly those from Amathous on Cyprus are one exception.[15] Here we can probably assume that the writers felt for some reason morally justified in having recourse to the extra help of the *defixio*. There are also some erotic *defixiones* in which the name of the sender appears.[16]

I shall introduce texts that carry all the obvious characteristics of the *defixio* but that also have particularities pointing to another kind of mentality. I shall limit myself at first to the Greek material (the Latin will follow later) and hope—on the strength of the material collected in *Defixionum Tabellae Atticae, Defixionum Tabellae,* and "Survey of Greek Defixiones" (texts found after the publication of the book collections)—to be able to lay claim to some degree of completeness.

THE BORDER AREA

The gods named in the *defixiones* invariably either belong to the dominion of death, the underworld, the chthonic or are reputed to have connections with magic. Kagarow (p. 59–61) names, in order of frequency, Hermes, Kore/Persephone, Hecate, Hades/Pluto, Ge, and Demeter. In addition, the Erinyes, the nymphs, Egyptian and oriental gods such as Osiris and Typhon, and many daemons appear. In principle they carry out tasks not as representatives of right or morality but on the strength of their dark nature. The only possible exceptions, namely the Erinyes, have in the period of the *defixiones* generally become daemonic Furies.[17] In this respect there is only one real exception, unmistakable because the name expresses the function clearly: Praxidike or the Praxidikai. These goddesses—most likely related to the Erinyes—are called upon for help only rarely,[18] clearly "to do justice." Particularly interesting is *DTA* 109, in which the Praxidikai are asked to restrain a victim: *φίλαι Πραξιδίκαι κατέχετε αὐτ⟨ό⟩ν*. The text concludes, "To you, Praxidikai and Hermes Restrainer, I shall, when Manes has fallen on hard times, bring an offer of rejoicing." Such a *votum* is, although not unique, still very exceptional in a *defixio*.[19]

If the Praxidikai have to carry out justice, as they clearly do, the sender obviously feels that he or she has been injured by someone. This sentiment is indeed expressed several times. Sometimes the injustice is explicitly mentioned. An unpublished tablet from Athens (first or second century A.D.)[20] curses "whoever gave a *pharmakon* to Hyacinthos." Likewise, an unpublished tablet in the Ashmolean museum (perhaps fourth century A.D.)[21] curses "whoever bewitched (*κατέδεσεν*) me, whether woman, man, slave, free, foreigner, townsman. . . . "

An Athenian *defixio* (second century A.D., *DT* 74) curses, (*καταγράφω κ⟨αὶ⟩ κατατ[ίθω]*) and wishes fever for, whoever "keeps and does not return" (*τοῦ κα⟨τά⟩σχοντος [κ⟨αὶ⟩ οὐκ] ἀποδ[όντος]*)[22] a certain object. At the end it also curses *Παῦλον λιθοξόον* / [. . .]*ο συνγνῶντα* ("Paulus, the stonemason, who has knowledge of the matter"). Practically identical in formula, time, and place, another tablet (*DT* 75) curses "the thief" (*τοῦ κλέπτ[ου]*). Sometimes more details are given. A tablet from Megara (*DT* 42, first or second century A.D.) gives on one side a curse of every conceivable part of the body in the typical manner of a *defixio*. On the other side the motive is explained. The text is quite mutilated, but we can clearly make out that the writer was (unfairly) accused of borrowing twenty denarii without returning them even though he had just repaid the debt. At other times we can gather from the terminology itself that the writer felt that he had been injured with regard to his rights. In the case of a second-century-A.D. tablet from Messana (*SEG* 4.47 = SGD 114) we should still, perhaps, hesitate; a woman is cursed who is called *ἁμαρτωλόν* ("criminal, sinful"), but this can often be an ordinary term of abuse, as she is also called *σκύζαν* ("lust"). Terms of abuse do not occur often in *defixiones*,[23] but we do encounter the term *ἁμαρτωλός*[24] once more on an Attic *defixio*, which contains an undeniable reference to the victim's guilt (*DTA* 103, fourth century B.C.): "To Hermes and Persephone I send this letter. Because I direct this (curse) against criminal people (*ἁμαρ[τωλο⟨ὺ⟩ς*), they must, O Dike, receive their deserved punishment (*τυχεῖν τέλο⟨υ⟩ς δίκης*)." Related to this text and to the text with the Praxidikai quoted above is the following curse from Centuripae on

Sicily (*SEG* 4.61 = SGD 115, first century A.D.): "Mistress, destroy Eleutheros. If you avenge me (ἐ⟨κ⟩δεικήσηs), I shall make a silver palm, if you eliminate him from the human race." The verb ἐκδικέω is a technical term in revenge texts on graves, and just like Latin *vindicare* it means, among other things, "to take revenge on someone on account of an injustice suffered." We should note that here again a *votum* is promised. A reference to revenge or punishment is usually a signal that the text is some sort of prayer for justice, for instance, in a text from Megara (*DT* 41, first or second century A.D.). It contains two operative verbs: καταγράφω (a familiar word in Greek *defixiones*) and ἀναθεματίζομεν, a unique term to which the *anathema* mentioned at the end of the text undoubtedly refers. It expresses the wish that the cursed person will moan and that his blood and flesh will burn. Finally, it directs the curse to "punishment and retaliation and revenge": [κατα]γρά[φ]ομεν, [εἰς] κολάσε[ις . . .] καὶ [ποι]νὴν καὶ [τι]μ[ωρ]ίαν. We are quite obviously in an atypical setting here; several elements betray a Jewish influence[25] and I think that in the quoted passages the punishments of hell are intended, just as the term κόλασις by itself also meant "hell."[26] If justice is looked for, it is in this case to be found in the hereafter.

Sometimes the writer only mentions that he has been "unfairly treated." *DTA* 102 (fourth century B.C., Attic) begins as follows: "I send a letter[27] to the daemons and to Persephone and 'deliver' to them Tibitis, daughter of Choirine, who has wronged me (τὴν ἐμ⟨ὲ⟩ ἀδικο⟨ῦ⟩σαν)." In *DTA* 120 (third century B.C., Attic) the victim is similarly designated: τὸ[ν] ἐμὲ ἀτιμο⟨ῦ⟩ντα ("who has treated me shamefully"). And on a very severely mutilated tablet (*DTA* 158, third century B.C., Attic) we read, το]ῖς ἀδικο⟨υ⟩μένοις . . . ἀ]δικο⟨υ⟩μέν[οις . . . ⟨κ⟩αὶ δίκ[η . . .].[28]

Another Attic tablet with the verb ἀδικέω presents us with an additional feature; *DTA* 98 (third century B.C.) curses Euryptolemos and Xenophon with terminology usually associated with *defixiones,* but it ends as follows: φίλη Γῆ, βοήθει μοι. ἀδικούμενος γὰρ ὑπὸ Εὐρυπτολέμου καὶ Ξενοφῶντος καταδῶ αὐτοὺς ("Dear Earth, help me. It is because I was wronged by Euryptolemos and Xenophon that I curse them"). We immediately feel a change in atmosphere. It is no longer an instruction to the god with an automatic result, but a flattering (φίλη!) request for help. And this request is supported by a reference to the injustice suffered. Perhaps more than argumentation is hiding in the phrase ἀδικούμενος γάρ. A fourth-century B.C. Attic tablet points in still another direction (*DTA* 100): "Saturos from Sounion and Demetrios and all of them as well, whoever else [is hostile] to me. I bind (καταδῶ) them, [I] Onesime. All of them, their persons and their acts against me I entrust to you, in order that you 'take care' of them, Hermes Restrainer, restrain their names, and all that belongs to them." In this first part of the text we find the traditional terminology of a *defixio,* particularly in the use of the verb καταδῶ. But it is remarkable that the writer makes herself known—Onesime. Then the tone changes abruptly:

Ἑ]ρμῆ καὶ Γῆ, ἱκετεύω ὑμᾶς τηρ⟨ε⟩ῖν
ταῦτα καὶ τούτους κολάζ⟨ε⟩τ⟨ε⟩
σῴζετε τὴ]ν μολυβδοκόπον

Hermes and Ge, I beg you to take care of all of this and punish them,
but [save her] who has "struck" the lead.

The verb *ἱκετεύω* means "to plead"; and as such, just like the phrase "help me" in the previous text, it is not exactly at home in the text of a traditional *defixio*. The key word here is *κολάζ⟨ε⟩τ⟨ε⟩*, which implies a justified punishment. What is of the greatest interest, however, is the closing passage, which (although it is mutilated) almost certainly asks that the writer of the tablet herself should suffer no harm. It seems as though she excuses herself for this indecent but unfortunately unavoidable device.

We encounter a similar tone in a very elaborate curse text from Athens (SGD 21, first century A.D.).[29] Someone curses and "deposits" (the verbs are *καταγράφω κὲ κατατίθεμε*) in the name of a number of gods of the underworld, including Hecate, the one who stole some articles of clothing mentioned in the text, as well as those who know about it but deny knowing it (*ἔτι κατατίθεμε κὲ τοὺς συνειδότας τῇ κλέψει κὲ ἀρν[ουμ]ένους*). Hecate, who is addressed directly several times (*ὦ δέσποινα Ἑκάτη*) is given the task "to cut the heart of the thieves or the thief." Despite these unique characteristics, the whole text is very similar to a traditional *defixio*. The author is clearly aware of this, for in the editor's version the first four lines we find:

> [. . . .]ες σ̣[έ]βου μὲ τὸν [κ]αταγρά̣-
> φοντ̣α κὲ τὸν ἀπολέ[σαντα] ὅτι οὐκ ἕ-
> κων ἀλλὰ ἀνανκαζ[όμεν]ος διὰ τ̣οὺς
> κλέπτας τοῦτο ποιεῖ.

Jordan has informed me privately that instead of [. . . .]ες σ̣[έ]βου μὲ, he reads *ἐξξερουμε* at the beginning of the first line. Interpreting this as *ἐξαιροῦμαι*, we can then translate:

> I make an exception for the one who is writing this *defixio* and thereby destroying the thieves, because he does not do this voluntarily but is forced[30] by the thieves.

It is as if the author contends that he does *not* belong in the collections of Wünsch and Audollent. But where does he belong? We will discover this forthwith with the help of a text that marks the transition between the traditional *defixio* and what we usually call prayer.

In 1957 a lead curse tablet was discovered on the island of Delos (SGD 58, first century B.C. or first century A.D.?) covered with writing on both sides. I shall discuss both sides, albeit not in the same sequence as the original editor.[31] Side B begins as follows:

> [Κύριοι] Θεοὶ οἱ Συκοναῖο[ι ? -] ΤΟΙΚΟΥΡΙ-
> —Κυρ]ία Θε⟨ὰ⟩ Συρία ΗΙ . . . ΤΟΙ . . . Συκο̣[να
> [ἐκδικ]ήσετε κὲ ⟨ἀ⟩ρετὴν γεν⟨ν⟩ήσετε·
> [κατα]γράφο τὸν ἄραντα, τὸν κλέ-
> ψαντα τὸ δραύκι⟨ο⟩ν· καταγράφο τοὺς
> συνιδότε⟨ς⟩, τοὺς μέρο[ς] λ̣α[βό]ντες.

Although without a doubt the characteristics of the traditional *defixio* predominate, we can once again detect a different kind of mentality in the opening lines:

> Lords gods Sykonaioi . . . , Lady goddess Syria . . . Sykona, punish, and give expression to your wondrous power.[32] I curse the one who took away, who stole, my necklace. I curse those who had knowledge of it, those who participated.

A recital of the specific parts of the body to be bound follows "from head to toe," identical to the lists found in traditional *defixiones,* and then we read the catchall phrase, "whether (the crook be) woman or man." We immediately recognize the verb *ἐκδικέω*, which (as we have seen above) signals the prayer for justice. It is this aspect of "prayer" that dominates the text of side A.

> Κύριο[ι] Θεοὶ οἱ Συκ⟨ο⟩ναῖοι Κ[--]
> [Κ]υρί⟨α⟩ Θε⟨ὰ⟩ Συρία ἡ Συκονα Σ̣[--]
> ΕΑ *ἐκδικήσετε καὶ ἀρετὴν*
> *γεννὴσετε κὲ διοργιάσετε*
> *τὸν ἄραντα, τὸν κλέψαντα τὸ δρ-*
> *άκι⟨ο⟩ν, τοὺ⟨ς⟩ συνιδότες, τοὺς μέ-*
> *ρ*[ο]*ς λαβόντες ἴδε γυνὴ ἴτε ἀ-*
> *νήρ.*

> Lords gods Sykonaioi, Lady goddess Syria Sykona, punish, and give expression to your wondrous power and direct your anger to the one who took away my necklace, who stole it, those who had knowledge of it and those who were accomplices, whether man or woman.

This text is nearly identical to the first except that instead of *καταγράφω* ("*I* curse") it says *διοργιάσετε* ("*You* must fulminate against"),[33] and the recital of the cursed parts of the body is missing. These two texts on one tablet show in a truly exemplary way the two possible appeals to the supernatural that were available to the victim of an injustice. Just like those who cannot blame their rivals for any wrong except their rivalry (e.g., in the context of circus, amatory, judicial, and commercial curses), the inscriber of this tablet, too, could avail himself of a traditional *defixio*. Unlike his colleagues, however, he could add references to his victim's guilt and his own innocence. Despite the fact that the text on Side B would (on formal grounds at least) be regarded as a typical *defixio,* the assertions of righteousness seem to diminish, if not neutralize, the negative connotations usually attached to this extreme form of black magic.

On the other hand, the injured party could totally refrain from the techniques and terminology of the *defixio* and appeal exclusively to the aid and might of the gods. The only "pure" illustration of this type is our final text (although the text from Centuripae cited above, p. 64 is certainly a strong possibility). All of the other texts are more or less hybrids of the traditional *defixio* and the standard prayer for justice that is the subject of the next section. Although it is conceivable to divide the material into two polar opposites—*defixio* and prayer for justice—there is, as we have seen, a whole spectrum of approaches that lie between them. Absolute distinctions, though sometimes indispensable for systematic definitions, are more-often-than-not blurred or even nonexistent in reality. Consequently, I do not plead for the complete elimination of the samples of our "border group" from the collections of the *defixiones,* provided that their specific peculiarities are duly recognized and appreciated. Just as elements usually associated with religious prayer tend to occur in the texts of the *defixiones,* so too we shall meet striking examples of curse terminology in the "pure" juridical prayers which follow.

In conclusion then, to the degree that the anticipated revenge against a guilty person is justified, we sometimes encounter in the boundary area between curses and prayers the following alien (i.e., nontraditional) elements in some *defixiones:*

1. the name of the author;
2. an argument defending the action, sometimes with a single term, sometimes with more elaborate detail;
3. a request that the act be excused or that the writer be spared the possible adverse effects;
4. the appearance of gods other than the usual chthonic deities;
5. address of these gods—whether because of their superior character or as a persuasive gesture—either with a flattering adjective (e.g., *φίλη*) or with a superior title such as *κύριος*, *κύρια*, or *δέσποινα*;
6. expressions of supplication (*ἱκετεύω*, *βοήθει μοι*, *βοήθησον αὐτῷ*) added to personal and direct invocations of the deity;
7. terms and names that refer to (in)justice and punishment (e.g., Praxidike, Dike, *ἐκδικέω*, *ἀδικέω*, *κολάζω*, and *κόλασις*).

We shall repeatedly encounter all these elements in the prayers for justice, the area of inquiry that lies just on the other side of the boundary between *defixio* and prayer.

THE PRAYER FOR JUSTICE

The person in antiquity who had suffered an injustice and had gone to the authorities in vain,—if indeed he had bothered to go at all[34]—had in fact only *one* authority at his disposal: he could lodge his complaint with the god(s). This did happen regularly in the form of prayers that I collect under the term *judicial prayer* or *prayer for justice*. Once again I shall give an overview of the Greek epigraphical material. I must disclaim any attempt here at a full treatment of two other special categories of prayer that I shall refer to no more than is necessary. These are the specific "prayer for revenge" and the so-called confession inscriptions, which are represented by a rather large number of inscriptions and are elsewhere described in detail, although a corpus of both would be desirable. I adduce, by way of an introduction, two often-quoted-and-studied texts. The first is one of the oldest Greek texts on papyrus, the famous curse of Artemisia from the temple of Oserapis in the Serapeum of Memphis (fourth-century B.C.).[35] It is too long to quote the whole in Greek here but I give a complete translation and sections of the Greek text wherever I deem them appropriate:

> O Lord Oserapis (*ὦ δέσποτ' Ὀσεράπι*) and you gods who sit enthroned together with Oserapis, to you I direct a prayer (*εὔχομ]αι ὑμῖν*), I Artemisia, daughter of Amasis, against (*κατά*) the father of my daughter, who robbed her of her death gifts(?) and of her coffin. Now, if he has done justice to me and to his children, then may that be just. Exactly

in the way that he did injustice (ἄδικα) to me and to my children, in that way Oserapis and the gods should bring it about that he not be buried by his children and that he himself not be able to bury his parents. As long as my accusation against him (καταβοιῆς) lies here, may he perish miserably, on land or sea, he and all his (possessions), through Oserapis and the gods who sit enthroned with Oserapis, and may he find no mercy (μηδὲ ἱλάονος τυχάνοι) with Oserapis nor with the gods who sit enthroned with Oserapis.

Artemisia placed this petition (κατέθηκεν τὴν ἱκετηρίην τα[ύ]την), begging (ἱκετύουσα) Oserapis to do justice (τὴν δίκην δικά[σαι) and likewise the gods who reign with Oserapis. As long as this petition (ἱκετηρία) lies here, may the father of the girl find no mercy in any way with the gods. Whoever takes away this petition (τὰ γράμματα) and does injustice to Artemisia, may the god punish him (τὴ⟨ν⟩ δίκην ἐπιθ[είη), . . . insofar as Artemisia has not ordered this to them . . . (a not-very-legible passage follows).

Here we have a real prayer for justice, requesting punishment of a guilty party, directed to powerful divine judges (of the underworld).[36] There are reminiscences of curse formulas,[37] but there is no coercion; the god is master, the human subservient. Evidently the prayer has been placed in the temple, clearly visible for everyone, with the risk that someone might take it away. That person will then also have to be punished by the god. Although it is not entirely clear from the text, it seems that Artemisia leaves open the possibility that she herself can (if she wishes) grant the order to remove the letter of supplication. Just like other, still-to-be-treated texts (in particular the next one), this supplication shows similarities with the worldly ἔντευξις formulas of the Ptolemaic, and especially the imperial, periods. It is likewise clear that aside from Greek influences, Egyptian (particularly demotic) culture plays an important part here; various demotic prayers for justice are known from as early as the sixth century B.C. down to the first few centuries A.D.[38]

The second text was found in 1899 on a lead tablet near Arkesine on Amorgos (SGD 60).[39] It was dated to the second century A.D. by Homolle, the first century A.D. by Bömer, and around 200 B.C. by Zingerle. Since the tablet has disappeared, linguists and papyrologists have the last word here, for the text shows unmistakable similarities with the petitions of the ἔντευξις type. In the narrative portion there is a description of how a certain Epaphroditos,[40] with the help of evil practices, incited the slaves of the writer to flee. With the exception of this passage I again give the translation of the whole text while quoting only the most important passages in Greek.

Lady Demeter, O Queen, as your supplicant, your slave, I fall at your feet (Κυρία Δημήτηρ, βασίλισσα, ἱκέτης σου, προσπίπτω δὲ ὁ δοῦλος σου). . . . Lady Demeter, this is what I have been through. Being bereft, I seek refuge in you: be merciful to me and grant me my rights (ἐγὼ ὦ ταῦτα παθὼν ἔρημος ἐὼν ἐπί σε καταφεύγω σοῦ εὐγιλάτου τυχεῖν καὶ ποῖσαί με τοῦ δικαίου τυχεῖν). Grant that the man who has treated me thus shall have satisfaction neither in rest nor in motion, neither in body nor in soul; that he may not be served by slave or by handmaid, by the great or the small. If he undertakes something, may he be unable to complete it. May his house be stricken by the curse for ever. May no child cry (to him), may he never lay a joyful table; may no dog bark and no cock crow; may he sow but not reap; . . . (?): may neither earth nor sea bear him any fruit; may he know no blessed joy; may he come to an evil end together with all that belongs to him. (Side A)

> Lady Demeter, I supplicate you because I have suffered injustice: hear me, O goddess, and pass a just sentence (λίτανεύω σε παθὼν ἄδικα, ἐπάκουσον, θεά, καὶ κρῖναι τὸ δίκαιον). For those who have cherished such thoughts against us and who have joyfully prepared sorrows for my wife Epiktesis and me and who hate us, prepare the worst and most painful horrors. O Queen, hear us who suffer and punish those who rejoice in our misery (ἐπάκουσον ἡμῖν παθοῦσι, κολάσαι τοὺς ἡμᾶς τοιούτους ἡδέως βλέποντες). (Side B)

Again we have a humble supplication from a submissive mortal ("your slave") to a sovereign goddess (here, even "queen")[41] who is asked to show her "mercy" and (here for the first time) to "hear"[42] the supplicant by avenging him and punishing the guilty. In this case the requested punishments recall curses more explicitly, not those that we normally encounter in the *defixio* but rather those of conditional self-cursing oaths or, more especially, prohibitive curses against possible grave-desecrators.[43] The main difference from the "borderline" *defixiones* of the previous section is that here the curses are not pronounced by the writer himself but rather placed in the hands of a goddess upon whose sovereign power the writer makes himself totally dependent even when he wishes that the goddess cast a κατάδε{ε}σμο⟨ς⟩ (= *defixio*) over the house of an enemy.

Besides the characteristics of traditional prayer noted above, these two prayers both request punishments that are irrevocable. The guilty must be punished for an irreparable damage, and the punishment serves exclusively as satisfaction for the sense of justice of the injured person; in short, it constitutes a request for *revenge*. Although not strictly belonging to it, both our texts recall a well-known category of prayer for which we can note several characteristics without giving an elaborate description: the prayer for revenge. Among the funerary inscriptions described by Strubbe (chap. 2), there is a category defined by (one of) two characteristics: (1) the Sun or another great god is invoked to avenge an injustice suffered; or (2) this supplication is symbolized by the depiction on the grave stele of two raised hands. Most of these inscriptions have been collected by Cumont in various publications, and a good overview can be found in Björck.[44] Since that time several new testimonia have come to light.[45]

The Sun, "who observes all things and hears all things" (Hom. *Il.* 3.277), is indeed the most qualified avenger,[46] but sometimes other superior and all-seeing gods are attested.[47] Often the Hypsistos Theos ("Greatest God") appears in a Jewish or Christian context. Likewise, the upraised hands are often, but not always, present. Thematically, however, all these texts have in common the fact that they beg the gods for retaliation, revenge, and justice and that they usually concern themselves with cases of abnormal and therefore suspicious death. Often the deceased is envisaged as an ἄωρος or βιαιοθάνατος, that is, someone who has died "before his fated time" or in a violent and unnatural manner.[48] As is typical in traditional, premodern societies, the inexplicable death—for example by a lingering illness—is frequently attributed to the evil practices of enemies,[49] who are suspected of using poison or magic spells. These revenge prayers aim at forcing the usually unknown perpetrator to atone for his or her crime.[50] Of the twenty-two pagan Greek and Latin texts collected by Björck all but three request revenge for manslaughter. In two of the exceptions[51] it is clearly a question of a security or *depositum* that the deceased had left and the present owner now (after the death of the real owner) refuses to relinquish. Here, as in the cases of manslaughter, it is a question of an irreparable loss.

The terminology is formulaic and shows a strong forensic influence. Terms with the element *ἐκδικε–* occur most often by far (the numbers refer to Björck's collection): *θεῖον φάος ἔκδικον ἔστω* (2); *ἐκδίκησον* (4, 18); *ἐκδικήσῃς τὸ αἷμα* (12);[52] *ἐκδικήσατε* (13); *ἐκδικήσειαν* (17); *este vindices* (16). Number 10 has *Tu indices eius mortem,* often read as *Tu* ⟨*v*⟩*indices* or *[u]t vindices.*[53] Other terms occur with the connotation of "prosecution": *μετέλθετε αὐτούς* (11) and *ἵνα με[τ]έλθῃ αὐτὴν ὁ Ἥλιος.*[54] The verb *μετέρχομαι* here literally means "pursue." Next in frequency is the verb *ζητέω*: *Ἥλιε, τὴν μοῖραν ζήτησο⟨ν⟩ ἐμὴν* (6) and *τὸ αἷμα . . . ζητήσεις τὴν ταχίστην* (12). I shall discuss below the several meanings of the verb *ζητέω* and its compounds. The following expressions are more distant variants of the same theme: *Ἥλιε βλέπε* (3) ("Sun, keep a watch") and also *Deus magnu oclu abet* ("The god has a large eye"), as is announced in an inscription from Rome;[55] *μὴ λάθοιτό σε ὁ ἐπίβουλος* (5); and *μὴ ἐκφύγοι τὸ κράτος τῆς θεᾶς* (14), in which it is asked that the evildoer or the crime not "escape"[56] the god. One can also "entrust" or "commit" the criminal directly to the god: *Sol tibi commendo qui manus intulit ei* (9, as well as 10).

In these short prayers we detect none of the characteristic features of a *defixio.* The situation is slightly different when we come to the related Christian revenge prayers, which are written on papyrus and can often provide much longer texts. In these texts there is often a detailed description of the diseases from which the guilty should perish, which recalls the prescriptions found on a number of *defixiones.*[57] Although in these Christian texts a (presumably) violent death is often the cause of the complaint, it is usually a question of activities described as "violations" or "crimes" committed against the plaintiff or else something more general, as, *ὅλα τὰ ἐναντία πέπονθα παρ' αὐτοῦ* (24) ("I have suffered all sorts of hostilities at his hands"). Here, too, revenge is regularly demanded instead of a simple redress of the injustice. In the Greek Christian texts we repeatedly see the verb *ἐκδικέω.* An appeal to the power of the god is also made: *ἵνα βλέπω τὴν δύναμιν* (24); compare *πολέμεσον αὐτοὺς τῇ σῇ δυνάμει.*[58] In the Coptic texts, which are larded with Greek terms, the judicial terminology is similarly dominant. The god is asked (I quote Björck's translation) "Recht zu schaffen." One time we read, "I bring this indictment (λίβελλος) against" (29), or "Führe meinen Prozess gegen" (27). Another text reads, "Bringe deinen Zorn (ὀργή)" (31). The sixth-century-A.D. papyrus now in Uppsala, which was the motivation for Björck's research, brings together several interesting elements: "Let Didymos and Severine be pursued (*κατιδιωχθήτω*). . . . Lord, quickly show them your might (δ[*εῖξον*] *αὐτοῖς ταχεῖαν τὴν δύναμιν σου*). . . . Let them come before your tribunal (*καταλαβέτωσαν τὸ βῆμα*). . . . I have submitted my affair to the Lord for punishment of the evil deeds (*ἐπιδέδωκ*[*α*] *τ̣ὰ̣ ἐ̣μ̣ὰ̣. . . . δεσπότῃ εἰς ἐκδίκησιν τῶν κ̣α̣*[. . . . .)." The verb *ἐπιδιδόναι* is a technical term for presenting petitions, such as the *libellos* mentioned in the Coptic texts.

Since the publication of Björck's collection, several Greek prayers for justice have been discovered in Egypt. I translate a text on an ostracon from Esna (*O. Cair.*

J. E. 38622) that although found in 1906, was published only in 1985:[59] "Claudius Silvanus and his brothers to mistress Athena against Longinus Marcus. Since Longinus—against whom we have often appealed to you because he was after our lives while we did no wrong, poor as we are—while he wins nothing with this, he still continues to be malicious against us, we beg you to do justice. We have already asked Ammon for help as well." In this case the prayer is directed to Athena (= Neith) and was probably placed in her temple, which lay underground in the necropolis.

Because it will be useful later on, I shall now cite from Björck (pp. 81–82) the four different meanings that the verb ἐκδικέω can have and that illustrate perfectly its judicial connotations: (1) claim a case, vindicate; (2) give someone legal aid or satisfaction, vindicate a person by taking up his cause, give satisfaction to a person (with a person as object); (3) punish a culprit, punish a crime, avenge (with a person or action as object); (4) act as avenger or as legal assistant (intransitive). In many aspects this verb corresponds closely to the use of the Latin verb *vindicare*.

We therefore note that the curse of Artemisia and the curse discovered on the island of Amorgos show strong thematic similarities with the prayers for vengeance, particularly with the subcategory that does not concern revenge for manslaughter itself but rather retaliation for some other injustice. In particular, they have in common that instead of redress, they ask for punishment and retribution. This also connects them to texts in the border areas between *defixio* and prayer, which all hold in common the necessity that the culprit suffer *deservedly* and that the offence be considered irreparable. We shall now introduce a group of texts in which a powerful divinity is called upon to help, but this time the prayer is used to put pressure on the culprit to redress the wrong.

During excavations at Cnidus in Asia Minor in the middle of the nineteenth century, C. T. Newton found thirteen lead tablets (*DT* 1–13)[60] that had apparently been placed in the temple of Demeter. As they are clearly formulaic, it suffices to cite one example, written by a woman named Artemis (*DT* 2):

> Artemis "dedicates" (ἀνιεροῖ) to Demeter and Kore and all the gods with Demeter, the person who would not return to me the articles of clothing, the cloak and the stole, that I left behind, although I have asked for them back. Let him bring them in person (ἀνενέγκα[ι] αὐτός) to Demeter even if it is someone else who has my possessions, let him burn, and let him publicly confess ([πεπρη]μένος ἐξ[αγορεύ]ων) his guilt. But may I be free and innocent of any offense against religion . . . if I drink and eat with him and and come under the same roof with him. For I have been wronged (ἀδίκημαι γάρ), Mistress Demeter.

In these texts,[61] dated by most specialists to the first or second century B.C., the plaintif is always a woman[62] who has been injured by a usually unknown, or at least unnamed, person (or persons). They concern (among other things) theft, the refusal to return a borrowed object, and slander. The plaintif "dedicates" (ἀνιερόω, ἀνατίθημι) the culprit to Demeter, Kore, and the gods with them. Sometimes there are two avenues open to the guilty parties: they can right the wrong—for example, by returning the stolen object, after which the affair is considered closed—or (if they do not elect to do this) Demeter must force them to come to the temple themselves (ἀναβαίνω) and to confess publicly their guilt (ἐξαγορεύω). In case of theft they must bring the stolen object to the temple (*ἀναφέρω*).[63] From these texts we can infer *how* the culprit is compelled by the goddess: he is described as *πεπρημένος* (*DT* 1A, 2A, 4A, 6A—and 6B and 7A?), a term that is sharply disputed (but I shall show elsewhere that it means "burned" in the sense of "afflicted by fever or illness"). At any rate, in *DT* 3 we find instead of this word the term *κολαζόμενοι*,[64] and in *DT* 8 there is mention of *τιμ[ω]ρίας τύχοι* and *πᾶ[σ]αν κόλασιν*. Elsewhere (*DT* 1) we read *μεγάλας βασάνους βασανιζομένα*,[65] "tormented by great agonies." The goddess therefore subjects the culprit to a painful illness and in this way forces him to confess and, if applicable, to surrender the stolen goods. There is still another formula that expresses the anger of the goddesses: *μὴ τύχοι εὐιλάτου Δάματρος* ("Let him not find Demeter merciful") and variants. Sometimes this seems to be a variant of the expression mentioned above, that is, the displeasure of the goddess forces the culprit to confession (*DT* 4A, lines 8–10; 8; and probably 5 and 6A). But elsewhere it is stipulated that the goddess must *remain* "unmerciful" and "implacable" even after the confession[66] (*DT* 1A; 4A, lines 4–6; 6B; and probably 12 and 13). These two alternative forms often seem to run somewhat together in the mind of the author.

Through this "consecration" to the goddess the culprit has entered a provisional taboo situation. He is cursed for the time being and belongs in one way or another under the control of the divine powers of the underworld. In order to protect themselves from "contagion" the writers in many cases added the proviso that the curse not strike the writer of the tablet, not even if he or she and the culprit are together accidentally—eating, drinking or living under the same roof[67]—things that can happen in a small community where the guilty party is anonymous. In addition the writer sometimes excuses the action with the expression, *ἀδίκημαι γὰρ Δέσπο[ι]να Δάματερ* (*DT* 2, "For I have been wronged, Mistress Demeter"; cf. *DT* 8, line 20). We have already encountered similar formulaes of excuse (see pp. 66–67). About the exact implications of the act of "dedicating" the victim I cannot digress at this point. It certainly does not mean that the culprit has fallen into a kind of holy slavery, as has been supposed, but rather that he has been "entrusted, committed" to the god, as we often find in the use of *(παρα)κατατίθεμαι*, *παραδίδωμι*, and similar verbs in the texts of the *defixiones*.[68]

With this, the culprit has become the "care" of the goddess, who now tackles the investigation and the prosecution and presides as the judge over an imaginary court. One could also "consecrate" other things to the deity besides the culprit; a bronze tablet (perhaps third century B.C.) found in Southern Italy around 1775 has often been compared with the Cnidian texts.[69] I shall quote only the second half:

> Kollura consecrates (*ἀνιαρίζει*) to the servants of the goddess the three gold pieces that Melitta received but does not return. Let her (Melitta) dedicate (*ἀνθ̣είη*) to the goddess twelve times the amount together with a *medimne* of incense according to the measure valid in the city. And let her not breathe freely until she has dedicated (*ἀνθείη*) it to the goddess. But if the writer of the tablet eats or drinks with her without knowing it,[70] then let her be unpunished, even if she finds herself under the same roof."

There are obvious parallels with the Cnidian texts; a prayer for divine justice containing elements of a "consecration" to the gods (here the verb is *ἀνιαρίζω*) pressures the (identified) victim to right the wrong she has committed. She is "not to breath freely" (*μὴ πρότερον δὲ τὰν ψυχὰν ἀνείη*)[71] until the stolen item is returned to the temple of the goddess (here the verb is *ἀνθείη*). The Italian tablet also ends with a very similar "protection clause." What is different is the extra, very heavy fine imposed upon the criminal by an overlap of the spheres of divine and human justice.[72] What concerns me especially is that here not the culprit but the stolen object is "consecrated" and is hence made the "god's affair." We encounter something similar in a recently published text of a bronze tablet, of unknown provenance but coming at any rate, from Asia Minor.

The tablet, which has been dated variously from 100 B.C. to 200 A.D. has a round hole in the middle of the top edge, presumably for attaching it to a surface for public display. I reproduce the text of the editio princeps:[73]

ἀνατίθημι μητρί σ̣ε θεῶν
χρυσᾶ ἀπ⟨ώ⟩λεσ⟨α⟩ πάντα ὥ-
στε ἀναζητῆσ⟨α⟩ι αὐτ-
ὴν καὶ ἐς μέσον ἐνε-
κκεῖν πάντα καὶ τοὺς
ἔχοντες κολάσεσθα-
ι ἀξίως τῆς αὐτῆς δυνά-
με⟨ω⟩ς καὶ μήτε αὐτ[ὴν]
καταγέλαστον ἔσεσθ[αι.]

I shall suggest as a correction that the reading *σ̣ε* in the first line is based on a misunderstanding, since the *ε* cannot be seen on the excellent photo published alongside; the putative *σ* is rather an unsuccessful start of the *θ* of the *θεῶν* that follows it.[74] My rejection of *σε* is supported by the fact that an addressed object is totally without parallel in such texts and by the syntactically impossible position between *μητρί* and *θεῶν*. Therefore I propose as translation,

> I consecrate to the mother of the gods the gold pieces that I have lost, all of them, so that the goddess will track them down and bring everything to light[75] and will punish the guilty in accordance with her power and in this way will not be made a laughingstock.

The text has a clear relation to the Cnidian curse as well as to the South Italian tablet. Here, too, the stolen object is consecrated to the goddess, who has to "track it down" and punish the guilty. However, this tablet also refers to another genre of inscriptions. We encountered the appeal to the miraculous power of the god in some prayers of revenge and in the lead tablet from Delos cited above. Its editor had not noticed that the *ἀρετή* mentioned there recalls inscriptions from Asia Minor, in which this term and the concept behind it are so pronounced that they are sometimes referred to as "aretalogies,"[76] although scholars usually prefer to call them "confession inscriptions." And in these texts we also encounter (aside from compounds of the verb *ζητέω* and an emphasis on punishments by the god) the warning, as in the bronze tablet just quoted, not to trifle with the god.

We have now considered the whole collection of Greek prayers for divine "legal aid" and have analyzed their characteristics. In the prayers in which there is no overriding demand for punishment or revenge but rather for clarification, confession, and (if possible) settlement of the dispute, the guilty is often "dedicated" to the deity for prosecution; at other times the stolen object itself is "consecrated" to the deity in hopes that it might be tracked down. We have saved one question until now, namely: What happened to the stolen property after the dispute was settled? Was it given back to the owner or did it remain not temporarily but permanently in the possession of the deity? There are no clear pronouncements about this. At Cnidus the thief first has the opportunity to return the stolen object to its owner and in this way can escape divine punishment, that is, he "burns" only if he fails to do this and "has to bring" the object to the goddess. The verb employed here (*ἀναφέρω*, "offer as sacrifice"!) seems to suggest that the object will indeed remain in divine possession, a situation that the term *ἀνθείη* in the South Italian text could also indicate. However, in this matter different practices could conceivably exist, as will certainly appear from the following discussion.

THE CONFESSION INSCRIPTIONS

Beginning in the last century steles were found in the northeastern area of Lydia (known as Maeonia or Katakekaumene) and in the adjacent area of Phrygia bearing similar texts, which were, as a group, different from the usual votive inscriptions. Since they often contain a kind of confession of guilt, they are generally called "confession steles".[77] The steles are often precisely dated and without exception come from the second and third century A.D. Although there exists great variation, they can generally be classified as praises for or aretalogies of the god, in which the *δύναμις* (power)[78] of the usually local divinity (e.g., the Great Mother, especially as Meter Leto; Men, with several epithets; Apollo, with epithets such as Lairbenos) is described and glorified. In addition we often encounter other elements. First there is an acclamation of the type "Great (is) Men" or "Great (is) Anaeitis," and so on. The reason for the erection of the stele is often a confession of guilt (*ὁμολογέω* or *ἐξομολογέω*), to which the author has been forced by the punishing intervention of the deity (*κολάζω*, *κόλασις*), often manifested by illness or accident. Sometimes the victim of the punishment has asked (*ἐρωτάω* or *ἐπερωτάω*) the deity the reason for the punishment and what he should do to propitiate the god; unsolicited divine commands also occur. By his confession and eventual reparation of the wrong the culprit appeases the god (*ἱλάσκομαι* or *ἐξιλάσκομαι*); the god therefore reveals his *dynamis* by both the punishment and the cure of the victim, and as homage to the god the story of the miracle is now written on a stele (*στηλλογραφέω*), sometimes at the command of the god but not necessarily so. The text often ends with a clear profession of faith: "And from now on I praise the god" (*καὶ ἀπὸ νῦν εὐλογῶ*) or sometimes, especially in Phrygia, a warning: "I warn all mankind not to disdain the gods, for they (i.e. mankind) will have this stele as a warning" (*παραγέλλω πᾶσιν μηδὲν καταφρονεῖν τῶν θεῶν ἐπεὶ ἕξει τὴν στήλην ἐξενπλάριον*).

These inscriptions often focus on transgressions of the type that we encountered in the tablets from Cnidus: theft, failure to return a deposit, and slander, especially with regard to allegations of poison or black magic.[79] Perjury, mentioned with some frequency,[80] forms a bridge of sorts between the transgressions against men and those against gods. For us it is interesting that in the category of profane transgressions the god is sometimes appealed to by the injured person in order to punish the guilty and in this way do justice. And in some cases it is also explicitly announced by what means the god has been brought to act. I quote a well-known confession inscription:[81] "To Men Axiottenos. Since Hermogenes, son of Glycon, and Nitonis, son of Philoxenos, have slandered Artemidoros with respect to wine, Artemidoros has given a *πιττάκιον*. The god has punished Hermogenes, who has propitiated the god, and from now on he will extoll (the god)." A *πιττάκιον* is a tablet covered with writing, which is presented to the deity. It is, among other things, a technical term for oracle questions placed before the gods in Egypt.[82] In another confession inscription (*TAM* 318) we are told how a woman, Tatias, stands accused (in a campaign of gossip) of poisoning her son-in-law, who had subsequently gone out of his mind. She reacted as follows:

> *ἡ δὲ Τατιας ἐπέστησεν*
> *σκῆπτρον καὶ ἀρὰς ἔθηκεν*
> *ἐν τῷ ναῷ, ὡς ἱκανοποιοῦ-*
> *σα περὶ τοῦ πεφημίσθαι αὐ-*
> *τὴν ἐν συνειδήσι τοιαύτῃ.*
> *οἱ θεοὶ αὐτὴν ἐποίησαν ἐν*
> *κολάσει, ἣν οὐ διέφυγεν.*
>
> Tatias drew up a scepter and placed *ἀραί* (curses) in the temple, as if to show that she was not guilty of the transgressions attributed to her,[83] although she was aware of her guilt. The gods subjected her to a punishment that she did not escape.

Apparently the poor woman died, and in the remainder of the text we are told how her relatives "unbound (*λύω*) the *ἀραί*" and successfully propitiated the gods.

The text contains many interesting elements, among which the following are of importance to us. There is a ritual opening of the judicial process by the "drawing up of a scepter."[84] We meet this and closely related expressions again several times. In particular some recently found texts make it clear that it is here a question of making visible the present power of the god. The *ἀραί* most likely contained a conditional self-curse that Tatias uttered in order to prove her innocence, a procedure similar to that which appears in a Cnidian tablet (*DT* 1). What is essential is that these *ἀραί* are clearly related to the *πιττάκιον* of the first text; the case is entrusted to the god in writing, and punishment is implored for the culprit. Perhaps we can find another example of such a "juridical prayer" in a confession inscription from Maeonia (*TAM* 362) that mentions "someone who has overthrown and removed the tablet (*π[ι]νακίδιον*)."[85] One way to appeal to divine justice (*ἐπεκαλέσατ[ο κατ᾽ αὐτοῦ τὸ]ν θεόν*, in the words of *TAM* 525) is therefore a written complaint in a temple. Now we have already become acquainted in detail with such petitions in the previous section. The similarity even goes so far that Artemisia feared that her "curse" would be taken away, something that seems to have happened in the Maeonian inscription under consideration. The Cnidian prayers provide closer comparison.[86] There is, to be sure, a time difference of at least one—and most likely two or three—centuries, but the dated confession inscriptions themselves prove precisely how persistently a religious practice can be maintained over two centuries, even to details of wording.

We could say that the Cnidian tablets form the opening to a legal proceeding, just like the ἀραί, the πιττάκιον, and the πινακίδιον in the confession inscriptions; while the confession inscriptions themselves describe the course and the conclusion of the whole lawsuit. Conversely, it also appears that such "divine justice" can function especially well where illness, misfortune, or even death are regarded in the first place as divine punishment and thus lead to serious self-examination for possible sins committed against gods or men. A series of short votive inscriptions from Philadelphia[87] illustrates perfectly the two possible reactions to illness and cure. On the one hand we find the usual type, εὐ[ξά]μενος . . . . . . εὐχὴν ἀνέ[θ]ηκα ("Having made a vow . . . I redeemed it"),[88] in which a vow for the cure of an illness is simply redeemed, but on the other hand there are texts like κολασθεῖσα το[ὺ]ς μασστο[ὺς] εὐχὴν ἀ[ν]έθηκα,[89] in which an illness affecting a woman's breasts is clearly seen as a punishment.

This relationship between the Cnidian tablets as complaints before a divine tribunal and the confession inscriptions as records of divine justice invites closer scrutiny of the confession inscriptions with regard to the question that we raised above: What was the fate of the possession obtained illegally by the culprit? As we have noticed, the wording of the Cnidian tablets and the related tablet from Southern Italy seemed to indicate that the object brought to the temple would in fact be the possession of the goddess(es) (which by itself would not necessarily have placed it outside human use).[90] Such an outcome, however, does not seem logical to every modern researcher: "Aber welcher Fluch! Wenn sie aber das Gestolene nicht wiedergeben, so sollen sie es der Göttin zustellen!" C. Wachsmuth wrote indignantly.[91] Some examples from the confession inscriptions demonstrate that this could indeed happen. According to an inscription from Koresa (Lydia)[92] an article of clothing was stolen from a bathing establishment: "The god was vexed with the man and after some time had him bring the cloak to the god. He openly confessed his guilt. Then the god ordered him, through the agency of an 'angel,' to sell the article of clothing and to publicize his miracles on a stele". One can of course contend that the exceptional nature of the command proves that *normally* the returned object will *not* belong to the deity. But there are other attestations. From the neighborhood of Kula (Maeonia) comes a confession inscription[93] whose legible part relates the gods' request that certain individuals return a sum of money that they are illegally withholding. They did *not* return it to its rightful owner, but instead "the sons of the culprits gave double the original amount to erect a stele"; thus it is somewhat comparable with the South Italian prayer (see p. 73). In an unpublished confession inscription from Usak[94] two people ask the *patrioi theoi* what they could have done wrong. Somewhere along the line this is made clear to them, for they tell us, "We deposited 100 denarii just as the *patrioi theoi* had demanded (καθὼς ἐπεζήτησαν)." Although there is no certainty here, I suspect that it concerns a sum of money acquired in a disreputable manner; again, this time, the money is not returned to the owner but is deposited in the temple.

Of course it also happens that the original owner gets the lost or stolen item back again. Clearly there existed a variety of practices; this will be confirmed further from the Latin material. At any rate, the very demand to erect a stele means an assessment on financial means. An inscription (*TAM* 327) begins as follows: "Anaeitis is great. Since Phoebus has sinned, the goddess asked for a *ἱεροποίημα*." This term, which occurs several times,[95] probably means a "gift to the deity" and is perhaps identical to the *στηλογραφία*. There is *one* text from Cnidus that although it has disappeared from scholarly discussion since Newton's edition, nevertheless offers a nice parallel to this.[96] In it we read that a woman dedicated *χαριστεῖα* and *ἐκτίματρα* to Demeter, Kore, and the gods with Demeter and Kore—in which the term *ἐκτίματρα*, otherwise unattested, must mean (using the analogy to terms such as *σῶστρα*, *λύτρα*, or *μήνυτρα*) something like "payment"—in this case in the form of high veneration and thanks—somewhat comparable to the *ἱεροποίημα* and praise of the confession stele.

Finally, a few words on the question of terminology. The verb *ἐπιζητέω* appears as one of the standard ingredients of the confession texts. This is significant since it has, like many of the terms discussed above, strong juridical connotations. I now summarize the results of a long and detailed analysis, the bulk of which I cannot present here. The verb, which in the confession texts nearly always expresses the action of the god(s), functions in roughly three divergent, yet nevertheless interrelated, ways:

1. When the direct object is inanimate, it means "to demand, to require something," for example, satisfaction, a *ἱεροποίημα*, a stele. Occasionally it means "to claim" an object that the gods regard either as their own property or as stolen property that has been entrusted to their care.
2. When the direct object is a person, the verb means "to pursue" a guilty party. Occasionally, we find a more concrete variant, such as the use of the verb *μεταβαίνω* in an inscription from Usak,[97] where a woman refused to pay a promised reward for child rearing: *καὶ οἱ θεοὶ μετέβησαν ἰς αὐτήν* ("And the gods chased after her"), easily comparable with the use of the verb *μετέρχομαι* in some prayers for revenge (see p. 71).
3. It also occurs without any object at all, in which case some object, animate or inanimate, is often easily supplied from the text. In some cases, however, it is used in the absolute sense of "to investigate the affair" or "to hold a judicial inquiry."

The investigation referred to in the last definition may be initiated by the injured party.[98] In *TAM* 317 several people have stolen pigs from Demainetos and Papias, and when they inquired into (*ζητέω*) the matter, the culprits denied it. The next step is to entrust the case to the justice of the god. This is exactly the same situation that occurs in Babrius' fable (no. 2), where a farmer loses his mattock and starts an inquiry (*ἐπιζητέω*) to ascertain whether it was stolen by certain individuals. When they deny it, he takes them to a temple in a nearby city to take an oath.

Our insight into the various meanings of (*ἐπι*)*ζητέω* can perhaps put an end to the long discussion[99] of another term in one very illuminating confession text (*TAM* 440) in which several gods are invoked with regard to a deposit of 40 denarii that Apollonios has left with Skollos. When Apollonios tries to get the money back, Skollos swears by the gods invoked at the beginning of the inscription (*τοὺς προγεγραμένους θεούς*) that the money had been returned at the agreed-upon time. Afterwards, because Skollos had broken his word, Apollonios *παρεχώρησεν τῇ θεῷ*. The gods punish Skollos with death and set up a judicial inquiry (*ἐπιζητέω* in its absolute sense). In the end Tatias, his daughter, redeems (λύω) the oath.

I have left the phrase *παρεχώρησεν τῇ θεῷ* untranslated. I cannot go into the lengthy debate over the use of the verb *παραχωρέω* here; it suffices to say that its semantic range includes "give way to," "leave or concede something to somebody," and especially (and to us most significantly) in legal parlance "surrender or cede a right or a legal claim to another person." As there is no explicit object for *παραχωρέω*, we may choose, as we do with *ζητέω*, to supply an animate or inanimate object; or we may understand it in an absolute sense. Just as we have seen the three possible functions of *ἐπιζητέω* we may conclude that the same three options are possible here. For *ἐπιζητέω* is precisely the divine response to the human act of *παραχωρέω*, the latter being a legal equivalent, of sorts, to the more hieratic terms used in the Cnidian and related tablets, such as *ἀνιερόω*, *ἀνιαρίζω*, or *ἀνατίθημι*, which could similarly have either the guilty party or the stolen property as objects. It is therefore quite unnecessary to single out one of the possible meanings. I think there is a good possibility that a larger concept of judicial cession is involved, that is, the plaintiff hands over the stolen property, the accused, and the entire case to the god(s) for a final decision. The act of cession may have been symbolically, if not legally, expressed by physically placing the *πιττάκιον* in the temple of the god; and it is possible, though in this case not demonstrable, that by this act the ceded object becames the property of the god.

At the end of our paraphrase of *TAM* 440 we also see the use of the verb λύω (loosen), which occurs in its various forms[100] rather frequently in the confession texts. It expresses the act of paying ransom in order to propitiate the god and assuage his wrath.[101] Therefore a curse in which the gods are called *ἀνεπιλύ[τους* ("unbending," lit., "unable to be loosened") or the "scepters" are termed *ἄλυτα* ("unable to be loosed") is particularly threatening,[102] since it no longer offers this last way out. The term λύτρον in the sense of "ransom," that is, "a means of escaping the consequences of sin" is commonplace in the New Testament, particularly in Pauline discourse,[103] which the otherwise purely pagan confession inscriptions recall in many respects. For us it is essential to stress the fact that in the case of a transgression against another person a simple settlement of the dispute may not be sufficient; additional "compensation" may also be necessary as atonement for an implicit wrong (or an explicit one, e.g., perjury) committed against the god. Such a λύτρον may have taken the form of a stele, in which case the terms λύτρον and *ἱεροποίημα* would seem to coincide.

CONCLUSION FROM THE GREEK MATERIAL

By way of the borderline *defixiones* we entered the domain of the judicial prayer. In doing so we noted that as an alternative to "taking the law into one's own hands" by means of a *defixio*, people could express their grievances against their fellow human beings in a prayer that submitted their complaint to the god. In practically all these prayers the deity is presented as a superior, majestic autocrat to whom human beings in all humility submit their cases, like in the Egyptian ἔντευξις to the monarch. The accuser transfers the whole lawsuit or the culprit or the stolen property (or all three at the same time) to the care of the god. In the explicit prayers of revenge there is principally a demand for retaliation against, and punishment of, the guilty party (κολάζω, ἐκδικέω, μετέρχομαι, καταδιώκω, or the Latin *vindicare*), to which end either the guilty person is "handed over" (κατατίθημι, κατατίθεμαι, or the Latin *commendare*) or the entire case is entrusted to the god (ἐπιδέδωκ[α] τ̣ὰ̣ ἐ̣μ̣ὰ̣, "I bring this *libellos,*" "führe meinen Prozess"). In the other judicial prayers as well as in the confession inscriptions, although the thought of revenge certainly plays a part, it is above all the restitution of, or indemnity for, the loss or damage (e.g., in the case of slander) that is demanded. Case, person, or stolen goods can be entrusted to the deity (the verbal forms vary, e.g., ἀνιερόω, ἀνιαρίζω, ἀνατίθημι, παραχωρέω), who then opens an inquiry, prosecutes the guilty, or claims the stolen property. The verb ἐπιζητέω is the typical term and is used in all these meanings, just as in the revenge prayers ἐκδικέω and *vindicare* possessed a comparable-though-not-identical spectrum of meanings. Likewise we encounter ἀναζητέω with the same meaning or simply ζητέω (e.g., with αἶμα or μοῖρα as a direct object)[104] in the biblical sense of "demand an account of."

The divine intervention appears in the form of illness, accident, or the death of the guilty and is seen either as an irrevocable punishment (κόλασις, τιμωρία, etc.) or as a conditional and temporary means of pressure (e.g., "May he not breathe easily until. . . ."), that is, a judicial torment (βασανίζειν and related words) by which the guilty is eventually brought to confess (ἐξαγορεύω, ὁμολογέω, or ἐξομολογέω)[105] and restitute or compensate for, what was owed (ἀνενέγκαι αὐτὸς παρὰ Δάματρα, etc.). Sometimes the stolen property is returned to the original owner, and sometimes it remains (whether according to an explicit agreement or not) property of the deity. The latter can be regarded as the result of an actual cession.

Finally, we must mention the physical presentation of these prayers for justice. Newton announced that he found the Cnidian tablets "broken and doubled up" (p. 382). Nevertheless he insisted that "they were probably suspended on walls as they are pierced with holes at the corners" (p. 724). Although some scholars claimed that these holes could not be found and that some tablets were covered with writing on both sides, many others, including C. Wachsmuth, J. Zündel, R. S. Conway, E. Ziebarth, and R. Wünsch [106] continued to believe that the tablets were placed in the temple in public view to be read by everyone. One continually detects the conviction (among modern scholars at least) that publication was necessary to give some real effect to the tablets. And indeed, there do occur formulas on the tablets that warn the thief, who thereby receives one last chance to remedy his transgression. Despite opposition by Audollent,[107] who thought that these prayers were buried just like real *defixiones,* Kagarow and Björck[108] argue for the publicity of these texts, not necessarily as "public advertisements" in the temple but possibly through the agency of a priest who received the complaint and then took the necessary steps to inform the accused. Zingerle[109] has gone very far—certainly too far—in his views about a *Priestergericht,* in which priests not only had control of the lawsuit but also carried out punishment.

Without going into details here, we can without effort establish that the Greek material does not point to one uniform procedure. Many of the judicial prayers from the "border area," just like the related *defixiones,* came from graves or wells or springs[110] and therefore were permanently hidden from human view. This is probably also true for the Amorgos tablet, just as it is also certain that some of the Coptic revenge prayers from Egypt were sent along with the dead in accordance with the practice of the ancient Egyptian "letters for the dead." But even where we are in a context of a temple, variations are possible. The reference to public fines in the bronze tablet from South Italy (see pp. 73–75) strongly suggests that it was a public document, and the bronze tablet from Asia Minor (see pp. 74–75) still has the hole from which it was originally suspended. The curse of Artemisia was certainly placed in the temple to be seen by everyone, just as the *πιττάκια* of the confession inscriptions presumably were, since the fear was expressed that they might be stolen. A demotic prayer for justice[111] says, "As for anyone in the world who will set this document on fire [to destroy] it, let him not escape from our plea." Still, some of these prayers were not available for everyone to read; a Coptic-Christian prayer for revenge (Björck no. 31) has the prescription, "He who opens this papyrus (*χάρτης*) will bring on his head what has been written on it: let it go on his head."

All this can and will be researched more systematically. For the time being, however, we can conclude that in some cases—among others the public prayers for revenge—the trust in the divine power alone sufficed as a motivation for the accusation. In temples prayers could be publicly displayed—as was also the case with oracle questions[112]—and could in this way bring the thief to repentance. But they were also offered closed,[113] just as today we still encounter "letters to heaven" in churches. In this case we may assume an occasional mediating role for the priests who represent the divine court of arbitration, a phenomenon that is still known to occur in many traditional cultures of our own day.[114] C. Wachsmuth[115] compared a church practice in Greece of his time in which the accusation, including the possible divine punishment, was read out loud by the priest. In my opinion the most obvious—though by no means exclusive—procedure may be that the injured party first tries to draw a confession from the suspected culprits[116] and then tells them explicitly that he is making a higher appeal to the god. The knowledge that the accusation now rests with an all-seeing, highest authority is sufficient to force the culprit to reconsider his deeds, especially when shortly thereafter he does not feel perfectly healthy. We shall see that the Latin lead tablets from England support this view to a great extent.

LATIN JUDICIAL PRAYERS

Augustine (*De Civ. D.* 6.10) borrows from Seneca a description of the people who resided in the temple of the Trias Capitolina in Rome: "One places names before

Jupiter, the other tells him the time, still another reads to him, another anoints him. There are women who arrange the hair of Juno and Minerva and hold a mirror for them. There are also those who hand them petitions and inform them about their lawsuits (*qui libellos offerant et illos causam suam doceant*)." Although the exact religious context of this curious fragment is not quite clear,[117] it does suggest that in the first century A.D. there existed, even at the center of the *imperium Romanum,* practices that provide a direct parallel to the Greek prayers for revenge and justice we have analyzed and that encourage us to search for similar texts in Latin epigraphical materials.

Let us return to the problematic text from Italica that opens this essay. The central question was how to explain the peculiar fact that a stolen object could be called the property of the goddess (*tuas res*) by the injured party. Whoever places this text in the category of the traditional *defixio* denies one excellent avenue for explanation; for if it is compared to the judicial prayers of Asia Minor, the difficulties dissolve immediately. The double usage of the verb *persequi* has its direct counterpart in the Greek verb ἐπιζητέω, and it occurs here in two constructions that occur in the confession inscriptions, namely in the sense of "tracking down" the stolen goods as well as "prosecuting" the culprit. It is then evident that in this Latin tablet, too, the injured party has "dedicated" (or perhaps better, "ceded" in the judicial sense) his or her stolen property to the goddess and that in this way it had become, legally speaking, the case or the property of the goddess (*tuas res*) just as we saw in the Greek material. The goddess was also addressed respectfully as *domina*. Now if this were our only Latin testimony to such prayers, there would be some ground for scepticism; for there are several lacunae in the text and the Latin is far from perfect. Not too much should be deduced from the one word *tuas*. But fortunately there are more examples. In a curse text from Corsica, called a *defixio* by the editor and much improved in Solin's publication we read,[118]

> [---] ụḷẹ vindica te. Qui tibi male f̣[aciet], q̣ụị [---]
> [--- v]indica te et si C. Statius tibi nocuit, ạḅ ẹọ vind[ica te ---]
> [--- persequa?]ris eum, ut male contabescat usque dum morịẹ[t]ụr ---
> cumque alis, et si Pollio conscius est et illum persequaris,
> ni annum ducat.

I prefer (for reasons that will become clear) to restore *f[ecit]* instead of *f[aciet]* in line 1. In lines 3–4 we should understand *quicumque alius* and in line 5 *ni* as *ne*. I translate,

> ——ule (probably the name of a god), avenge yourself. Whoever has done you harm . . . avenge yourself on him, and if C. Statius has injured you, avenge yourself on him [. . . persecute] him in order that he may waste away horribly until he dies. And whoever else—for instance, if Pollio—is an accomplice, persecute him as well, so that he won't live out the year.

Without a doubt, elements of the standard terminology of *defixiones* are present, like *tabescat* and *ni annum ducat,* but Solin remarks, "Wenn ein Gegner des den Fluch aussprechenden der Gottheit Schaden zufügt, soll diese sich von jenem befreien. Dergleichen habe ich auf keiner antiken Fluchtafel gefunden. Der verfluchende identifiziert sich sozusagen mit der Gottheit, die ihm helfen soll." The text, however, immediately loses its enigmatic character when we classify it with the judicial prayers, a classification that seems warranted by the terms *vindica, conscius,* and *persequaris*. The injured party has ceded his case to the goddess, who thereby becomes a "party" to the lawsuit herself and will act as *vindex*. Hence my preference for the restoration *male fecit* (has done harm) instead of *male faciet* (will do harm).

We find added support for such interpretations in several Latin "curse tablets" (for the most part from England), of which some have been known for a long time, while others have been found very recently and are published in the latest issues of *Britannia*.[119] In 1964 the great scholar of the Latin *defixio,* R. Egger, discussed a curse tablet from Wilten-Veldidena near Insbruck that had been published in 1959 by L. Franz:[120]

Side A

Secundina Mercurio et
Moltino mandat, ut siquis * XIIII
sive draucus duos sustulit, ut
eum sive fortunas eius infi-
dus Cacus sic auferat quo-
modi ill⟨a⟩e ablatum est id quod
vobis delegat, ut persecuatis
vobisque deligat, ut

Side B

persicuatis et eum
aversum a fortunis ⟨s⟩u-
is avertatis et a suis prox-
simis et ab eis quos caris-
simos abeat, oc vobis
mandat, vos [e]um cor[ipi]a-
tis.

Egger made excellent use of the extant Latin and Greek *defixiones* and other related texts from England and shed light on several troubling aspects of the text. I do believe, however, that we can make this text yield its full meaning only by setting it against the background of the (Greek) judicial prayers. I first give a translation:

> Secundina charges Mercurius and Moltinus that whoever has stolen 14 denarii or two necklaces,[121] that the perfidious Cacus take him away or his possessions, just as they (her possessions) have been taken away from her, the very things that she transfers to you to track down. And she also assigns you to persecute him and separate him from his possessions and from his fellow men and from those who are dearest to him. With that she charges you; you have to catch him.

The parallels to the Greek judicial prayers strike us immediately. Mercurius and Moltinus have to take over the tasks of the earthly judge. The task or transfer is expressed by *mandare* and *delegare*. The latter of these words is the exact Latin equivalent of the Greek verb *παραχωρέω*, meaning "to assign, confide, commit, entrust any thing to a person (for attention, care, protection, etc.)" and "to make over either one who is to pay the debt or the debt itself," being, therefore, the term for judicial cession. It is now up to the gods to *persequi* (with a double object) first the stolen object that is ceded (*id quod vobis delegat ut persecuatis*) and next the thief himself or possibly the whole case *(vobisque deligat ut persicuatis),* all exactly as we saw in the case of the Greek equivalent *ἐπιζητέω*. The prosecution here similarly involves punishment as well, in this case by a sort of "isolation" of the guilty. I doubt that *ut eum coripiatis* means "ihr aber sollt ihn vor Gericht bringen." I should prefer "seize," "catch," perhaps "take into custody."

What happens to the stolen object? Egger thought that the thief, in Secundina's expectation, would contritely report to the gods and carry out restitution of the stolen object. Although this is not impossible, it seems equally likely that revenge and punishment of the guilty is the main purpose of the prayer. Either way, the god takes over the task of punishing the guilty and demanding an account.

In his day Egger could refer to three tablets from England that, however different they may have been, still were somewhat related to the tablet from Austria. Of these, the tablet from Lydney Park (Gloucestershire) has been known the longest:[122]

Devo
Nodenti Silvianus
anilum perdedit
demediam partem
donavit Nodenti
Inter quibus nomen
Seniciani nollis
petmittas sanita
tem donec perfera⟨t⟩
usque templum No-
dentis.

To Divus Nodens. Silvanus is missing his ring. He has given half (of its worth) to Nodens. Let no one in the group to which Senecianus belongs live in good health until he brings (the object) to the temple of Nodens.

Here we have a mechanism that corresponds in detail to that of the Cnidian tablets; pressure is exerted on the thief in the form of an illness (cf. πεπρημένος). The phrase *perferat usque templum Nodentis* corresponds to ἀνενέγκαι αὐτὸς παρὰ Δάματρα, except that here the writer suspects in what social circles the thief could be found. And there is a prearranged agreement with the deity concerning the stolen property; the god receives half of the value, which must mean that the original owner receives back the ring if it ever shows up. This is confirmed by other texts.

A tablet from the third or fourth century A.D.[123] found in Kelvedon, Essex confirms this picture. The text of R. P. Wright in the editio princeps (*Journal of Roman Studies* 48 [1958]: 150, no. 3) reads as follows:

quicumque res Vareni in
volaverit si mulrer si mascel
sangu⟨i⟩no suo solvat
et pecuni⟨a⟩e quam exesuerit
Mercurio dona et Virtuti s⟨acra⟩.

Egger improved the interpretation suggested by the editor in several respects, especially in his reading of the final line as *Mercurio dona⟨tur⟩ et Virtuti s⟨emis⟩*. He was mistaken in one detail only; he understood the phrase *sanguino suo solvat* to mean *soll persönlich zahlen,* that is, "let him pay in person."[124] Apart from the fact that such a notion cannot be expressed by the Latin here, later discoveries among the British lead tablets make it clear that *sanguine suo* can only mean "with his own blood," that is, "with his health" or "with his life." Thus, I would translate this tablet as follows:

Whoever stole the property of Varenus, whether man or woman, let him pay for it with his own blood. From the money that he (the thief) will pay back, half is given to Mercurius and Virtus.[125]

From these texts and others discussed below, it appears that the guilty must personally atone for their crime with physical suffering, illness, and so on. In addition, some form of restitution is sometimes demanded for the stolen object, which then benefits (at least partially) the deity as well. In short, just as in the Cnidian tablets, the criminal is "punished" with physical discomfort (there the term πεπρημένος is used, here *sanguine suo*) either to force him to restore the stolen property or (failing that) to provide real—and therefore permanent—punishment for the crime.

Several recently published tablets from England demonstrate quite clearly that this is the correct interpretation of the phrase *in sanguine suo*. I shall first cite the ending of a rather unusual specimen from Bath (*Britannia* 12 [1981]: 378, no. 9). Uricalus and several members of his family have sworn an oath at the spring of Dea Sulis in which the stipulation occurs, *Quicumque illic periuraverit deae Suli facias illum sanguine suo illud satisfacere* ("Whosoever committed perjury will give satisfaction [or will atone] to the Dea Sulis for it with his blood"). This text comes very close to the confession inscriptions in which there is a question of atonement for perjury, in particular, the one in which the term ἱκανοποιοῦσα (= *satisfacere*) occurs.[126] The most important observation, however, is that *sanguine suo satisfacere* can mean nothing else here except "atone *with* his own blood" (therefore with illness, in particular with fever, and perhaps even death).[127] Before discussing some recently discovered texts in which all this is expressed more explicitly, I first want to cite a puzzling tablet from Essex[128] (*Britannia* 4 [1973]: 325, no. 3) that has not really been understood:

Side A	*Side B*
Ḍiọ Ṃ(ercurio) dono ti⟨bi⟩	Dono tibi
negotium Et-	Mercurius
{t}ern⟨a⟩e et ipsam	aliaṃ neg [o-
nec sit i⟨n⟩vidi⟨a⟩ me⟨i⟩	tium N VIN
Timotneo san-	. . . .
gui[n]e suo	NII[. . . . . .
	MIN[. . .]S NG[
	SVO

The editors R. P. Wright and M. W. C. Hassall translate,

> To the god Mercury, I entrust to you my affair with Eterna and her own self, and may Timotneus feel no jealousy of me at the risk of his life-blood. (Side A)
>
> I entrust to you, O Mercury, another transaction. . . . (Side B)

At the end of side B there follows another instance of *sanguine suo*. There are numerous grammatical and spelling mistakes: *Dio* instead of *Deo, Timotneo* instead of *Timotheo, aliam* instead of *alium*. This encourages me to suggest an alternative interpretation. Eternus is a name that occurs rather often in Roman Britain and has just recently appeared again, this time on a curse tablet.[129] I have long wondered whether *negotium Eternae et ipsam* might not mean "Eterna's trade (shop, business, store) and Eterna herself," as we find them cursed regularly in Greek *defixiones*[130] (e.g., τὰ ἔργα, ἐργασία, πρᾶξις, etc.). I prefer an interpretation that the editors give with regard to *alium negotium* on side B, especially because of the parallels to other tablets from Britain and (in particular) to the Greek judicial prayers; we have here in an exemplary way the entrusting to the god of the "lawsuit itself and of the (guilty) person." However, as we have seen, in such prayers the author often does mention his name. Besides, the proposed translation of *nec sit invidia mei Timotneo* is somewhat artificial. I therefore propose to interpret as follows: "And don't let there be envy on my account (*mihi* must be understood), Timotheus."[131] The last-named is the author, who asks the god that his action not be to his disadvantage. Is this not the same stipulation that we found in the tablets from Cnidus and the related one from South Italy? If this is so, then *sanguine suo* (perhaps also on side B) must be regarded as a syntactically freestanding expression, "With his blood!" (sc., "The guilty party must pay"). This is not without parallels in related Greek texts. One of them we have already cited under the prayers for revenge: *τίς αὐτὸν ἠδίκησε τῷ ἦ αἷμ[α]*.[132] Although this does not prove my interpretation, I believe it does show that such a reading is possible and indeed—in the context of judicial prayers—more plausible than what has been previously suggested.

The profit of our comparison of the Greek and Latin texts will become even more striking when we apply our method to a well-known "*defixio*" that has often been discussed without producing any convincing interpretation. It concerns a tablet found in the amphitheater at Caerleon (the ancient Isca Silurum, encampment of the *legio II Augusta*). The editor, R. G. Collingwood, read, *Domna Ne- / mesis do ti-/bi palleum / et galliculas / Qui tulit non / redimat n[isi fusa] sanguine / sua*. Practically all of the previous interpretations of this tablet begin with the assumption that we have here a traditional *defixio* using an appropriate kind of sympathetic magic,[133] namely, that the person gives the clothing of an enemy to the goddess Nemesis in order that the owner will get them back only by paying with his life. In order to elicit this reading from the tablet, however, one must understand the word *tulit* to mean "he has worn" or "he has brought" and *redimat* to mean "may he get them back," both of which are strained, if not impossible, translations of the Latin.

If, however, we take these verbs in their normal meaning as they appear on other British lead tablets (i.e., *tulit* [= *abstulit*], "he has stolen" and *redimat,* "may he atone or redeem") we immediately see that this text is another example of juridical prayer. The translation is then very straightforward: "Mistress Nemesis, I give you (*my*!) cloak and shoes. Whoever has stolen them will not atone for it unless with his life, with his blood." It is therefore a question of cession to the goddess by which the stolen object has become her possession (cf. *tuas res* from the Italica text). This is clearly a prayer for revenge and for punishment of the guilty. *Redimere* is comparable to the Greek verb *λύω*, in the sense of "ransom, redeem, buy off, atone for."

Since I first arrived at this interpretation long ago, it has been happily confirmed by texts from Bath (*Britannia* 14 [1983]: 336, nos. 5 and 6). Number 5 reads,

ẹxẹcro qui involaver-
it qui Deomiorix de ḥos-
{i}pitio suo perdiderit quicumque rẹ⟨u⟩ṣ deus illum-
inveniat sanguine et
vitae suae illud redemạt.

I curse (him) who has stolen, who has robbed Deomiorix from his house. Whoever is guilty, may the god find him. Let him pay for it with his blood and his life.

Number 6 reads,

> Minervae
> de⟨ae⟩ Suli donavi
> furem qui
> caracallam
> meam invo-
> lavit si seṛvus
> si liber
> si ba
> ro si mulier
> hoc donum non-
> redemat nessi
> sangu⟨i⟩nẹ suo.

> I have given to Minerva, the Dea Sulis, the thief who has stolen my hooded cloak, whether slave or free, whether man or woman. He is not to redeem (i.e., in the sense of "buy off") this "gift" unless with his own blood.

The phrase "to pay with his own blood" occurs twice. The second text, in particular, is interesting because here the thief is delivered to the goddess (as in the Cnidian tablets), and he can only redeem himself from this "transfer" (*donum*)[134] with his blood.

I shall now cite some (parts of) recently discovered Latin texts to show to what extent they correspond to Greek—in particular Cnidian—texts, partially also to show how the Greek material can contribute here to the interpretation. The first is a tablet from Uley (*Britannia* 10 [1979]: 343, no. 3). As I have done before, I shall begin with the excellent translation and interpretation of the editors, M. W. C. Hassall and R. S. O. Tomlin:

Side A

> Commonitorium deo
> Mercurio [written over *Marti Silvano*] a Satur-
> nina muliere de lintia
> mine quod amisit ut il-
> le qui hoc circumvenit non
> ante laxetur nis{s}i quando
> res ssdictas ad fanum ssdic-
> tum attulerit si vir si mu-
> lier si servus s[i] liber

Side B

> Deo ssdicto tertiam p-
> artem donat ita ut
> ex{s}igat istas res quae
> ssta⟨e⟩ sunt.
> Ac a quae perit deo Silvano
> tertia pars donatur ita ut
> hoc ex{s}igat si vir si femina, si serv-
> us si liber[. . .] E[. . .]TAT.

> A memorandum to the god Mercury (erased: Mars Silvanus) from Saturnina a woman concerning the linen cloth she has lost. Let him who stole it not have rest until he brings the aforesaid things to the aforesaid temple, whether he is man or woman, slave or free. (Side A)

> She gives a third part to the aforesaid god on condition that he exact those things which have been written above. A third part from what has been lost is given to the god Silvanus on condition that he exact this, whether (the thief) is man, woman, slave or free. . . . (Side B)

This is an official complaint, called *commonitorium* here and on another tablet (*Britannia* 13 [1982]: 400, no. 4), *petitio*. In harmony with such a technical term is the vaguely official style in *ssdictus* (cf. the use of the term *προγεγραμένοι* in the confession inscription on p. 78, which concerned the dispute between Apollonios and Skollos).[135] Saturnina asks that the unknown thief "have no rest until" (*non ante laxetur*) he returns the stolen cloth, a stipulation that recalls not only the language of some *defixiones* (e.g., *ne quis solvat nisi nos qui fecimus* [*DT* 137] or the use of *ἀναλύω* in the Greek *defixiones*)[136] but, above all, the expression in the South Italian prayer (see pp. 73–74): "Let her not breath freely until. . . . " The thief can only "be free," "be liberated" when he has brought the stolen object to the temple, as in the Cnidian tablets. Saturnina in advance pledges to the god a third of the value on condition that he goes after these things; here as elsewhere the Latin *exsigat* is used in a manner identical to the Greek verb *ἐπιζητέω*.

A tablet from Bath (*Britannia* 12 [1981]: 371, no. 6) has a similar usage: "I have given to Dea Sulis the six silver pieces that I have lost. The goddess will claim them back (*exactura*) from the names listed below." The names then follow. The verb *exsigatur* occurs again on a tablet from Bath (*Britannia* 13 [1982]: 403, no. 6) that begins by dedicating some lost money to the same goddess: *Deae Suli Minervae Docca / dono numini tuo pecuniam quam* [. . . . *a]misi id est* (*denarios quinque*).

The following tablet from Uley (*Britannia* 10 [1979]: 342, no. 2) is also very interesting.

Side A	*Side B*
Deo Mercurio	habeant nis{s}i
Cenacus queritur	[[nis{s}i]] repraese [n-
de Vitalino et Nata-	taverint mihi iu-
lino filio ipsius d[e	mentum quod r[a-
iumento quod erap-	puerunt et deo
tum est. Erogat	devotionem qua[m
deum Mercurium	ipse ab his ex-
ut nec ante sa-	postulaverit.
nitatem	

Cenacus complains to the god Mercury about Vitalinus and Natalinus, his son, concerning the draught animal that was stolen. He begs the god Mercury that they will not have good health until they repay me promptly the animal they have stolen and (until they pay) the god the "devotion" that he himself will demand from them.

Again we perceive the full judicial setting of the Cnidian tablets. Here we see the use of the actual legal term *queri* ("to make a complaint before the court"). The last sentence contains a proviso unique for British lead tablets; the thief, forced by "the court," must not only give back the stolen animal or pay its value (*repraesentare* can mean both), but in addition he has to give a *devotio* to the god. Here the term can not, of course, have had the usual meaning (i.e., "a curse, imprecation, etc.") but must be something like a devotional act or gift by which penance is done. This may refer to a part of the value of the object (which elsewhere, however, is invariably promised by the original owner) but also to the *ἱεροπόημα* mentioned in some confession inscriptions that accompanied the secular settlement of the case and was also requested by the god in a dream.

I cannot cite here all the known texts from Britain. Several offer some further interesting details. One curse (*Britannia* 14 [1983]: 338, no. 3) reads, *si servus si liber t⟨a⟩mdiu siluerit vel aliquid de hoc noverit* and further on, *is qui anilum involavit vel qui medius fuerit,* that is, in addition to the actual thief, the tablet curses any accomplices or people who know about the crime. We have met stipulations of this sort several times in Greek and Latin texts discussed above (cf. the term *conscius* or, in Greek, *συγγνῶντα, τοὺς συνειδότας, τοὺς συνιδότες*). Consequences for the blood can also be expressed in a more concrete way. Another tablet (*Britannia* 14 [1983]: 334, no. 2) wishes that the thief of a bronze vessel *sangu⟨in⟩em suum in ipsmu* [*i.e., ipsum*] *aenmu* [*i.e., aenum*] *fundat* ("may spill his blood into the vessel itself"). Another variant occurs in one of the more intriguing texts, in which a Christian is named explicitly. I give the text as presented by the editors (*Britannia* 13 [1982]: 404, no. 7).

seu gen(tili)s seu C-
h⟨r⟩istianus quaecumque utrum vir
ụṭrụṃ mulier utrum puer utrum puella
utrụm sẹṛvụs utrum liḅer mihi Anniaṇ
o ma⟨n⟩tuṭeṇe de bursa mea s⟨e⟩x aṛgente[o]s
furaverit tu d[o]mina dea ab ipso perexi[g]-
e[. . . eo]ṣ si mihi pẹṛ [f]ṛaudem aliquam INḌẸP̣-
ṚẸG̣[.]STVM dederit nec sic ipsi dona sed ut sangu-
inem suum ẸPVṬES qui mihi hoc inrogaverit

There can hardly be any disagreement about the interpretation of the first six lines. I point to the deferential apostrophe *domina dea,* which we noted in related Greek and Latin texts. *Perexigere* is a unique intensification of the more common *exigere*. The editors, however, are at a loss about what to do with the last three lines: "The syntax is obscure and the text (if correctly transcribed) probably corrupt. The purpose is apparently apotropaic (to make any counter-spell by the thief rebound upon him?), and *eputes* possibly conceals a verb equivalent to *solvat*." As for the reading *indepreg*[.]*stum,* "it is uncertain whether to read something like *indeprehensum* ("undiscovered") or to separate *inde* ("thence") from an obscure technicality like *pr⟨a⟩egestum* ("previous action"), *pr⟨a⟩egustum* ("foretaste"), etc."

I believe that we make further progress by adducing some comparable Greek texts, especially those from Cnidos. I shall begin with the verb *irrogare,* which means "to impose, inflict, ordain a punishment or penalty upon somebody." How could the injured party himself ever deserve punishment? I think that the solution is hidden in one of the conjectures of the editors with regard to *indepreg*[.]*stum,* namely, *indeprehensum.*[137] However, this sentence becomes clear only in comparison with the Cnidian practice. Several excuse formulas were to safeguard the writers from the wrath of the gods. In particular they had to emphasize that they themselves were not guilty and were forced to this appeal to divine justice. We have interpreted *nec sit invidia mei Timotneo* in the same way. If the complaint is not somehow justified, the accuser himself must risk divine retaliation. This would a fortiori be the case if nothing had been stolen at all. That is why the authors of the Cnidian tablets first give the guilty an opportunity to right the wrong. Divine help is invoked only when human endeavors fail. Invoking divine help against false accusations occurs repeatedly on confession inscriptions. Since we know that already in Augustan times *donare* occurs generally with the meaning of *condonare* ("to forgive, to pardon"), the following translation of the last three lines seems best to me:

> If through some deceit (i.e., on the sly) the thief has given it back to me so that it remained undiscovered, do not in any way pardon him, but may you destroy (or drink, or may he vomit)[138] his blood, whoever has inflicted this guilt upon me.

Thus, although of course discussion is still possible about the details of the individual readings and interpretations, it is by now clear how helpful the Greek material can be for the explanation of related Latin texts.

FROM THE GREEK EAST TO THE LATIN WEST: CONCLUSION

We must conclude that it is worthwhile to break the language barrier. The differences between the *defixio* and the judicial prayer that we have observed in the Greek material, seem to occur in exactly the same way in the Latin tablets.[139] Indeed, only when this distinction is fully appreciated can some Latin texts that previously resisted analysis now be fully understood. The differences in mentality that form the background of the two categories is similar in both the Latin texts and the Greek. In the traditional form of the *defixio* there is more-often-than-not an anonymous person who desires to harm an enemy without any argumentation or justification for the action; the daemons or gods who carry out the curse are manipulated, rather than persuaded. In the judicial prayer, however, an individual, often giving his or her name, supplicates the god(s) in a subservient way (*domina,* etc.) and asks for divine assistance in the form of retaliation for an injustice suffered. In this context there is abundant use of formulaic language closely imitating that used in the secular courts of law. Although once in a while a great god is invoked (Iupiter O.M., for example), these prayers are generally directed to local deities—or gods from the Roman pantheon identified with local deities—for instance, spring nymphs, Divus Nodens, Mercurius, Neptunus, and, especially in Bath, the Dea Sulis (Minerva). The British tablets, insofar as it is possible to check, were not placed in a temple open to public view, but rather they were folded or rolled, possibly pierced with a nail (so at any rate unable to be read by outsiders) and buried or thrown into a spring; at Bath dozens of tablets manipulated and deposited in this way were recovered from the hot spring and the Roman reservoir excavated beneath the floor of the King's Bath.[140] We can therefore conclude that in these cases the trust in the power of the god was so dominant that publication of the complaint in order to bring the thief to repentance was deemed superfluous. This does not, however, preclude the possibility that the injured persons may have also mentioned the accusation in some more public forum in such a way that the culprits were made aware of their indictment before the god(s).

I have left one problem until now that I hope to be able to go into elsewhere in more detail: How do we explain the striking similarities between the texts of such far-flung regions, cultures, and periods (e.g., the Cnidian tablets, of hellenistic date, and the English tablets, dating mostly from the third or fourth century A.D.)? In this case spontaneous generation is certainly not to be rejected out of hand as a possible explanation: a certain shared religious mentality[141] in which the dependence and subservience of mankind to the superior power of the god predominates might indeed lead independently to prayers of a judicial nature. Nor is the belief that illness is punishment from the god limited to a small number of cultures. Elsewhere[142] I have argued that the more specific, detailed, and exceptional the elements shared by comparable customs and formulas, the greater the likelihood of a derivation and the smaller the probability of an independent development. In the case of the judicial prayer I would prefer to believe that borrowing and transmission has taken place because I consider the similarities so striking as to make spontaneous development far less likely. We should note that such borrowing among the *defixiones* can sometimes be demonstrated, just as the South Italian judicial prayer was also undoubtedly related to the Cnidian prayers. Sample books or professional formularies may have played a role.[143] Particularly in England, we have to take into account the strong influence of international migration that often resulted when Roman soldiers from far distant parts of the empire finished their military duty there and opted not to return to their native lands.

I conclude with an exceptionally interesting example of borrowing in a judicial prayer from Spain. Besides the prayer from Italica, which we took as our starting point, there exists a prayer, already known for a long time, on marble bricked into the wall of a water basin near Emerita.[144] It reads,

dea Ataecina Turi-
brig. Proserpina
per tuam maiestatem
te rogo obsecro
uti vindices quot mihi
furti factum est; quisquis
mihi imudavit involavit
minusve fecit [e]a[s res] q(uae) i(nfra) s(criptae) s(unt)
tunicas VI, [p]aenula
lintea II, in[dus]ium cu-
ius I. C . . . m ignoro
i . . . ius

Goddess Ataecina Turibrigensis Proserpina, by your majesty I ask, pray, and beg that you avenge the theft that has been done to me. Whoever has changed (*immutavit,* or replaced?), stolen, pilfered from me the things that are noted below (*quae infra scriptae sunt*): 6 tunics, 2 linen cloaks, an undergarment. . . .

The respectful language is striking. The verb *vindicare* corresponds again to *ἐκδικέω* in the Greek revenge prayers. In reaction to a paper I gave in Paris in April 1985 on this subject, Patrick Le Roux informed me that in a well in Baelo, Spain a lead tablet was found in 1970 showing a clear relation to the one just cited. He was kind enough to send me the text and allowed me to use parts of it in advance of full publication.[145] The text invokes the goddess Isis Myrionymos[146] and "commits a theft" to her [*tibi conmendo furtum*]. The goddess is addressed as "mistress" (*domina*) and asked *per maiestate⟨m⟩ tua⟨m⟩* to pass sentence on this theft. This time the verb is not *vindicare,* but *reprindere* (= *reprehendere,* "to convict, pass judgment on"). All this is clearly similar to the usual practice in judicial prayers. There is however one element unique in the whole Latin collection: *fac/tuo numini maes-/tati exsemplaria.* It is paradoxical that neither the term *exemplarium* nor anything similar occurs in any other Latin judicial prayer[147] but that it does occur several times in Greek confession inscriptions from Asia Minor—clearly, as happens more often, as a loanword from Latin—and in a formulaic expression: "I warn all mankind not to disdain the gods, for they (i.e. mankind) will have the stele as a warning (ἐξενπλάριον)" (see p. 75) Moreover, the phrase *ut tu evide⟨s⟩ immedi/o qui fecit autulit* ("that you publicly punish[?] whoever did it, [whoever] stole it") recalls the phrase ἐς μέσον ἐνεκκεῖν in the text from Asia Minor (see pp. 74–75). It seems clear to me, because of these similarities, that these texts from Spain cannot have originated independently from the texts from Asia Minor. In this way a welcome bit of information in Paris about Spain possibly provides a link between Greek Asia Minor and Roman Britain.

Finally, what about the problem of "magic" and "religion"? As the reader will have noticed I have opted for a cautious use of the term *magic.* We have observed that many *defixiones* in the traditional sense of the term display clear characteristics of black magic. At the same time, however, we also observe a shift here and there. As Faraone remarks, elements of prayer may intrude. In and of itself this fact does not necessarily exclude such texts from the category of magic. Invocations to gods and daemons of the underworld in prayerlike formulas often occur in (particularly later) magical texts in order to encourage divine cooperation, and they may readily be included in a definition of magic. Indeed, we often descry elements of coercion. As soon as aspects of supplicatory prayer turn up, however, we notice that these are restricted to the texts that also display other "atypical" elements, for instance, the invocation of Olympian as opposed to chthonic gods, the use of deferential titles and formulas, excuses for the disturbance, and so on. We concluded that all these elements were characteristic of utterances to which in their most ideal form no one would deny the label of "religious prayer." There are, then, two complications: first, it seems better to see prayer and *defixio* as two opposites on the extreme ends of a whole spectrum of more or less hybrid forms; second, the terms *magic* and *religion,* which may be applied to each of these extremes, tend to lose their distinctive force as one approaches the middle ground of this spectrum. This does not—at least in my mind—imply that we need to abandon altogether the use of these terms. On the contrary, it should provoke our interest and encourage us to document and explain the conditions and the circumstances that foster the blurring of the boundaries. We have seen in our case that the essential criterion for the definition of judicial prayer should be sought in the legitimation and motivation of the wish, that is, we should ask ourselves whether the wish was justified according to some unwritten laws of public morality or whether the action was a legitimate one (i.e., as an act of rightful retaliation). In this situation and only in this situation a person could and did resort to divine aid by means of a judicial prayer, in addition to, or as an alternate to, magical *defixiones.* It is obvious, however, that what we distinguish as magical and religious attitudes correspond closely to coercive or performative attitudes and supplicative or negotiative attitudes.

What is perhaps most interesting is that the "manipulative" aspects predominate in the traditional *defixiones* found in Greece proper, whereas we find supplicative elements in areas where Greek culture was imported at a later period and where for centuries prior very different social and political forces had exercised their influence on the culture and mentality of the inhabitants. A strongly monarchical ideology has deeply influenced religious perceptions here; for the common man one of the chief tasks of the distant king and his more-approachable subordinates was the administration of justice. The fact that the prayer for justice employed the official language of a royal petition is significant. It appears that in these regions people had a choice of options when it came to interacting with the supernatural; the fact that in the case of a justified complaint they so often opted for the deferential judicial prayer instead of the traditional *defixio* speaks volumes about their belief in divine power and its direct involvement in human affairs.[148]

Notes

The following oft-cited works will be referred to by the author's last name and date of publication only, or by the abbreviation given in square brackets:

A. Audollent, *Defixionum Tabellae* (Paris, 1904). [*DT*]
G. Björck, *Der Fluch des Christen Sabinus,* Papyrus Upsaliensis 8 (Uppsala, 1938).
F. Cumont, "Il sole vindice dei delitti ed il simbolo delle mani alzate," *Mem. Pont. Acc.,* 3d ser., 1 (1923): 65–80.
R. Egger, *Römische Antike und frühes Christentum,* vol. 1 (Klagenfurt, 1962/63).
P. Herrmann, *Ergebnisse einer Reise in Nordostlydien,* DenkschrWien 80 (Vienna, 1962): 1–63.
———, *Tituli Asiae Minoris* vol. V, pt. 1 (Vienna, 1981). [*TAM*]
D. R. Jordan, "A Survey of Greek *Defixiones* Not Included in the Special Corpora," *GRBS* 26 (1985): 151–97. [SGD]
E. G. Kagarow, *Griechische Fluchtafeln,* Eos Suppl. 4 (Leopoli, 1929).
K. Latte, *Heiliges Recht: Untersuchungen zur Geschichte der sakralen Rechtsformen in Griechenland* (Tübingen, 1920).
K. Preisendanz, "Fluchtafel (Defixion)," *RAC* 8 (1972): 1–29
H. Solin, *Eine neue Fluchtafel aus Ostia,* Comm. Hum. Litt. vol. 42, pt. 3 (Helsinki, 1968)
W. Speyer, "Fluch," *RAC* 7 (1969): 1160–1288
F. S. Steinleitner, *Die Beicht im Zusammenhange mit der sakralen Rechtspflege in der Antike* (Munich, 1913).
H. S. Versnel, ed. *Faith, Hope, and Worship: Aspects of Religious Mentality in the Ancient World* (Leiden, 1981).
———, "'May he not be able to sacrifice . . . :' Concerning a Curious Formula in Greek and Latin Curses," *ZPE* 58 (1985): 247–69
R. Wünsch, *Defixionum Tabellae Atticae,* IG, vol. 3, pt. 3 (Berlin, 1897). [*DTA*]
E. Ziebarth, "Neue attische Fluchtafeln," *GöttNachr.* 1899:105–35
J. Zingerle, "Heiliges Recht," *JOAI* 23 (1926) 5–72

1. J. Gil and J. M. Luzón, *Habis* 6 (1975): 117–33; *L'année épigraphique* 1975, no 497. The editors of *AEpigr* interpret *abutor* "avec le sens superlatif et non d'abus," but this does not make the text clear to me.

2. I hope to devote a more detailed study to this question, in which I shall return to

several important questions that at present are not under discussion. In this connection I would greatly appreciate it if readers would call my attention to any testimonia I may have overlooked. Many texts, such as more than two hundred tablets from England, are still awaiting publication. There are about 80 from Bath (*Britannia* 16 [1985]: 322) and about 200 from Uley (*Britannia* 10 [1979]: 342, n. 11).

3. Ziebarth 1899, Steinleitner 1913, Zingerle 1926, Björck 1938, and Latte 1920.

4. See n. 59.

5. Jordan 1985, 151.

6. For details see Faraone's essay (chap. 1, p. 3–4).

7. See the collection of formulas in Audollent, *DT,* pp. 483–86, and note that practically all these instances of direct instructions to the gods or daemons date from the period of the Roman Empire. Earlier instructions to the gods are the exceptions, not the rule.

8. It is significant that apart from a few instances of ἐπικαλοῦμαι and ἀξιῶ (without exception dating from imperial times) all of the (few) instances of verbs connoting the idea of "imploring" in the index to *DT* (e.g., *rogo, obsecro, oro, peto,* etc., which indeed belong to the genuine language of prayer) are Latin and therefore late. The Greek verb ὁρκίζειν and others, of course, belong to another category. For vows in *defixiones* see below, p. 00.

9. This is the usual classification. Kagarow (1929, 50 ff.) distinguishes between five categories separating curses against evildoers in general from curses against magic incantations. Cf. Faraone above, p. 24 n.23

10. E. C. Banfield, *The Moral Basis of a Backward Society* (New York, 1958), 83. His view is shared by many other modern anthropological studies on circum-Mediterranean societies. For the able protector, see J. Davis, *People of the Mediterranean* (London, 1977), chaps. 3 and 4. He explains success in social competition mainly in terms of honor, which in turn will give the successful competitor access to limited resources.

11. Among many passages, see, e.g., Pl. *Rep.* 364B: "If he wants to hurt an enemy, he will damage both the righteous and the unrighteous by means of incantations and spells," a statement that clearly denounces the "amoral" aspects of this type of black magic.

12. R. and E. Blum, *The Dangerous Hour* (London, 1970), 34. Their book reports many such examples of the marked ambivalence concerning magic in general and curses in particular.

13. Björck 1938, 112.

14. See SGD 162 and 164 for two examples.

15. T. B. Mitford, *The Inscriptions of Kourion* (Philadelphia, 1971), nos. 127–42. Cf. Th. Drew-Bear, "Imprecations from Kourion," *BASP* 9 (1972): 85–107. For the correct provenance (Amathous, not Kourion) see P. Aupert and D. Jordan, *AJA* 85 (1984): 184.

16. *DT* 231, 260, 261, 270, 271; SGD 151–53, 155, 156, 158–61. Jordan (SGD, pp. 186 and 191) also mentions two unpublished examples. SGD 91 is a curious and extraordinary piece.

17. On Erinyes in defixiones, see Wünsch, *DTA* vi and xxi; Audollent, *DT* lxi and xciii. Combination of Erinyes with other gods occurring in *defixiones:* B. C. Dieterich, "Demeter, Erinys, Artemis," *Hermes* 90 (1962): 124–48. In general: A. Dieterich, *Nekyia* (Leipzig, 1893), 54ff.; L. R. Farnell, *The Cults of the Greek States,* vol. 5 (Oxford, 1904), 437ff.; P. Robin, *Le culte des Erinyes dans la Gréce classique* (Paris, 1939); R. E. Wüst, "Erinys," *RE* Suppl. VIII (1956): 82ff.; R. Gladigow, "Jenseitsvorstellungen und Kulturkritik," *ZRGG* 26 (1974): 289–309, esp. 299ff.

18. Kourouniotes, *Ephemeris Epigraphica* 1913:185 (= SGD 14 [Athens, third century B.C.]): πρὸς τὰς Πραξιδίκας; the same formula is in SGD 62 (Athens, third century B.C.?). A votive inscription at Volos has Πραξιδίκαις Μεγαλοκ[ῆς . . .] (B. Helly, *Gonnoi,* vol. II [Amsterdam, 1973], no. 204). A *defixio* from Cyrene (third century B.C.) identifies Praxi-

dike with Kore: Πραξιδίκα κώρα μεγαλήτο⟨ρος⟩ Ἀγλαοκάρπου; see C. Pugliese Carratelli, *ASAtene* 23/24 (1961/62): no. 193; C. Gallavotti, *Maia* 15 (1963): 450–54 (= SGD 150). Cf. also Wünsch, *DTA vii*. For the goddess Dike see below. In a late bilingual word list mention is made of Ἐκδικήσεις as *Ultrices* for which see G. Goetz, *Hermeneumata Pseudositheana*, Corpus Glossariorum Latinorum 3 (Leipzig, 1892), 237.

19. The expression is *εὐαγγέλια θύειν* ("sacrifice at the occasion of good tidings"); cf. Xen. *Hell.* 1.6.37. Cf. *PGM* IV.2094: *τέλεσον δαῖμον τὰ ἐνθάδε γεγραμμένα· τελέσαντι δέ σοι θυσίαν ἀποδώσω*. There are only a few more such votive formulas in the *defixiones*. Cf. for the moment Kagarow 1929, 40 and D. R. Jordan, *Hesperia* 54 (1985): 243.

20. Jordan, SGD, p. 158.

21. Jordan, SGD, p. 197; cf. *DTA* 46.

22. The stolen object may even be a *παρθένος* (young girl). See Audollent, *DT* 74. SGD 38 mentions a tablet from the Athenian Agora (inv. IL 1722) from the 3d cent. A.D. that reproduces this text against thieves.

23. We occasionally find phrases like: *τὴν γλῶτταν τὴν κακήν*, etc. (*DTA* 84) or *τὸν κύσθον τὸν ἀνόσιον* (*DTA* 77). There is a slight possibility that apparent professional indications like *πορνόβοσκος* (brothel keeper or pimp) (SGD 11) may simply be a form of ridicule or abuse. A special type of *defixio* is based on the use of abuse, e.g., *DT* 155: *τὸν δυσσεβὴν καὶ ἄνομον καὶ ἐπικατάρατον κάρδηλον* and *DT* 188 *τόνδε τὸν ἄνομον καὶ ἀσεβῆ*. These are *diabolai*, which accuse the opponent of evil deeds against the god in order to provoke divine wrath against him. Cf. S. Eitrem, "Die rituelle ΔΙΑΒΟΛΗ," *SO* 2 (1924): 43–58. For this reason one should not include *DT* 295 into category 4, as Audollent does, since this accusation—*tibi commendo quoniam maledixit partourientem*—is undoubtedly a *diabolē*. Cf. Eitrem, ibid., 57.

24. The term *ἁμαρτωλός* occurs frequently in funerary curses of Asia Minor. See Strubbe above, p. 48.

25. Inter alia, the Greek term *ἀνάθεμα*, which is typically Jewish (Speyer 1969, 1240 ff.).

26. A. Cameron ("Inscriptions Relating to Sacral Manumission and Confession," *HTR* 32 [1939]: 158 states clearly that *κολάζω* is "the term for the divine punishment"; Th. Homolle, *BCH* 25 (1901): 422, n. 1. "châtiment divine, à l'expiation imposée même au delà de la vie"; Steinleitner 1913, 96, n. 3: "typische Ausdruck für Hölle und Höllenstrafen." Cf. F. X. Gokey, *The Terminology for the Devil and Evil Spirits in the Apostolic Fathers* (Ph.D. diss., Catholic University, 1961), 89f.; *TWNT* III, 817. References to punishment in hell or netherworld do occur now and then in *defixiones*. A tablet from Athens has [*εἰς τὸ*] / *δαιμόνιον σκοτεινὸν καὶ καταχθονίων καὶ* / [[ ]] *εἰς ὅλους τοὺς θεοὺς* / ΠΑΣΑΓΗΣ (= *πάσ⟨ης⟩ γῆς*?) *πέμπω δῶρον* (SGD 54 [Jordan's text] after R. P. Austin, *BSA* 27 [1925/26]: 73). Perhaps something of the kind is intended in a tablet from Apamea against charioteers who are supposed to see daemons from their doors: *δέμ[ο]νας ἀώρο[υ]ς, δέμονας βιέους, Ἡφέστου πῦρ* . . . (Jordan, SGD, p. 193).

27. On this *Jenseitskorrespondenz* see Preisendanz 1972, 7–8 and Faraone above, p. 4.

28. Perhaps *DT* 14 from Phrygia belongs to this category: *γράφω πά[ν]τ[α]ς τοὺς ἐμοὶ ἀντία π[ο]ιοῦντας μετὰ τῶν [ἀ]ώρων*. The context does not refer to the juridical atmosphere normally indicated by terms like *antioi* or *antidikoi*. Cf., further, *DT* 92 from the Thracian Chersonesos (third century B.C.): *Αἶσα ἀναιροῦσι κά[δι]κοῦσι*.

29. *SEG* 30.326 (= SGD 21), originally published by G. W. Elderkin (*Hesperia* 6 [1937]: 382–95) together with another *defixio* that contains a *diabolē* and the phrase *ναὶ Κύριε Τυφὼς ἐκδίκησον* (. . . . . .) *καὶ βοήθησον αὐτῷ*. See, for some emendations, *BE* 1938, no. 23 and D. R. Jordan, *Glotta* 18 (1980): 62–65, whose text I follow.

30. It is no accident that this term is also a *sermo technicus* in official petitions in Egypt; see below, p. 80.

31. SGD 58; Ph. Bruneau, *Recherches sur les cultes de Délos à l'époque hellénistique et à l'époque impériale* (Paris, 1970), 650ff. Cf. J. Triantaphyllopoulos, *Mélanges helléniques offerts à G. Daux* (Paris, 1974), 332–33. The editor makes an occasional slip, not noticed by L. Robert. Lines 10–12 of side B read καταγράφο τοῦ ἄραντος ṬΑΟΙΔΕ\, τὰ ἀνανκε͂α αὐτοῦ, and he translates, "Je dévoue . . . de celui qui a emporté, ce qui lui est nécessaire." ṬΑΟΙΔΕ\ is almost certainly τὰ αἰδοῖα and τὰ ἀναγκαῖα is surely not "what is necessary" but a variant of αἰδοῖα.

32. The editor follows a suggestion made by A. Plassart by translating, "Donnez naissance à votre pouvoir." He does not realize, however, that this corresponds closely with the terminology of the so-called confession texts from Lydia and Phrygia (see below, p. 75). Cf., on the meaning of *aretē,* for instance, Y. Grandjean, *Une nouvelle arétalogie d'Isis à Maronée* (Leiden, 1975), 1–8.

33. The use of the imperative διοργιάσετε is related to the expression θεοὶ κεχολωμένοι found in many funerary curses. Cf., e.g., a curse from Halos in Phthiotis (third century B.C.): . . . . ἕξει δὲ ὀργὴν μεγάλην τοῦ μεγάλου Διός. See, on these texts, Versnel 1985, 259ff. and Strubbe above, chap. 2.

34. Failure of human efforts is a *topos* in reports of divine healing miracles; see O. Weinreich, *Antike Heilungswunder,* RGVV, vol. 8, pt. 1 (Giessen, 1909) 195ff. Cf. a confession text ἀφελπισθοῦσα ὑπὸ ἀνθρώπων (Steinleitner 1913, no. 19). I do not know any comparable reference to failing human justice in *defixiones* or juridical prayers. A funerary curse containing μὴ δυνάμενος ἄγειν τὸ πρᾶγμα (Zingerle 1926, 54ff.) is of a different nature.

35. Fr. Blass, *Philologus* 41 (1882): 746ff.; K. Wessely, 11. *Jahresber. d. Franz-Joseph Gymnasiums* (Vienna, 1885); Wünsch, *CIA* III, app. XXXI; Preisigke, *Sammelbuch* I, 5103; Wilcken, *UPZ* I, no. 1; Gerstinger, *WS* 44 (1924/25): 219, with a new reading adopted by Wilcken, *UPZ,* 646ff.; *PGM* XL; Steinleitner 1913, 102; Björck 1938, 131ff. Some remarks: W. Crönert, in *Raccolta di scritti in onore di G. Lumbroso* (Milan, 1925), 470–74; R. Seider, in *Festschrift zum 150 jährigen Bestehen des Berliner Aegyptischen Museums* (Berlin, 1974), 422–23.

36. Not only because Oserapis is a god of the netherworld but also since gods sitting together as judges are typical for the images of underworld and hereafter; E. Rohde, *Psyche,* vol. I, 10th ed. (Tübingen, 1925), 310–11. Cf. the instances at Cnidus and elsewhere (below, p. 72) and SGD 164: Εὐλαμων, μετὰ τῶν σῶν παρέδρω[ν].

37. S. Eitrem (see n. 23), 43: "Ein Gebet kann doch in einen Fluch hinübergleiten wie z.B. in dem Rachegebet der Artemisia." On the differences between this type of prayer and the genuine curse, see Kagarow 1929, 22ff., 49ff.; Björck 1938, 112ff.; P. Moraux, *Une defixion judiciaire au musée d'Istanbul,* Mém. Ac. Roy. Belg. vol. 54, pt. 2 (Brussells, 1960): 4–5.

38. Below, p. 81.

39. Th. Homolle, *BCH* 25 (1901): 412–30; *IG* XII.7², p. 1; R. Wünsch, *BPhW* 25 (1905): 1081; Latte 1920, 81, n. 54; Zingerle, 1926: 67–72; Björck, [1938]: 129–31; Versnel 1981, 32; H. W. Pleket, in *Faith, Hope and Worship,* ed. H. S. Versnel, 1981, 189–92. I have given a recent treatment of the text in Versnel 1985, 252ff., to which I refer for the details.

40. In a letter Jordan suggests to me that perhaps *epaphroditos* should not be capitalized and interpreted as "a certain charming fellow." This may well be true. As we have seen and shall further observe, names of thieves and the like are mentioned if known, but the curse is usually the specific refuge for those who do not know their opponents.

41. Zingerle 1926, 67–72 was the first one who drew attention to the similarity with the ἔντευξις, and he was followed by Björck 1938, 60 ff. Indeed, many of the expressions of

this prayer (and the one of Artemisia) have exact parallels in the *ἔντευξις*: *βασιλεῦ; ἐπί σε καταφεύγω; τοῦ δικαίου τύχω; ἱκέτις; προσπίπτω; ἀδικοῦμαι*. Later collections and studies confirm this: Maria T. Cavassini, in "Repertorium Papyrorum Graecarum Quae Documenta Tradant Ptolemaicae Aetatis," *Aegyptus* 35 (1955): 299–334 ("Exemplum vocis *ἔντευξις*"); O. Gueraud, *Enteuxis* (Cairo, 1931); J. L. White, *The Form and Structure of the Official Petition,* SBL Dissertation Series 5 (Missoula, 1972). On *ἀδικοῦμαι* as a stereotyped element of the *ἔντευξις* see also W. Schubart, "Das hellenistische Königsideal nach Inschriften und Papyri," *APF* 12 (1937): 7. Zingerle could have also included the term *εὐίλατος*, which is used in reference to kings or emperors: *P. Petr*. 2, 13, 19; *UPZ* 109, 6; cf. Versnel 1985, 260–61.

42. Björck (1938, 137) points out that the same can be found in magical papyri. *PGM* LI has *παρακαλῶ σε, νεκύδαιμον* (. . . . . .) *ἀκοῦσαι τοῦ ἐμοῦ ἀξιώματος καὶ ἐκδικῆσαί με*.

43. See Versnel 1985, 254; Strubbe above, chap. 2.

44. The main collections and discussions are F. Cumont, "Il sole vindice dei delitti ed il simbolo delle mani alzate," *Mem. Pont. Acc.* ser. 3, vol. 1 (1923): 65–80; idem, "Nuovi epitafi col simbolo della preghiera al dio vindice" *Rend. Pont. Acc. Arch.* 5 (1926/27): 69–78; idem, *Syria* 14 (1933): 392–95; Björck 1938, 24ff.; Cf. also G. Sanders, *Bijdrage tot de studie der Latijnse metrische grafschriften van het heidense Rome,* Verhandelingen Kon. Vlaamse Acad. Wet. Kl. Letteren 37 (Leiden, 1960), 264ff.; F. Bömer, *Untersuchungen über die Religion der Sklaven in Griechenland und Rom,* vol. 4 (Wiesbaden, 1963), 201–5. On the symbol of the raised hands see Strubbe above, p. 42.

45. L. Robert, *BE* (1965), no. 335 and (1968), no. 535; D. M. Pippidi, *"Tibi commendo," RivStorAnt.* 6/7 (1976/77): 37–44; D. R. Jordan, "An Appeal to the Sun for Vengeance (*Inscr. de Délos* 2533)," *BCH* 103 (1979): 522–25.

46. See F. J. Dölger, *Die Sonne der Gerechtigkeit* (Münster, 1919), 90ff.; F. Cumont, *Afterlife in Roman Paganism* (New Haven, 1922), 133ff.; J. Bidez, *La cité du monde et la cité du Soleil chez les Stoiciens,* Mém. Ac. Roy. Belg. 26 (Brussels 1932); R. Pettazzoni, in *Hommages à J. Bidez et F. Cumont* (Bruxelles, 1949), 245–56.

47. In the collection of Björck: Serapis (1); Theos Hypsistos (11, 12); Hosios Dikaios (13); Hagne Thea (14); *Manes vel Di Caelestes* (16); *οἱ θεοί* (17). Cf. E. Schwertheim, *Inschriften von Kyzikos und Umgebung* (Bonn, 1980), no. 522 for *Δίκη καὶ* Ζεῦ Πανεπόψιε.

48. J. H. Waszink, "Biothanatoi," *RAC* II (1954): 391–94 and Pippidi (see n. 45). On *ἄωροι* see the literature in Bömer (see n. 44), 202–3.

49. K. Meuli, "Lateinisch 'morior'—deutsch 'morden'," *Gesammelte Schriften* vol. I (Basel, 1975), 439–44.

50. L. Robert, *Collection Froehner,* vol. 1, *Inscriptions grecques* (1936), 55–56: "Poison et magie jouaient un grand rôle et en fait et dans les imaginations. Il serait intéressant d'en relever les traces dans les épitaphes grecques." He gives some examples to which, indeed, many more could be added. Cf. also Zingerle 1926, 18–19 and Latte 1920, 68, n. 18. The word δόλος is frequently attested, see L. Robert, *BCH* 101 (1977): 49, on δόλον *πονηρόν*. I wonder whether in *Inscriptions de Délos,* ed. P. Roussel and M. Launey (Paris, 1937) no. 2533—*κ*]*αὶ ἔι τις α*[*ὐτῇ βλάβη*]*ṿ*(?) *ἐπεβούλευσ̣εν*, according to the conjectures by Jordan (see n. 45)—we should not prefer δόλο]*ṿ* to *βλάβη*]*ṿ*. This nearly always implies that the culprit is unknown, e.g., *τίς δὲ τούτους ἠδίκησε ἐνκεχαρισμένος ἤτω εἰς αὐτὰ τὰ νέκυεια* (*MAMA* vol. VII, p. 402 = S. Mitchell, *Regional Epigraphic Catalogues of Asia Minor* [Oxford 1982] II, 362); *τίς αὐτὸν ἠδίκησε τῷ ἢ αἷμ*[*α*] (N. P. Rosanova, *VDI* 51 [1955]: 174–76). I find it difficult to follow Jordan (see n. 45), 522, n. 2, who makes *τίς* interrogative (*τίς αὐτὸν ἠδίκησε;*). My colleague Strubbe has transcribed another funerary

text in the museum of Afyon where τίς has unmistakably the function of ὅστις. Cf. also *BE* (1959): 273 and (1980): 341. C. Naour ("Inscriptions du Moyen Hermos," *ZPE* 44 [1981]: 18, no. 1) gives another unequivocal example. W. Peek (*ZPE* 42 [1981]: 287–88) refers to it as "ein seit dem Hellenismus ganz gewöhnlicher Sprachgebrauch," (he gives yet another example). Cf. G. Petzl, *ZPE* 46 (1982): 134. Naturally, the name of the murderer is rarely mentioned. Björck 1938, no. 20 (*Atimeto liberto, cuius dolo filiam amisi*) and no. 21 (*Acte libertae*) belong to the rare exceptions.

51. Björck 1938, no. 13 (= S. Mitchell, [see n. 150]: 242) and no. 14. In no. 17 (from Nabataea) mention is made of κακολογούντων, which leaves the option of death caused by charms or slander. In no. 18 δειράντων may imply bodily violence, although δείρομαι may also be "Nachteil erleiden" (F. Preisigke, *Wörterbuch der griechischen Papyruskunde* [Berlin, 1856–1924] s.v.).

52. This expression tallies with the marked Jewish character of the famous inscription from Rheneia.

53. Cumont 1923, no. 5B: *tu* ⟨*v*⟩*indices;* followed by Björck and Bömer (see n. 44), 203; Jordan (see n. 45), 524, n. 8, suggests *[u]t vindices,* provided that the stone proves to have been damaged.

54. *Inscriptions de Délos* no. 2533. Cf. Jordan (see n. 45).

55. *CIL* 6. 34635a; Cumont 1923, no. 7 with related texts.

56. Cf. *CIG* 3.5471 (= *IG* 14.254): μὴ λάθοιτο τὸν θεόν, and Cumont 1923, 74, no. 9: Κύριε Ἥλιε [ἤ τι]ς(?) κλαπῇ σε μὴ λάθοιτο.

57. Björck 1938, 58: "Das Verweilen bei der besonderen Art von Heimsuchung die die Gegner treffen soll, eignet weniger dem Rachegebet als dem Schadenzauber und Fluch."

58. On the wall of a monastery in Nabataea: Le Bas-Waddington, *Voyage archéologique en Grèce et Asie Mineure,* vol. 3 (Paris, 1870), 2068.

59. In this translation I have tried to preserve the faulty syntax of the Greek, which provides a good example of the deficient and often somewhat breathless language characteristic of many of these texts. Cl. Gallazzi, "Supplica ad Atena su un ostrakon da Esna," *ZPE* 61 (1985): 101–9. I thank W. Clarysse for having drawn my attention to this text before it was published. Gallazzi refers to two related texts from the same area, of which only a poor transcription remains, published by B. Boyaval (*Chron. d'Égypte* 55 [1980]: 309–13). One of them has ἀξιούμεθα ἡμᾶς κρίνεσθαι μετ' αὐτῶν καὶ βεβοηθῆσθαι ὑπὸ ὑμῶν (. . . . . .) ἐγκαλοῦμεν ὑπὸ τούτων τῶν καταράτων καθ' ἡμέραν ὑμῖν. For a *repeated* prayer for justice compare a text from England (*Britannia* 15 [1984]: 339, no. 7): *iteratis* [*pre*]*c*[*i*]*bus te rogo ut* . . . . Gallazzi neglects to place this prayer in the context of the prayers collected by Björck and denies any similarity with the curse of Artemisia because it contains *una lunga esecrazione* (p. 103), surely an insufficient argument. Several authorities quoted by him do compare demotic prayers for justice with the Artemisia text: G. R. Hughes, "The Cruel Father," in *Studies J. A. Wilson* (Chicago, 1969), 43–54; J. Quaegebeur, in *Schrijvend Verleden,* ed. K. R. Veenhof (Leiden, 1983), 263–76, esp. 272ff.

60. C. T. Newton, *A History of Discoveries at Halicarnassus, Cnidus, and Branchidae,* vol. 2 (London, 1863), nos. 81ff. For older discussions see literature in *DT,* p. 5. The texts can also be found in H. Collwitz and F. Bechtel, *Sammlung der griechischen Dialekt-Inschriften,* vol. 3, pt. 1 (Göttingen, 1899), 234ff.; *DT* 1–13; Steinleitner 1913, nos. 34–47; *DTA,* pp. xff.; some of them in *SIG,* 3d ed., 1178–80. Discussions in Zingerle 1926; Latte 1920, 80; Björck 1938, 121ff.; M. P. Nilsson, *Geschichte der griechischen Religion,* vol. 1, 2d ed. (Munich, 1955), 221f. There are some textual conjectures in Kagarow 1929, 52.

61. I shall rigorously restrict my comments to the themes directly bearing on my investigation.

62. Note that most of the *defixiones* from the sanctuary of Demeter and Kore at Corinth are directed against women (N. Bookidis, *Hesperia* 41 [1972]: 304; R. S. Stroud, *AJA* 77 [1973]: 228–29). Among the Cnidian curses there is one conditional self-curse.

63. The verb *ἀναβαίνω* is a technical term for "to go up to god or temple." Cf. *UPZ*, vol. I, p. 42, no. 3: *ἀνάβασιν εἰς τὸ ἱερόν*, and *TWNT*, vol. I, pp. 517f. It also occurs in confession texts: *TAM*, vol. V, p. 1, no. 238; *MAMA*, vol. IV, pp. 283, 289. The verb *ἀναφέρω* is the technical term for "offering a present to the god." See *TWNT* vol. IX, 62ff. Cf. a recently published funerary text: *ἀμπέλους τῷ Διεὶ ἀνάφερε* (G. Petzl, *EA* 6 [1985]: 72).

64. This forms a distant parallel to the term *κολαζόμενος* in a confession text from Eumeneia: Th. Drew-Bear, "Local Cults in Graeco-Roman Phrygia" *GRBS* 17 (1976): 261, n. 54: ". . . a present participle of which our text thus furnishes the first attestation."

65. The term also in a magic papyrus: *P. Berlin* 10587. That it may imply a feverish illness is demonstrated by an unpublished tablet from Carthage or Hadrumetum (SGD, pp. 186–87), from which L. Robert, *JSav*. (1981): 35, n. 1, quotes: *καὶ ἐπιθυμίᾳ πυρούμενοι τὰς ψυχὰς, τὰς καρδίας, τὰς σπλάνγχνας αὐτῶν βασανιζόμενοι ἐπὶ τὸν τῆς ζωῆς μου χρόνον*. On *βάσανος* as juridical torture see *RAC* 8 (1972): 101ff.

66. On the consequences of this prayer see Versnel 1985. There is a marked inconsistency in the sequence of what is wished for in lines 4–6 and 8–10 of *DT* 4A. This, however, is not unparallelled in this type of text. Cf. SGD 163 (Hebron, third century A.D.): *βάλεται* (= *βάλετε*) *αὐτὸν ἐπὶ κάκωσι⟨ν⟩ καὶ θάν[ατον κ]αὶ κεφαλαργίας*; and a *defixio* mentioned below, n. 139, where a thief must die and bring back a stolen vessel (in this order).

67. On the implications of this formula see Latte 1920, 55, 64f., 75, n. 40 and addendum; idem, "Schuld und Sünde in der griechischen Religion," in *Kleine Schriften* (Munich, 1968), 9; and W. Speyer 1969, 1165 and 1181. The formula had a long life: it is found in a different context in magical texts against illness of the sixteenth century (F. Pradel, *Griechische und süditalienische Gebete, Beschwörungen und Rezepte des Mittelalters,* RGVV 3 [Leipzig, 1907], 22, line 11): *μὴ συμπιῇς, μὴ συμφαγῇς, μὴ συγκομηθῇς, μὴ συναναστῇς μετὰ τοῦ δούλου τοῦ θεοῦ*. I know of only one similar curse from antiquity: D. R. Jordan, *Hesperia* 54 (1985): 223–24, no. 7.

68. *DTA* 100; *DT* 74–75. The terms are frequently used in the Sethianic curses. Cf. in particular Jordan (see n. 67), 241; idem, *AM* 95 (1980): 236–38; SGD 112. Cf. also the *defixio* (above, page 65) *αὐτοὺς* (. . . .) *σοι παρακατατίθεμαι τηρεῖν* ("I entrust them to you that you may keep watch over them" [as over a deposit]).

69. See for literature Audollent's comments on *DT* 212. The text also appears in *SEG* 4.70, *IG* 14.644, and with a commentary in V. Arangio-Ruiz and A. Olivieri, *Inscriptiones Graecae Siciliae et Infimae Italiae ad Ius Pertinentes* (Milan, 1925), 165ff.

70. This may seem strange, since the author knows the culprit by name. There are several possibilities. The author may accidently find herself in Melitta's company through no fault of her own, or she may simply be using a prescribed formula that is inappropriate for her present situation.

71. I interpret the phrase in this way, although the syntax suggests that the culprit herself is the subject. I recognize here a situation opposite from, e.g., *στρέβλωσον αὐτῶν τὴν ψυχὴν καὶ τὴν καρδίαν ἵνα μὴ πνέωσιν* (*DT* 241, line 14). In a tablet from Hadrumetum we read, *καὶ μὴ ἀφῇς τὴν ψυχήν* (SGD 147).

72. The combination of divine and secular penalties is well known: Zingerle, *Philologus* 53 (1894): 347ff.; Latte 1920, 80, n. 53; J. Merkel, "Ueber die sogenannten Sepulcralmulten," in *Göttinger Festgabe für R. von Ihering* (Göttingen, 1892), 79–134. On the duplication or multiplication of fines, see Zingerle 1926, 36; W.-D. Roth, *Untersuchungen zur Kredit-*

παραθήκη im römischen Ägypten (Ph.D. diss., Philipps-Universität zu Marburg 1970), 91–95 and 99.

73. C. Dunant, "Sus aux voleurs! Une tablette en bronze à inscription grecque du Musée de Genève," *MusHelv* 35 (1978): 241–44.

74. The text is not without difficulties. L. Robert (*BE* [1980]: 45) would prefer a full stop after *θεῶν* and then a separate statement: "I have lost all my gold." This does not seem very likely since *ὥστε* must depend on some previous wish or command. Nor do I follow him in maintaining the subjunctive *ἀναζητησῆι* of the text—not because *ὥστε* could not take a subjunctive but because *αὐτήν* must be the subject-accusative belonging to this verb. Clearly *αὐτήν* cannot be the object of *ἀναζητησῆι* ("search for her") since the unknown thieves are in the plural and the goddess (*αὐτήν*) is in the subject-accusative in the rest of the text.

75. The phrase *ἐς μέσον ἐνεκκεῖν πάντα* means "make known the truth." This might be a slight argument for maintaining *indices* in the juridical prayer (see n. 53) "that you make known (the cause of) his death," but I still prefer the other solutions.

76. For this reason V. Longo, *Aretalogie nel mondo greco* I: *Epigrafi e papiri* (Genova, 1969), 158–66, included five confession texts in his collection. The term most commonly used is *δύναμις* or *δυνάμεις*. See the survey in E. Varinlioglu, "Zeus Orkamaneites and the Expiatory Inscriptions," *EA* 1 (1983): 83, n. 42. It has invaded late magical texts: *ἀξιῶ καὶ παρακαλῶ τὴν σὴν δοίναμην* (SGD 189). Cf. the prayer for revenge *P. Upsal.* 8 (above, p. 71): *δεῖξον τὴν δύναμίν σου*, which appears in a different form in line 15 as [δ]*εῖξον δ' ὡς τὸ πάροιθε θεουδέα θαύματα σεῖο*. These *θαύματα* are indeed identical with what generally are called *ἀρεταί*. In one confession text we read *καὶ ἐνέγραψα τὴν ἀρετήν* (*TAM* vol. V, pt. 1, no. 264), where *ἀρετή* = *δύναμις*. For an analysis of *δύναμις* as divine power to do miracles, see F. Preisigke, *Die Gotteskraft der frühchristlichen Zeit* (Leipzig, 1922); J. Röhr, *Der okkulte Kraftbegriff im Altertum* (Leipzig, 1923); H. W. Pleket, "Religious History As a History of Mentality: The 'Believer' As Servant of the Deity in the Greek World," in Versnel 1981, 178–83.

77. In his famous 1913 collection and commentary F. S. Steinleitner counted seventeen examples from Maeonia, four from other parts of Lydia, and twelve from Phrygia—a total of thirty-three inscriptions in all. Since then dozens have been found—especially lately—and several (for example, the specimens in the museum of Usak) have not yet been published, and others have been published in very scattered studies. For Maeonia there is a recent edition in the volume published by P. Herrmann, *Tituli Asiae Minoris,* vol. 5, pt. 1 (Vienna, 1981), abbreviated throughout this essay as *TAM.* A new edition with commentary of all the material is a pressing need. The most important older collections are Steinleitner 1913; W. H. Buckler, "Some Lydian Propitiatory Inscriptions," *BSA* 21 (1914–16): 169–83; Zingerle 1926; *MAMA* vol. IV, nos. 279–90. There are discussions of several particular aspects in Cameron (see n. 26); O. Eger, "Eid und Fluch in den maionischen und phrygischen Sühne-Inschriften," in *Festschrift P. Koschaker,* vol. 3 (Weimar, 1939), 281–93. For a fundamental survey of the religious mentality see J. Keil, "Die Kulte Lydiens," in *Anatolian Studies Presented to W. M. Ramsay* (1923): 239–66. On the confession see R. Pettazzoni, *La confessione dei peccati,* vol. III, pt. 2 (Bologna, 1936). There have been many new discoveries: Herrmann 1962, 1–63; L. Robert, *Nouvelles inscriptions de Sardes* (Paris, 1964), 23–31; Drew-Bear (see n. 64), 260ff.; G. Petzl, "Inschriften aus der Umgebung von Saittai," *ZPE* 30 (1978): 249–58; H. W. Pleket, "New Inscriptions from Lydia," *Talanta* 10/11 (1978/79): 74–91; Chr. Naour, "Nouvelles inscriptions du Moyen Hermos," *EA* 2 (1983): 107–22. There are several recent studies on specific gods connected with the confession texts, some of them with new material: E. N. Lane, *Corpus Monumentorum Religionis Dei Menis* vols. I–IV (Leiden, 1971–1978), with a good introduction in vol. III (1976), 17–38; I. Diakonoff, "Artemidi Anaeiti anestesen," *BABesch* 54 (1979): 139–75; E. Varinlioglu (see n. 76); P. Herrmann and E. Varinlioglu,

"Theoi Pereudenoi," *EA* 3 (1984): 1–17; H. Malay, "The Sanctuary of Meter Phileis near Philadelphia," *EA* 6 (1985): 111–25; K. M. Miller, "Apollo Lairbenos," *Numen* 32 (1985): 46–70; P. Herrmann, "Men, Herr von Axiotta," in *Festschrift F. K. Dörner* vol. I (Leiden, 1978), 415–23; and now H. Malay and G. Petzl, "Neue Inschriften aus den Museen Manisa, Izmir und Bergama," *EA* 6 (1985): 60ff.; P. Herrman, "Sühn- und Grabinschriften aus der Katakekaumene im archäologischen Museum von Izmir," *Anz. Österr. Ak. Wiss.* 122 (1985): 249–61. Since this essay was submitted in 1986, several new and important studies here appeared that could not be incorporated here.

78. On the *δύναμις* see note 76 above. The *στῆλαι* are *μαρτυρίαι*; see Versnel (1981) 60ff.

79. There is an exemplary list of sins in a sacral law from Philadelphia concerning a private sanctuary (second or first century B.C.): *SIG,* 3d ed., 985; *LSAM* 20. The religious climate is very similar to that of the confession *στῆλαι* (M. P. Nilsson, *GGR* [see n. 60], vol. 2, 2d ed. [Munich, 1961], 291) and to that of Christian communities, see S. C. Barton and G. H. R. Horsley, "A Hellenistic Cult Group and N. T. Churches," *JAC* 24 (1981): 7–41.

80. For *ἐπιορκέω* or *παρορκέω* see E. Varinlioglu (see n. 77), 78 and 81 and Drew-Bear (see n. 64), 265ff.

81. *TAM,* no. 251, where there are references to literature on the term *πιττάκιον*; cf. also Strubbe, above, p. 45.

82. *ἐξέλθῃ τὸ πιττάκιον* (*POxy* 8.1150 = *PGM* VIIIb). That this is formulary is demonstrated by a recently published oracle: *ZPE* 41 (1981): 291.

83. This is my interpretation of *ἱκανοποιοῦσα*, on which I cannot expand here. Cf. *satisfacere* in a British tablet on page 85.

84. For the discussion on the meaning of these *σκῆπτρα* cf. Naour (see n. 77), 119–20 and Strubbe above, pp. 44–45).

85. On the word *πινακίδιον*, see Chr. Habicht, *Altertümer von Pergamon,* vol. VIII, pt. 3, *Inschriften des Asklepieions,* ad no. 72.

86. The correspondence was noticed for the first time by Ziebarth 1899, 122ff. and has been explored by Steinleitner 1913, 100–104; Zingerle 1926, 19f.; Björck 1938, 112ff. Cf. Eger (see n. 77), 288ff.; Pettazzoni (see n. 77), 74-76 and 141, n. 96.

87. Malay (see n. 77). I quote from his nos. 2 and 9 respectively.

88. On the formulaic aspects of similar *εὐχή* dedications cf. Robert (see n. 77), 35, n. 4; Naour (see n. 77), 108; Varinlioglu (see n. 76), 79.

89. For a most unequivocal expression of this see, in a recent dedication to Men Axiottenos, *πεποσχότα . . . ὑπὸ* Μηνός ("suffering . . . through Men"), which has been edited by G. Manganaro (*ZPE* 61 [1985]: 199 ff.). The terminology *κολασθεὶς εἰς* or *ἀπό* plus part of the body is formulaic. For a recent survey of pictures of parts of the body in confession texts, cf. Naour (see n. 77), 109. In general see F. T. van Straten, "Votive Offerings Representing Parts of the Human Body (the Greek World)," in Versnel 1981, 105–51.

90. It was possible, for instance, to give an object to the god and keep the usufruct, cf. A. Cameron (see n. 26), no. 1, Edessa no. 10. Actually the sacral *manumissio* is an example of this principle, so that *εἶναι αὐτὴν τῆς θεοῦ* actually means *ἐλευθέραν εἶναι*; see Cameron, ibid., 149. For comparable examples from the Roman world see P. Veyne, "*Titulus praelatus:* offrande, solennisation et publicité dans les ex-voto gréco-romains," *RA* (1983), 296f.

91. C. Wachsmuth, "Inschriften von Korkyra," *RhM* 18 (1863): 5.

92. *TAM,* no. 159, where one finds references to the discussion on *ἄγγελος*. On dream commandments in general see F. T. van Straten, "Daikrates' Dream: A Votive Relief from Kos and some other *kat' onar* Dedications," *BABesch* 51 (1976): 1–38.

93. Herrmann 1962, 57 and *TAM,* no. 510.

94. It is the same inscription from which Petzl ([see n. 77], 257, n. 41) and P. Herrmann (*TAM,* no. 159) quote another passage.

95. *TAM,* no. 328 (ἐπεζήτησεν ἱεροπόημα); nos. 320 and 321 (ἀπέδωκαν τὸ [ἱε]ροπόημα); no. 322 (ποήσαντες τὸ ἱεροπόημα)—all from the temple of Anaeitis and Men Tiamou at Kula. Cf. Diakonoff (see n. 77). For the discussion on the meaning of the term cf. Robert (see n. 77), 30, n. 4 and Lane (see n. 77), III:18f.

96. Newton (see n. 60), 716: "this word may mean 'atonements' or 'sin-offerings'." For ἴατρα demanded by the god see Wörrle in Chr. Habicht, *Altertümer von Pergamon,* vol. 8, pt. 3, *Die Inschriften des Asklepieions,* 184ff.

97. Herrmann/Varinlioglu (see n. 77), no. 9. I think that the translation "die Götter zu Iulia übergegangen sind" (Varinlioglu) is mistaken and that the views of Herrmann and Petzl, "gingen auf sie los," should be preferred. In ibid., no. 7 τοῖς συνεπερχομένοις has the comparable meaning of "to attack together." Cf. the new text from Esna (see p. 71–72) with the verbs ἐπέρχομαι and προσέρχομαι.

98. Cf. Keil (see n. 77), 38: ἐπιζητέω πρός τινας ὁμολογίαν. Finally, ἐπιζητέω can take the meaning of (ἐπ)ερωτάω in the function of "consulting the god on the sin that causes illness." Thus in H. W. Pleket (see n. 77), no. 13: Γλυκία Ἰουλίου τοῦ Ἀγρίαι κολασθεῖσα ὑπὸ τῆς Ἀναείτιδος τῆς ἐγ Μηρῷ τὸν γλουθροῦν ἐπ̣ιζητήσασα ἀν̣[έθ]ηκεν.

99. Ziebarth 1899, 122, cf. K. Buresch, in *Aus Lydien: Epigraphisch-geographische Reisefrüchte,* ed. O. Ribbeck (Leipzig, 1898), 112; Buckler (see n. 77), 179; Eger (see n. 77), 282. Cf. Lane (see n. 77), 31. Ziebarth saw in this the presentation of a judicial complaint on a πιττάκιον, comparable to the Cnidian petitions, in which he is followed by Steinleitner (p. 100): "Apollonios machte der Meter Artemis Platz, d.h. machte sie zu Vertreterin seiner gerechten Sache." Buckler agrees: ". . . a cession of Apollonios. This was probably a curse inscribed on a πιττάκιον and placed before the goddess' shrine." Here the verb is clearly interpreted in its absolute sense or at best with, as its hidden object, the lawsuit itself. Yet O. Eger, who with this one exception follows the footsteps of his predecessors, interprets, "Apollonios überantwortete (παρεχώρησεν) deshalb der Göttin den ihr durch seinen Eidbruch bereits verfallen Skollos" and therefore sees the guilty as object. With this he particularly reacts to Zingerle (1926, 36), who had described the action as a real *Zessionsverfahren,* in which the right to claim is transferred to a third party, in this case the deity, exactly as on the bronze tablet from South Italy. The result would then be that the sum would in the end become temple property.

100. For the discussion on these terms see Herrmann 1962, 47f. and *TAM,* no. 255.

101. Eger—(see n. 77)—(p. 283) compares Polyb. 6.58.4: τηρεῖν τὴν πίστιν καὶ λύειν τὸν ὅρκον; cf. Liv. 22, 61, 4 (*iure iurando se exsolvisset*) and ibid 8 (*religione se exsolvisset,* "redeem oneself from the obligations of an oath"). Cf. Speyer 1969, 1191.

102. See Versnel 1985, 261f. and Strubbe above, p. 45.

103. A. Deissmann, *Licht vom Osten,* 4th ed. (Tübingen, 1923), 277ff. Cf. on the psychology H. S. Versnel "Self-sacrifice, Compensation, and the Anonymous Gods," in *Le sacrifice dans l'antiquité,* Entretiens Hardt 27 (Genève, 1981), 135–85. Cf. an inscription from Jerusalem (seventh century A.D), *BE* (1960) no. 416: ὑπὲρ λύτρου τῶν αὐτοῦ ἁμαρτιῶν.

104. Björck 1938, nos. 6 and 12. On the use of ζητέω in this sense see *TWNT* vol. II, 897.

105. The terms alternate in literary texts: R. Merkelbach, "Fragment eines satirischen Romans: Aufforderung zur Beichte," *ZPE* 11 (1973): 89–90.

106. C. T. Newton (see n. 60); C. Wachsmuth (see n. 91); J. Zündel, "Aegyptische Glossen," *RhM* 19 (1864): 481–96; R. S. Conway, "The Duenos Inscription," *AJP* 10 (1889): 445–59; Ziebarth 1899, 126; Wünsch, *DTA* xiib.

107. Audollent, *DT*, p. cxvi and 5.

108. Kagarow 1929, 22ff.; Björck 1938, 123ff.

109. Zingerle 1926. His theory was severely censured, e.g., by Eger (see n. 77) and L. Robert (*BE* [1978]: 471, no. 434). Cf., further, Strubbe above, p. 44 with n. 102.

110. For lead tablets in wells cf. Jordan (see n. 19), 207 nn. 3–5 and 210, n. 7.

111. It is discussed by Hughes and Quaegebeur (see n. 59).

112. Cf. J. D. Ray (*JEA* 61 [1975]: 181–188), who refers to Amm. Marc. 19.12.3 concerning the oracle of Bes at Abydos: *chartulae seu membranae, continentes quae petebantur, post data quoque responsa interdum remanebant in fano*. Cf. also Lucian *Alex*. 19. For more general remarks on these practices see Versnel 1981, 32ff. Sometimes oracular questions come very near to prayers for justice; a tablet found at Dodona asks, ἔκλεψε Δορκίλος τὸ λᾶκος (H. W. Parke, *The Oracles of Zeus* [Oxford, 1967], no. 29). Cf. Versnel 1981, 6.

113. The emperor Trajan sent a closed letter to the oracle of Jupiter Heliopolitanus (Macrob. *Sat*. 1.23).

114. Cf. Gallazzi (see n. 59), 105–6.

115. Wachsmuth (see n. 91), 569.

116. In Egypt we have clear evidence of oracles with this purpose; see B. Kramer, "*POxy* 12.1567: Orakelfrage," *ZPE* 61 (1985): 61–62 for earlier literature.

117. For the discussion see Versnel 1981, 30, n. 118. The complex of guilt feelings, penalties by the gods, confession, and redemption is attested for Rome, too: R. Reitzenstein, *Die hellenistischen Mysterienreligionen* 3d ed. (Leipzig, 1927; repr. Darmstadt 1966), 137ff.; L. Koenen, "Die Unschuldsbeteuerung des Priestereides und die römische Elegie," *ZPE* 2 (1968): 31–38; R. Merkelbach (see n. 105), 81–100; P. Frisch, "Ueber die lydisch-phrygischen Sühneinschriften und die 'Confessiones' des Augustinus," *EA* 2 (1985): 41–45.

118. G. Moracchini-Mazel, *Les fouilles de Mariana (Corse)*, vol. 6, *La Nécropole d'I Ponti* (Bastia, 1974), 18f. (first or second century A.D.); H. Solin, *Arctos* 15 (1981): 121–22.

119. They are being published by M. W. C. Hassall and R. S. O. Tomlin, but many tablets from Uley and Bath are still unpublished. For a short introduction see M. W. C. Hassall, "Altars, Curses and Other Epigraphic Evidence," in *Temples, Churches, and Religion: Recent Research in Roman Britain*, ed. W. Rodwell (London, 1980), 79–89; and R. Tomlin, "Curses from Bath," *Omnibus* 10 (1985): 31–32. On the temple and cult of Minerva Sulis at Bath, see I. A. Richmond and J. M. C. Toynbee, "The Temple of Sulis-Minerva at Bath," *JRS* 45 (1955): 97–105; H. J. Croon, "The Cult of Sul-Minerva at Bath," *Antiquity* 27 (1953): 79–83; B. Cunliffe, "The Temple of Sulis Minerva at Bath," *Antiquity* 40 (1966): 199–204; P. Salway, *Roman Britain* (Oxford, 1981), 686ff.; B. Cunliffe, *Roman Bath Discovered* (Oxford, 1971), 27ff.; M. J. Green, *The Gods of Roman Britain* (n.p., 1983), 31, 43, 52; M. J. T. Lewis, *Temples in Roman Britain* (Cambridge, 1966), 57–61. On Uley see A. Ellison, "Natives and Christians on West Hill, Uley," in *Temples*, etc., ed Rodwell, (London, 1980): 305–20.

120. L. Franz, *JOAI* 44 (1959): 69ff.; Solin 1968, no. 12.

121. Egger interpreted *draucus* as "a head of cattle," but Jordan points out to me in a letter that this is actually a transliterated Greek term meaning "necklace," which also appears in the diminutive (δραύκι⟨ο⟩ν) on the Delian tablet discussed above on p. 66–67.

122. *CIL* VII.140; *ILS* 4730; *DT* 106; *RIB* 306. The common interpretation, "among those who are called Senicianus," seems wrong to me. The interpretation in my text is closer to Latin syntactical rules and *nomen* is generally "the person" in this type of text. R. G. Goodchild ("The Curse and the Ring," *Antiquity* 27 [1953]: 100–102) connects this tablet with a golden ring with the name Senecianus that was found in a fourth-century Christian context.

123. Solin 1968, no. 18. The archaeological context is third-to-fourth-century-A.D., although the writing seems earlier.

124. He derived this suggestion from a third, well-known tablet discovered in England, which I cannot treat here *in extenso,* cf. E. G. Turner, *JRS* 53 (1963): 122ff. (= Solin [1968, no. 21], who dates it to c. 200 A.D.). In it Jupiter is required to "haunt," (*exigat*) the mind and intestines of the person who has stolen some *denarii* from Canus Dignus. Then there follows, *ut in corpore suo in brevi temp[or]e pariat. Donatur deo ssto decima pars eius pecuniae quam [so]luerit* ("that in his own person [or perhaps lit. "with his own body"] in a short time he may balance the account. The above-named god [*ssto = s(upra) s(crip)to*] is given a tenth part of the money after he [i.e., the thief] will have repaid it."). Note that *exigo* even more than the customary *persequor* exactly parallels the Greek verbs *ἐπιζητέω* ("claim, reclaim, demand") and *ἐκζητέω* ("require an investigation," "require an accounting," "exact an amount" or [in an absolute sense] "start an investigation," "punish"). Note also that *pario* and *solvo* approach the meaning of the Greek verb *λύειν* as used in the confession texts, i.e., "redeem, atone for, buy off."

125. The editor translates "from the money that the thief had 'consumed' " (i.e., the verb is some form of *exedere*), which is hardly possible. Egger assumes that *exesuerit* is a slip of the pen for something like *solverit*. Could it not be *exsolverit?*

126. *TAM* 318 (see above, p. 76); cf. Björck 1938, no. 25, where I suggest reading *ποιῶ αὐτὴν ἱκανόν μου, Μέσα.*

127. The editors have recognized this (see p. 375, n. 24).

128. From a pit or well (third/fourth century A.D.).

129. On a pewter plate from the hot springs at Bath, (*Britannia* 16 [1985]: 323, where one can find further references).

130. E.g., SGD 11, 13, 15, 40, 46, 52, 69, 73, 94, 124, 170, 177; cf. Faraone above p. 11.

131. Or maintain *mei* as objective genitive followed by a dative.

132. See n. 50. Since this text is also disputed, I cite several other curse formulas. L. Robert (*CRAI* [1978]: 280ff.) published a funerary curse from the Karayü valley in Pisidia. It contains the following wish for the potential graverobber or vandal (I quote the version proposed by H. W. Pleket in *SEG* 28, no. 1079; I owe this reference to Strubbe): *τέκνα / τέκνοις αἵματι καὶ θανάτοις ἀποδώσο̣υσ[ι]*. In his commentary Robert connects the verb *ἀποδίδωμι* to the committed crime and interprets, "He will pay for this crime with his blood and with many dead." Threatened punishments of many or even a thousand deaths do occur (see L. Robert, ibid., 281, n. 36), but for us it is most significant that *sanguine suo solvat* recurs here literally in a Greek curse. Robert points to a related tombstone curse from Philomelion, Phrygia (*MAMA* VII, 199; *CIG* III, 3984; cf. L. Robert, *Hellenica* 13 [1965]: 97–98), which reads, *ὅς ἂν τούτῳ τῷ μνήματι κακῶς ποιήσει, οἴκῳ, βίῳ, τῷ σώματι αὐτοῦ* ("Whosoever harms this tombstone, [he will pay for it] with his house, his life [and] his body"). Here we are dealing with what Robert calls a "carcasse d'une malédiction traditionnelle," in which the added asyndetic exclamation "with his house, his life, and his body" is sufficient to express that these are the objects with which the guilty must pay or atone.

133. R. G. Collingwood (*JRS* 17 [1927]: 216 and *Archaeologica* 78 [1928], 158, no. 10) correctly understands *tulit* as *abstulit* and *redimat* as "buy back" and interprets it as meaning that the thief or the owner will buy back the objects placed with Nemesis only with his death. This isn't very logical because it concerns stolen objects. A. Oxé (*Germania* 15 [1931]: 16ff.) improved the text by reading *n[i vita] sanguine suo* instead of *n[isi fusa] sanguine sua*. His translation reads, "Wer sie brachte, möge sie wiedererhalten nur mit seinem Leben, mit seinem Blute." He imagines that someone has left his clothes at the wardrobe of a public bath and that in the meantime an enemy buried the tablet with the wish to Nemesis that the owner can get the clothing back only by paying with his blood. Without commenting on the precise

situation, Preisendanz (*APF* 11 [1935]: 155) judges "wahrscheinlich handelt es sich um Verfluchung eines Gladiators durch seinen Todfeind, der die Kleider und Schuhe des anderen der Domna Nemesis schenkt unter Bedingung seines Todes." Egger (*Wien. Jahresh.* 35 [1943]: 99ff. = Egger 1963, 281–83) generally agreed with these suggestions. Without the benefit of the recent discoveries we have discussed—and reading *sanguinei sui*—he translated, "Herrin Nemesis! Ich übergebe dir Mantel und Schuhe, wer sie getragen hat, möge sie nur dann zurückerhalten, wenn sein Rotfuchs umkommt" (idem 1963, 281–83). He is followed by Solin 1968, no. 20 and *RIB* 110, no. 323. It had escaped him that H. Volkmann (*ARW* 31 [1934]: 64) had at least made a start towards a better solution by understanding *redimere* as *culpam redimere* ("büssen"). Whoever chooses to translate this verb as "to get back" (this is not at all its usual meaning) has to conclude that it is a question of sympathetic magic in which the clothing of a person by consecration to a deity draws the owner along as well. It is, however, notable that although the burial of clothing, hair, nails, etc. is widely known in Greek magical texts, allusions to it on magic *defixiones* are rare. There is one highly dubious—at any rate abnormal—instance in *DT* 210, and there is a new tablet published by Jordan (see n. 19), 251f. in which the hair of the victim is given to the daemon. He mentions a few other instances. See H. Solin, "Tabelle plumbee di Concordia," *Aquilea Nostra* 48 (1977): 146–63, esp. 149 for the translation of *tulit* as "he has taken away." We need not say more about the bay (horse), cf. *Britannia* 14 (1983): 352, n. 12.

134. A tablet from Aylesford (Kent) has *donatio;* cf. *Britannia* 17 (1986): 428, n. 2.

135. For the term *ὑπογεγραμμένοι* on a gold tablet and a collection of parallels, see D. R. Jordan, *AJA* 89 (1985): 164–65, and idem (see n. 19), 241 and 252. On this "official" language cf. Hassall (see n. 119), 87.

136. Cf. Deissmann (see n. 103), 259; *καταδέσμους ἀναλύσεις* (*PGM* IV.2177). Cf. Preisendanz 1972, 6f.; Speyer 1969, 1191. Cf. SGD 170: *τούτων μηδεὶ[ς] θεῶν λύσιν ποιήσαιτο.*

137. E. Courtney suggests *idem regestum* (heaped back upon), which would yield a similar interpretation.

138. *epotes*(?) or, as F. G. Naerebout suggests, *eructet*(?). After this essay had been sent off to the publisher, Tomlin informed me that he now prefers to read ⟨*r*⟩*eputes,* i.e., "reckon the stolen coins with his blood," which seems to solve the riddle quite well.

139. We also find ourselves in the "borderland" now and then in the texts of the Latin tablets. A text from England of the usual juridical type has strong reminiscences of the traditional *defixio,* especially in its detailed series of stipulations (*Britannia* 15 [1984] 339 no. 7): *nec illis* [*p*]*ermittas sanit*[*atem*] *nec bibere nec ma*[*n*]*ḍ*[*u*]*care nec dormi*[*re*] [*nec nat*]*ọs sanos habẹ*[*a*]*ṇt* . . . . . .The usual text of a *defixio defigo Eudemum.* This victim must die as soon as possible, and yet *infra dies nove*⟨*m*⟩ *vasum reponat.* Apparently it is a case of theft, therefore it has the tone of an appeal to justice (R. Egger, "Eine Fluchtafel aus Carnuntum," *Der römische Limes in Österreich* 16 [1926] 136–56 = idem 1963, 81–97); Solin 1968, no. 6). Cf. also R. Marichal, *CRAI* (1981): 41–51 for a lead tablet from Montfo (Gallia Narbonensis): *Qomodo hoc plumbu non / paret et decadet, sic deca / dat aetas, membra, vita / bos, grano, mer eorum qui / mihi dolum malu fecerunt,* etc. It is magic, but the victim is guilty. Cf. *CIL* vol. XIII, 11340: *ut me vindicetis de ququma;* see R. Egger, ibid., p. 87 and Solin (see n. 133), 149.

140. Cf. Cunliffe (see n. 119), 27ff.

141. Cf., on this mentality, Pleket (see n. 76), 152–92.

142. Versnel 1985.

143. On these "handbooks" see Björck 1938, 134; Jordan (see n. 19), 211 and 233f. On handbooks for magic spells and rituals, see Faraone above p. 4.

144. *CIL* II.462 (with a very adventurous conjecture by Mommsen); *DT* 122; J. Vives, *Inscripciones Latinas de la España Romana* (Barcelona, 1971), no. 736. The terminology shows influences of legal language.

145. It has been published in J. H. Bonneville, S. Dardaine, and P. LeRoux, *Fouilles de Belo: Les inscriptions* (Paris, 1987–88).

146. Thus I explain *Isis Muromem* of the text.

147. One Latin *defixio* belonging to the Sethianic texts (*DT* 142) has *ut omnes cog*[*n*]*osc*[*ant*] *exempl*[*um e*]*or*[*um*].

148. I should like to thank Jeannette K. Ringold, who translated this essay from the Dutch original. I am also very grateful to David Jordan and Christopher Faraone for their numerous helpful comments.

4

Incantations and Prayers for Salvation on Inscribed Greek Amulets

Roy Kotansky

The use of magic for protection and deliverance from diseases must have been widespread from the earliest times.[1] The sufferer had recourse to healing through prayers and offerings, rites of incubation, and any number of rituals performed by itinerant holy men who adhered to that heterodox and often arcane aspect of religion known as "magic."[2] Apart from the more empirically minded doctors, a wide variety of practitioners from herb-gatherers to midwives could be sought for remedies in the classical period.[3] As an infection festered or a fever lingered, even the sternest critics of traditional or "superstitious" remedies turned to the application of amulets. In his lost *Ethics,* Theophrastus questions whether a man's character changes when the circumstances of his life change, and he reports how the "freethinking" Pericles, sick with the plague, had been prodded by his womenfolk into wearing an amulet (Frag. L21 Fortenbaugh); Diogenes Laertius tells a similar deathbed tale about the philosopher and notorious atheist Bion, who at the end of his life dons amulets and renounces his former attacks "against religion" (εἰς τὸ θεῖον, 4.54; cf. 4.56–57). Both anecdotes, whether historically accurate or not, present a plausible picture of competing cures and "second opinions" as a disease worsens and seem to set the use of amulets squarely within the sphere of traditional beliefs.

Amulets were in demand for every imaginable situation in life.[4] Although they could often serve to introduce desirable qualities such as love, wealth, power, or victory,[5] amulets were usually used to cure medical complaints (both injuries or illnesses) and to thwart the daemonic influences often held responsible for disease. Etymologically speaking, the Greek terms for amulet (περίαμμον and περίαπτον) are derived from the verb περιάπτειν ("to tie on") and refer to an object or material that is "attached to" or "tied on" to a person.[6] A phylactery (the English word comes directly from the Greek φυλακτήριον, formed from the verb φυλάσσειν, "to protect") is a type of amulet used more specifically to protect an individual or community from some impending calamity or plague.[7] Amulets could be organic substances or simple compounds; as we shall see, their application was often accompanied by spoken prayers or incantations. In later times the large number of amulets containing written prayers and incantations allows us to document the gradual transition from the "unlettered" practice of oral magic to the full-fledged literary compositions found in the Greek magical papyri. Simple, uninscribed amulets are difficult-if-not-impossible to identify; even when they carry some telltale symbol or design, they remain silent about their specific purpose or the source of their efficacy. Those, however, that are inscribed with texts (no matter how brief) provide information about the ancient medical and religious contexts of their use and will be the focus of our inquiry here.

THE COMBINED USE OF INCANTATION AND AMULET

As far as the use of written charms is concerned, presumptive antecedents can be traced to ancient Egyptian and Near Eastern rituals with which early Greek traders may have had contact.[8] This does not, however, mean that such amulets were taken over directly from oriental prototypes; magic is indigenous to every culture, and the employment of magic incantations and the like for apotropaic purposes already occurs in the earliest Greek writings. We cannot be certain just how and when written charms came into regular use among the Greeks (and later the Romans), although it is clear that in the very early periods healing words or other incantations often accompanied the protective or therapeutic act. The locus classicus is a passage in Homer describing how the sons of Autolycus stop the bleeding of Odysseus' boar wound by binding it with an incantation (*Od.* 19.457–59):[9]

ὠτειλὴν δ' Ὀδυσῆος ἀμύμονος ἀντιθέοιο
δῆσαν ἐπισταμένως, ἐπαοιδῇ δ' αἷμα κελαινὸν
ἔσχεθον

And the wound of noble, god-like Odysseus they bound up skilfully, and checked the black blood with a charm. (trans. A. T. Murray)

Here, the "skillful binding" of the leg, although applied with a magic utterance, appears to modern readers to serve the practical medical function of staying the hemorrhage by use of a tourniquet; the widespread testimony, however, to the popular ancient belief that knots could bind the flux sympathetically cannot be overlooked.[10]

One of the earliest mentions of Greek amulets and the pronouncement of spells appears in Pindar, where we read of the adult Asclepius, who was taught medical lore by Chiron the Centaur. The passage seems to give a detailed description of early fifth-century B.C. medical practice (*Pyth.* 3. 47–54):[11]

τοὺς μὲν ὦν, ὅσσοι μόλον αὐτοφύτων
ἑλκέων ξυνάονες, ἢ πολιῷ χαλκῷ μέλη τετρωμένοι
ἢ χερμάδι τηλεβόλῳ,
ἢ θερινῷ πυρὶ περθόμενοι δέμας ἢ
χειμῶνι, λύσαις ἄλλον ἀλλοίων ἀχέων
ἔξαγεν, τοὺς μὲν μαλακαῖς ἐπαοιδαῖς ἀμφέπων,
τοὺς δὲ προσανέα πί-
νοντας, ἢ γυίοις περάπτων πάντοθεν
φάρμακα, τοὺς δὲ τομαῖς ἔστασεν ὀρθούς·

> And whosoever came to him [sc. Asclepius] afflicted with natural illnesses or with their limbs injured by grey bronze or stones far-slung or with their bodies ravaged by summer's fever or winter's chill, these he frees from the bonds of every sort of pain, tending some with gentle incantations, giving others soothing *φάρμακα* to drink or attaching *φάρμακα* to their limbs from every side, and still others he cures by incisions.

Here Pindar presents us with a compact description of Asclepius' threefold medical methodology (incantations, *φάρμακα*, and surgery), wherein he subdivides the second category into *φάρμακα* drunk as potions and those tied on as amulets (*περιάπτειν*). The *φάρμακα* (drugs or charms) that are attached as *περίαπτα* were perhaps applied with an incantation (*ἐπῳδή*) in a way similar to that described in Homer;[12] but the important fact here is that at this early stage incantations remained separate from the amulet.

One is reminded of the well-known incantation in Cato's *De agri cultura* 160, which is probably considerably older than the second- or third-century B.C. text in which it is preserved. It is part of a recipe for healing dislocated or fractured bones that also involves the "sympathetic" gesture of cleaving a reed in two and then rejoining the pieces while brandishing an iron knife and uttering the apparently nonsensical incantation, MOTAS VAETA DARIES DARDARES ASTATARIES DISSUNAPITER.[13] The pieces of reed are then attached (*adligare*) to the limb in question, perhaps as a practical splint of sorts but more likely in order to place the symbolically rejoined reed in direct contact with the damaged joint or bone(s), that is, as a sort of primitive amulet. For the following days there is an additional charm for recitation: HUAT HAUT HAUT ISTASIS TARSIS ARDANNABOU DANNAUSTRA.

The first explicit reference to an amulet applied with an incantation occurs in Plato's *Charmides* (155e–156e). Socrates reveals the recipe for a headache amulet that he once learned from the Thracians while serving on a military campaign in the area, probably at Potidaea or Amphipolis (cf. *Ap.* 28e). As is often the case with Plato,[14] the account serves to introduce a dialogue on the meaning and definition of something more important than a mere detail of medical lore; nevertheless, it does preserve a fascinating folkloristic belief probably contemporary with the dialogue's composition if not indeed known personally to the historical Socrates. The charm is described as follows (155e): "So I told him that the thing itself was a certain leaf (*φύλλον*), but there was a charm (*ἐπῳδή*) to go with the remedy; and if one uttered the charm at the moment of its application, the remedy made one perfectly well; but without the charm there was no efficacy in the leaf" (trans. W. R. M. Lamb). The cure, therefore, is only effective when the leaf is applied with the requisite incantation; unfortunately Socrates does not give the text of the incantation in the course of the dialogue.

Later writers also report that the recitation of incantations or magic formulas could, with the application of a special material, rectify fractures or heal other medical problems. Lucian, for example, in a way reminiscent of the *Charmides,* also describes the process of enchanting an amulet with powerful words. During a revealing discussion about the efficacy of amulets and incantations (*Philops.* 7–8), the question is raised about the value of applying external remedies to ailments that have internal causes. Those who are defending the use of amulets explain that oftentimes objects employed for the cure of rheumatism in the feet, such as a weasel's tooth or lion's whiskers, are only effective if one knows how to use them with a suitable incantation (*εἴ τις ἐπίσταιτο αὐτοῖς χρῆσθαι μετὰ τῆς οἰκείας ἐπῳδῆς ἑκάστῳ* [*Philops.* 7]).[15] A little later in that same dialogue we learn that the disease is believed to be driven away by uttering magic words (*Philops.* 9), and a story is told about a Babylonian magus who healed a certain gouty Midas by chanting an *ἐπῳδή* and binding a fragment of a tombstone of a deceased virgin to the sufferer's foot (*Philops.* 11)![16]

These examples suggest that incantations and amulets were often used in tandem for healing.[17] As in the case of the *defixiones* (discussed by Faraone in chapter 1), the verbal incantation and the material used in the attendant gesture (e.g., the leaf applied to the head) seem, with the introduction of a written language, to merge according to some natural law of economy. As a result, a new and more sophisticated type of amulet begins to appear, as the words of incantations, formerly only spoken, are now engraved directly onto the amulet itself.[18]

EARLY INSCRIBED AMULETS: THE *EPHESIA GRAMMATA*

There is no sure way of knowing when the first written charms were employed by the Greeks; and much of our understanding of such texts must remain hypothetical, since the discoveries of actual inscribed amulets from the classical period are few. The rather scanty evidence, of course, does not provide an accurate picture of what probably was a widespread practice. Amulets written on perishable material (e.g., leather, wood, wax, and the like) would not have survived intact. A similar problem attends the texts engraved on more durable materials: tablets of gold or silver were reused because of their innate value, while those of cheaper metals such as lead were regularly recycled on account of their pliability and noncorrosive nature.[19]

Let us look at some of the early evidence for texts written on amulets. The use of rings as amulets, possibly engraved, is mentioned by the comic poet Antiphanes (Frag. 177 Kock), a contemporary of Demosthenes. The passage simply states that someone purchased from Phertatos for a drachma a ring for digestive pains. We read nothing about what the ring was made of, whether it was a simple band or held a carved stone, nor is anything said of an inscription. Such rings used as amulets must have been common. When Aristophanes has the "Just Man" ignore the threats of Karion, a treacherous sycophant, he seems to allude to rings that actually carried engraved texts for use as amulets (*Plut.* 883–85):

ΔΙ. *οὐδὲν προτιμῶ σου. φορῶ γὰρ πριάμενος*
τὸν δακτύλιον τονδὶ παρ' Εὐδάμου δραχμῆς.
ΚΑ. *ἀλλ' οὐκ ἔνεστι "συκοφάντου δήγματος."*

Just Man: I fear you not, for I wear a ring that
Eudamos sold me for a drachma.
Karion: But it is not inscribed, *For an informer's bite.*

Here, too, the ring's price and supplier is named. Karion's witty reply suggests that there was a market for inscribed rings that protected the wearer from the bites of dangerous insects or animals; Bonner believes that a designation such as *σκορπίου δήγματος* (for scorpion bite) or the like must have been in common enough use as part of the actual inscription to warrant such an allusion.[20]

But to find detailed evidence for the early use of engraved amulets we must turn elsewhere. The *Ephesia grammata,* mystic letters allegedly incised on the famous cult statue of Artemis of Ephesus,[21] were often used in apotropaic rituals, both verbally and as parts of inscribed texts. The text of the incantation, traditionally given in ancient sources as *ασκιον, κατασκιον, λιξ, τετραξ, δαμναμενευς, αισιον* (or *αισια*) shows that (like the charms recorded by Cato) they were not comprehensible to later ancient writers.[22] Menander describes them as "evil-averting spells" (*ἀλεξιφάρμακα*) spoken as one walked in a circle around newly wed couples.[23] Plutarch (*Mor.* 706e) reports that the "Ephesian letters" could be uttered to expel daemons and that Croesus supposedly recited them to save himself from being burned alive on the funeral pyre. But were the Ephesian letters also used as written charms, that is, engraved on objects to be used or worn as talismans? A story (albeit recorded in late sources) does tell of an Ephesian who by wearing the letters tied onto his ankle repeatedly defeated his Milesian rival in boxing; as soon as the amulet was detected and removed the man was soundly defeated.[24] One is immediately reminded of the many *νικητικά* (victory charms) recorded in the Greek magical papyri and useful in a variety of agonistic settings.[25] In *PGM* IV.2145–50, in particular, we find a multipurpose talisman that employs Homeric verses (*Il.* 10.521, 564, and 572) engraved on an iron tablet; one of the virtues of this magic tablet is that a contestant who carries it will remain undefeated (lines 2159–60).

Mention of the use of inscribed *Ephesia grammata* comes from an early source as well. A fragment of Anaxilas, a fourth-century B.C. comic poet, reads (Frag. 18 Kock),

> *ἐν σκυταρίοις ῥαπτοῖσι φορῶν*
> *Ἐφεσήϊα γράμματα καλά.*
>
> Carrying about the excellent Ephesian
> letters in little stitched hides.

Like the reference to the Ephesian boxer given above, we do not know on what material the *grammata* were engraved or in what manner they were carried, though the passage suggests a sort of leather pouch (similar to the metal tubes worn by both Greeks and Romans, which often enclosed an inscribed metal tablet or papyrus).[26] A lead tablet actually inscribed with the *Ephesia grammata* allegedly from Phalasarna in Crete dates securely to the fourth century B.C. (it is roughly contemporary with Anaxilas); it had been folded over about six times into a small mass only three or four centimeters wide[27] and had evidently been used as a protective charm, perhaps for an individual, though one cannot rule out protection for a household or sanctuary. The rather long text scratched into the tablet is mostly composed of hexameters that clearly served as an *ἐπῳδή* to ward off some general malady or plague on Crete in the fourth century B.C.[28]

Although the text is fragmentary (most present editions of the opening line, for instance, must be rejected), the apotropaic language of the piece is evident throughout. Some sections are hymnic in nature (like the hexametric hymns of the *Papyri*

Graecae Magicae) and invoke deities appropriate for healing or success: Zeus Alexikakos (line 3), Herakles Ptoliporthos (line 3), Iatros, Nike, and Apollo (line 4). The formula used at the beginning of the tablet seems to address some type of evil directly: "I bid . . . flee from our homes"[29] followed by the repeated command φεῦγε (Flee!) addressed to "wolf" and "dog" and other unidentifiable entities, all probably designations for daemons. In addition to the appearance of the Ephesian letters (lines 5, 9–11, and 15ff.) towards the end of the spell there seems to be a request for protection against the magic operations of others.[30] If worn about a person, as its compact size would suggest, the folded tablet could have been enclosed in a σκυτάριον ῥαπτόν like that described by Anaxilas and suspended from the neck by a thong.[31]

AMULETS WITH ἐπῳδαί IN THE ROMAN EMPIRE

The texts of the extant amulets from the classical era are few, but for the period of the Roman Empire the situation is considerably different. Sometime between the hellenistic period and the height of the Roman Empire the manufacture of inscribed amulets began to flourish. Among the extant magical papyri, one document in particular, known as the Philinna papyrus (*PGM* XX) and of relatively early date (first century B.C.), gives instructions for using ἐπῳδαί in apotropaic contexts.[32] These tattered fragments of a medicomagical handbook[33] preserve two incantations (ἐπῳδαί) composed in hexameters[34] that are clearly antecedents of some of the inscriptions found on amulets of the later Roman period. The first is to be used against inflammation and is entitled *An incantation of the Syrian woman from Gadara*. Employing some sort of "sympathetic magic," it briefly describes an initiate to a mystery religion (μυστοδόκος) who is set aflame on a mountaintop and subsequently doused with water:[35]

[The most majestic goddess' child] was set
Aflame as an initiate—and on
The highest mountain peak was set aflame—
[And fire did greedily gulp] seven springs
Of wolves, seven of bears, seven of lions,
But seven dark-eyed maidens with dark urns
Drew water and becalmed the restless fire. (trans. E. N. O'Neil)

Despite the obscure references to initiation and predatory animals, the nature of the incantation is fairly straightforward: just as the immolated μυστοδόκος is subsequently doused with water, so too will the bodily inflammation of the patient be extinguished.

These types of spells are known as *historiolae*—short stories recounting mythological themes that sympathetically persuade the sufferer's illness to cease.[36] Another *historiola* occurs on a silver phylactery found at Carnuntum. The spell, dated to the third century A.D. on archaeological grounds, contains a description of an encounter between Artemis of Ephesus and Antaura the mermaid; the text on the tablet reads as follows:[37]

πρὸς ἡμικράνι⟨ο⟩ν· Ἀνταύρα
ἐξῆλθεν ἐκ τῆς
θαλάσσης, ἀναβόησεν ὡς
ἔλαφος, ἀνέκραξεν ὡς βοῦς.
ὑπαντᾷ αὐτῇ Ἄρτεμις Ἐφεσ[ία]·
"Ἀνταύρα, πο[ῦ]
ὑπάγ{ι}εις ἡμικρ[άνιο]ν;
[μ]ὴ οὐ[κ ε]ἰς τὰν . . . ."

For migraine headache. Antaura came out of the ocean; she cried out like a deer; she moaned like a cow. Artemis Ephesia met her: "Antaura, where are you bringing the headache? Not to the . . .?" (here the text breaks off).[38]

In this case, Artemis' presumed interdiction of the disease will hopefully be reenacted in the body of the person who wears the amulet. Another tablet (discussed below) contains a similarly short mythological story for the cure of epilepsy, and possibly also for headache.[39]

The second spell of the Philinna papyrus is the headache remedy of its namesake, the mysterious Philinna of Thessaly. This short hexametric spell cures by using a "flee" formula:

φεῦγ' ὀδύνη κεφαλῆς· φεύγει δὲ [λέων] ὑπὸ πέτραν,
φεύγουσιν δὲ λύκοι, φεύγουσι δὲ μώνυχες ἵπποι
[ἱέμενοι] πληγαῖς ὑπ' [ἐμῆς τελέας ἐπαοιδῆς].

Flee, headache, [lion] flees beneath a rock,
Wolves flee; horses flee on uncloven hoof
[And speed] beneath blows [of my perfect charm]. (trans. E. N. O'Neil)

The same "flee" formula appears on the much earlier Phalasarna tablet (see pp. 111–12), where a presumably daemonic dog and wolf are similarly put to flight. But this is not the only indication of the antiquity of this particular kind of ἐπῳδή; a similar formula, quoted by Pliny (as a hexameter) as part of a ritual cure for *impetigo*, seems to be a product of the classical period or earlier:[40] φεύγετε κανθαρίδες· λύκος ἄγριος ὔμμε διώκει ("Flee beetles, a fierce wolf pursues you"). The incantation is repeated while the infected area is touched with a special stone.

As the last example shows, the use of unengraved materials as amulets continues unabated in the Roman period side by side with the talismans and phylacteries that carried texts. Chapters 24–32 of Pliny's *Historia Naturalis* attest well to the situation in the first century of the common era. Numerous folklore remedies are described, but among the literally thousands of "magical" remedies (both herbal and mineral) we find only a few recipes that employ an incantation by itself; for instance, Attalus is reported to have uttered DUO to avert scorpion sting (28.5.24), and elsewhere a formula is given for protection while one picks a powerful herb (24.116.176).[41] Pliny also gives an example of an incantation to be uttered simultaneously with the application of an amulet; the healer, while fasting and applying nine herbal knots as an amulet, recites the *carmen*, "Fasting I give a cure to a fasting patient" (24.118.181). Inflammations are to be treated with a plant called *reseda* while the following chant is repeated three times: "Reseda, allay diseases; dost know, dost know, what chick here uprooted thee? May he have neither head nor feet" (27.106.131). Elsewhere to cure superficial abscesses, a fasting, naked woman is to touch a patient's back and say, "Apollo tells us that a plague cannot grow more fiery in a patient if a naked maiden quench the fire" (26.60.93), a charm that recalls the incantation against inflammation in the Philinna papyrus quoted above. Thus throughout antiquity we find the continued use of the therapy of the spoken charm in tandem with the application of the uninscribed amulet.

Elsewhere Pliny records a somewhat early example of an incantation to be written on papyrus as an amulet; Marcus Servius Nonianus, consul in 35 A.D., cured his ophthalmia by engraving the Greek letters PA on a slip of papyrus and tying it around his neck (28.5.27).[42] In addition to papyrus and lead, amulets were fashioned from a wide variety of other writing media as well. Magical texts (often containing just symbols or very short spells) are often inscribed on small, semiprecious stones that are then set into rings and necklaces or otherwise simply carried in an individual's clothing.[43] Slips of gold or silver foil (*lamellae*) inscribed, like the Artemis and Antaura amulet quoted above, with apotropaic prayers and incantations, are often described in the recipes of the magical papyri and the late medical writers.[44] Although these tablets have often been neglected in the past, or at least not fully appreciated, they will provide the focal point of our discussion of inscribed amulets of the later period.[45]

THE MAGIC *LAMELLAE*

Gold and silver magic *lamellae*[46] have been unearthed in every corner of the Roman Empire and are usually inscribed with protective charms similar to those found in the magical papyri. Their existence testifies to a popularity at least as widespread as that of the gemstones, though due to their often fragile condition (and for the other reasons discussed above), not nearly as many *lamellae* as gemstones have survived. References, however, in both the *Papyri Graecae Magicae* and a variety of other ancient literary sources show that these inscribed amulets were recommended more frequently than the gemstones for healing and other magic operations. Also, the so-called Fayum portraits, which sometimes picture women and children wearing the telltale tubular capsules, provide evidence for their regular use among Egyptians in the Greco-Roman period.[47]

The so-called Orphic *lamellae* come foremost to mind when discussing possible prototypes for these tablets.[48] They date to the late classical period (c. 400–330 B.C.; the "Cretan" group [B3–B8 Zuntz] is somewhat later) and do not, at first, seem to have been amulets in the conventional sense described above; rather, they were usually placed unrolled on a corpse as a form of phylactery that protected the dead person from either the terrors of the afterlife or the equally feared cycle of rebirths (metempsychosis).[49] There is, however, some physical evidence that "Orphic" tablets could indeed be used as traditional amulets as early as the classical period. The tablet labeled C by Zuntz was found with one of the more "standard" texts (A4) in 1897 (in the "Timpone Grande" tumulus). Carelessly written and uncertainly interpreted, the text seems to contain a hymnic address of some sort. More importantly, though, this tablet (A4) was folded "like an envelope" inside of tablet C, which itself had been folded over nine times from right to left in the manner of amulets.[50] There is also the curious case of the fourth-century-B.C. gold tablet from Petelia (B1)[51] that was apparently found enclosed in a tubular necklace dating to the second or third century A.D., either carefully handed down from one generation to the next or disinterred at a later date and reused, this text was clearly employed as a conventional amulet.[52]

Finally there is the gold tablet written for a certain Caecilia Secundina (A5), which "has every appearance of having been rolled up in a cylinder similar to that which contained the Petelia tablet."[53] The Greek text of six lines contains portions of the standard "Orphic" formulas found in various forms on the other extant *lamellae* and in this regard would belong wholly to this exclusive class of inscriptions were it not for the surprising fact that it is inscribed in a cursive hand six hundred years removed from the dates of the other known pieces and carries the name of its bearer, Caecilia Secundina. Thus in these three examples (out of the dozen or so extant "Orphic" *lamellae*) the fact that the tablet had once been rolled up or folded suggests that at some point an "Orphic" tablet could have been used, like the Phalasarna tablet, as a personal amulet. One might speculate, then, that the widespread use of the gold and silver phylacteries was indeed patterned after the "Orphic" *lamellae,* that is, that the protection of the recently dead from the dangers of the underworld may have been, or gradually became, a desideratum for living folk as well.

If it cannot be irrefutably demonstrated that the "Orphic" gold leaves served as precursors to the use of inscribed gold (and silver) *lamellae* as amulets, their antecedants may lurk in special types of inscribed gold amulets excavated from other areas in the circum-Mediterranean basin. One group, in particular, seems to be an excellent candidate: the prophylactic inscriptions of Punic-Phoenician origin, disinterred primarily from tombs of the seventh to fifth century B.C. in Carthage and Sardinia; like our amulets of a later date, they too, were enclosed in tubular capsules.[54] Most of the debate about these amulets has centered on the configuration of the suspension capsules enclosing the *lamellae;* worn perpendicularly, rather than horizontally like the later Greco-Roman capsules, the tubes usually terminate at the top in a sculpted representation of an Egyptian deity or animal (e.g., Osiris, Bastet, Sokhit, a swan's head, etc.). More important for our study, however, are the long rolled-up strips of gold (and occasionally silver and papyrus) found within. Although they are mostly inscribed with hieroglyphic and animal figures, and in some cases odd monstrosities, their prophylactic function can be readily perceived. For example, one strip of gold foil measuring twenty-eight by twenty-four centimeters and covered with approximately 250 different Egyptian figures, carries two Punic inscriptions: "Protect and guard Hilletsbaal, son of Arisatbaal" and "Guard and protect Hilletsbaal, son of Asi." The function, style, and general character of the piece is readily comparable to those of the magic *lamellae* of the Roman period.[55]

Another little-known category of gold *lamellae* that has some bearing on the discussion of both the "Orphic" and magic tablets is represented by a number of thin sheets of inscribed gold foil dating from the second century B.C. (and later) found primarily in tombs from Palestine.[56] Each tiny tablet, usually in the shape of a *tabula ansata,* carries a terse, formulaic expression found often on local steles, that is to say, *θάρσει*, NN, *οὐδεὶς ἀθάνατος* ("Take courage, NN, no one is immortal!").[57] Judging from the content alone and from the fact that generally such tablets were not enclosed in capsules, it would seem that these were also laid with the deceased as "passports for the dead," just like the "Orphic" gold leaves. But again, as with the "Orphic" pieces, a clearly magical—and hence amuletic—exception can be discerned in this group; a gold band found in a tomb "à Fîq, dans le Gaulan"[58] evidently served a peculiar talismanic function. In the shape of typical funerary headbands, with terminal holes for affixing it to the forehead, the piece is hardly remarkable except that it is engraved with four magic inscriptions made up of "characters" or perhaps a cipher of magic letters.[59] Evidently, protective magic was sought in the realm of the dead as well. At the end of this essay I will return to this blurring of the distinction between protection in the present life and in the hereafter again.

THE SOCIAL CONTEXT OF THE INSCRIBED AMULETS OF THE ROMAN PERIOD

Although early Greek literary texts (quoted at the beginning of this essay) often describe the use of *ἐπῳδαί*, we know few details about their actual content or purpose. With the Greek *lamellae* of the Roman period, however, a wide variety of texts came into use, allowing us to observe the context in which they operated. These charms seem to be primarily concerned with health, and they aim at either curing existing diseases or preventing them in the future. The *voces magicae,* the long strings of magical *logoi,* and especially the series of vowels[60] that occur so often on amulets can all be regarded as *ἐπῳδαί* in the broadest sense of the word. The habit of engraving Homeric verses on gemstones, papyri, and metal phylacteries is also widely attested.[61] We also have other examples of verses inscribed on amulets, usually dactylic hexameters (such as the *φεῦγε* charm from Pliny) or iambic trimeters of uncertain authorship, which occasionally seem to preserve scraps of early liturgical material like that found on the Phalasarna lead amulet.[62]

In the examples of early Greek amulets and magic practices discussed above we noticed an interest in recording the specific malady for which the charm was written. Socrates' leaf was used strictly as a headache remedy; and the inscription on Eudemos' ring, *συκοφάντου δήγματος* (*For an informer's bite*), suggests that specific charms against bites from other, more literal predators were in circulation.[63] On the later *lamellae* spells are often headed by such descriptions, usually with *πρός* plus the accusative, as witnessed by the Artemis and Antaura phylactery from Carnuntum (whose inscription begins with the rubric *For migraine*) and by the two charms in the Philinna papyrus. Thus during the Roman Empire the treatment of diseases with amulets seems to have required the proper diagnostic identification of the ailment, and we find that the texts found on amulets often indicate the specific diseases for which they are written. Some of the complaints addressed in the magical *lamellae* are discussed below; a study of the charms given by medical writers, the papyri, and the texts of gemstones would necessarily expand this list.[64]

The spells given in the magical papyri and those preserved on the extant amulets often do not differentiate between the specific ailment afflicting the patient and the daemonic influence held responsible for the disease.[65] Descriptions of bodily ailments do occur, but they are usually made with special references as well to the malignant, preternatural influence behind the disease's manifestation. The most common expression for such an influence is "evil spirit" or simply "spirits."[66] In addition to the general protections against daemons and spirits, a number of the magical phylacteries like those used to combat headache describe specific maladies.

A gold charm in Latin found near Picenum at Ripe San Ginesio, though not fully published, reads, *ad oculo⟨rum⟩ dolorem* (*For eyeache*).[67] Far more significant is a gold *lamella* from Tyre employing a Christian trinitarian formula to cure ophthalmia: "In the name of God, and of Jesus Christ, and of the Holy Spirit, RABA SKAN OMKA LOULA AMRI KTORATH ĒNATHA BATHAROURAK . . . (for the one) praying the great name IAO divert the inflicting ophthalmia and do not allow any more attacks of ophthalmia."[68] Except for the few lines of *voces magicae* (which may, indeed, be Aramaic), this phylactery can in every sense be understood as an inscribed Christian prayer; the believer prays both for deliverance from a current medical condition and for the prevention of a relapse of the disease. Among the scattered remedies preserved in Marcellus Empiricus' *De Medicamentis* we also read of the manufacture of a gold phylactery for eye problems.[69] Elsewhere Marcellus gives recipes that recommend the use of magical *lamellae,* not to speak of other types of amulets and incantations.[70]

A silver *lamella* from Ságvár (Tricciana), Hungary, whose text has only been partially published,[71] reads "SESENGENBARPHARANGES, the great and pitying(?) and unconquerable name," in Greek, followed by the name *Romulus* written with Latin characters; it is apparently a cure for swelling.[72] Although none of the extant magical recipes for curing this particular disease recommends the making of a *lamella, PGM* XXIIa. 15–17 requires writing a certain Homeric verse (*Il.* 4.141) as a general amulet for the sufferer of elephantiasis. In 1901 Homolle[73] published a putatively lead inscription from Amorgos that was designed to "banish" a tumor—medically, a type of swelling; D. R. Jordan, on the basis of the medical nature of the text, suggests that the piece (now lost) may have been silver, and was presumably identified as lead in error because of the metal's tarnished color.[74]

Although epilepsy, the so-called sacred disease, appears rarely in the recipes of the *Papyri Graecae Magicae,*[75] at least three metal phylacteries (one unpublished) mention the disease. The first is a gold phylactery acquired in Damascus, which, after a lengthy invocation of standard angelic and divine names, reads (lines 12–19):

> [- - -] κύριοι ἀρχάν-
> γελοι θεοὶ καὶ θ⟨ε⟩ῖοι χαρα-
> κτῆρες ἀπελάσατε
> πᾶν κακὸν καὶ πᾶσαν
> ἐπίλ]ηψιν καὶ πᾶσαν
> κεφά]λγιαν ἀμπ-
> c. 4]ου ἣν ἔτεκε[ν] Ορη-
> c. 4] κτλ.

> O Lord archangels, gods, and divine "characters," drive away all evil and all epilepsy and every headache(?) from [. . .]os, whom ORE[. . .] bore . . . .[76]

Epilepsy, a chronic condition with intermittent attacks, would require that the sufferer wear the amulet at all times. A second example, in the J. Paul Getty Museum,[77] begins with a variation of a traditional prayer: "O God of Abraham, God of Isaac, God of Jacob, God of our ⟨Fathers⟩." Then follows the request, "I implore you (sing.), Lord, Iao, Sabaoth, Eloaion, etc. protect Aurelia from every evil spirit and from every epileptic fit and seizure." A third example, an unpublished silver phylactery in the Walters Art Gallery, Baltimore (inv. no. 57.1961), conjures the "spirit of the sacred disease epilepsy (*πνεῦμα ἱερᾶς νόσου ἐπιληψίας*)" to depart.

Phylacteries for the treatment of fever appear regularly in the papyri.[78] Among the *lamellae* we find in a bronze phylactery from Acrae and a silver phylactery (formerly in the Louvre) references to a number of different ailments.[79] There is no expression describing gout on any of the preserved magic *lamellae,* but the use of a magic *logos*[80] on a gold tablet from Brindisi[81] suggests the possibility that the phylactery served to protect its bearer from this disease. The evidence comes from a fortunate citation in the collection of veterinary medical and magical recipes known as the *Hippiatrica* (*Hippiatr. Paris.* 440, p. 63 Oder and Hoppe):[82] "*For gout.* Write these 4 names on a tin *lamella* with a stylus that has not been filed down, and on a Sunday bind (the amulet) on the foot of the patient, then again in 36 days on the 36th day, which falls on a Sunday (untie it). And these are the things to be written: *χεντιμα τεφῆκεν τέφρα γλύκαινε.*" The inscription of the gold *lamella* from Brindisi bears the same formula, though in a slightly garbled form:

> ΧΕΝΤΕΜΜΑ
> ΤΕΦΡΕΙΧΕΝ
> ΤΕΦΡΑΙΣ [.
> ΒΛΥ [. . . . .

It is an intriguing possibility that this tablet and the recipe from the *Hippiatrica* came from a common source.

The comparisons between literary and epigraphical sources concerning the treatment of gout do not end here. Alexander Trallianus (vol. II p. 581 Puschmann) reports that if a particular line of Homer's (*Il.* 2.95) is engraved on a gold *lamella* while the moon is in Libra or Leo, a person afflicted with gout will recover. It seems no small coincidence that the only Homeric verse found engraved on a magic *lamella* is this very same verse written neatly in two lines on a gold tablet of the third century A.D.:

> *τετρήχει δ' ἀγορή, ὑπὸ δὲ στεναχίζετο γαῖα*
>
> And the place of gathering was in turmoil,
> and the earth groaned beneath them.[83]

Still another remedy for gout employing a magic *lamella* occurs in the demotic magic papyri.[84] In addition, a silver tablet is used to treat gout in a late Coptic collection of magicomedical recipes whose original text must have been Greek, judged from the untranslated titles preserved in the manuscript.[85] A gemstone in the

Hermitage collection also treats gout. This text is of particular interest because it treats the disease by the *φεῦγε* formula that we have found as early as the fourth-century-B.C. Phalasarna tablet. The sardonyx pictures Perseus holding a *harpē* and the Gorgon's head as he flys through the air; on the reverse the inscription reads: *φύ[γε] ποδάγρα, [Π]ερσεύς σε διώχι* ("Flee, Gout, Perseus is chasing you.")[86]

Like the phylacteries designed for general protection from many kinds of danger, some amulets list a number of medical complaints from which the bearer seeks protection. By listing the various afflictions, the practitioner seems to be safeguarding his health by precluding the possibility of any harm coming his way. One silver amulet in particular seems to address a host of disorders for a certain Syntyche. The spell, after summoning the "great and holy name of the living Lord God Damnamanaios and Adonaios, and Iao and Sabaoth," conjures "all spirits" (*πάντα τὰ πνεύματα*), "every falling sickness" (*πᾶν πτωματισμόν*), "every hydrophobia" (*πᾶν ὑδροφόβαν*), the "evil eye" (*τὸν βάσκανον ὀφθαλμόν*), and what is apparently a reference to a violent daemonic attack (*πᾶσαν ἐπαποστολὴν βιαίαν πνευμα[τικὴ]ν*).[87] A few of the *lamellae magicae* also adopt general prayers for protection but with no particular description of the affliction. For example, a gold phylactery found in Segontium (Caernarvon, Wales) contains more than twenty lines, some of which is actually Hebrew text written with Greek letters, a surprising feature in a text found so far from the Eastern portions of the empire; besides the simple *ἀεί* (ever), the only Greek portion reads simply *διαφύλαττέ με, 'Αλφίανον* ("Protect me, Alphianus").[88]

PRAYERS FOR PROTECTION ON INSCRIBED AMULETS

We have discussed above the use of metrical incantations, apparent "nonsense" words, and other *voces magicae* as inscriptions on amulets. Finally, we shall turn to the texts that contain "prayer formulas" that aim at a more general kind of protection. As in the prayer formulas found on the *defixiones* discussed by Faraone (chap. 1), the texts on the *lamellae* are usually very laconic, sometimes preserving only an invocation of the god(s) and the request in the imperative. We have seen that imperatives are employed in two different ways in the texts of the inscribed amulets: some contain "performative" incantations in which the disease/daemon is directly addressed (e.g., the *φεῦγε* formula) and some are simple prayers that use an imperative to bid the deity to take action (e.g., *ἀναχώρησον*, bidding a god "banish" a disease, is used in the fragmentary silver phylactery from Antiochia Caesarea, mentioned above p. 117 with n. 66).[89] The latter type will concern us now.

As befits their apotropaic purpose, many of the texts on amulets employ verbs compounded with the Greek preposition *ἀπο–*. The gold phylactery from Rome[90] employs the imperative *ἀπάλλαξον* ("Take away!"), but it is too fragmentary to tell us what sort of disease is concerned. The occurrence of this verb in texts on other amulets suggests that the disease is of the acute, intermittent type, such as fever or headache.[91] Epilepsy is treated in a gold phylactery purchased in 1924 at Damascus and discussed above (pp. 117–18);[92] the use of the imperative form of the verb *ἀπολαύνειν* ("to drive away, to expel") suggests the driving out of an already sedentary and chronic ailment.[93] The use of the form *ἀπόστρεψον* ("Turn aside!") on the Christian gold *lamella* from Tyre underscores the difficulty of differentiating between prayers for prevention and those for deliverance, as the line between the two is often thinly drawn: *ἀπόστρεψον τὴν ἐπιφε[ρ]ομένην ὀφθαλμίαν καὶ μη[κέ]τι ἐάσῃς ὀφθαλμ[ίας τι]ν' ἐνβο[λὴν γένεσθαι]* ("Divert the inflicting ophthalmia and do not allow any more attacks of ophthalmia.")[94]

The imperative *σῴζετε* ("Save, protect!") addressed to a deity is a very common plea on amulets. In the text of a bronze phylactery found in excavations in the area of Mazzarino (near Syracuse), after a long invocation of angel names and other deities (including Artemis), we find it in combination with the incantatory *φεῦγε* formula: "Flee from Judah and every evil . . . in the glory of the Holy God. . . . [Pro]tect THYBES(?) him who bears your holy law, Judah."[95] Since the actual purpose of this spell is not made explicit in the text, we cannot know for sure what therapeutic category this text belongs to. But even in the better preserved examples, the meaning of *σῴζειν* is often ambiguous. A silver tablet found in Beroea, in Macedonia, addresses a string of magic names (AKRAMMACHAMARI, BARBATHIAOTH, ABLANATHANALBA, IAO, etc.) and ends "Lord angels, save Euphēlētos to whom Atalanta gave birth!"[96] We do not know what sort of danger Euphēlētos faced, but we cannot wholly rule out the possibility that salvation in the broadest sense is intended and that this spell, like the "Orphic" gold leaves of centuries before, asks for salvation from eternal punishments or from a disease thought fatal enough to make the wearer concerned about his own personal salvation following his imminent demise.

The use of the verb *σῴζειν* is not as common on pagan amulets as it is on Christian ones. Two gold tablets published by M. Siebourg[97] contain very short prayer requests using similar language but are conspicuous for their lack of magical names or symbols. One of these reads, in part, "Zeus-Serapis, have mercy!"[98] The second tablet, broken off at the beginning reads, ". . . TE, Abba, Father, save (me), have mercy (on me)!" The latter shows a Christianizing tendency, if it is not wholly Christian. Chronic sufferers, like those afflicted by epilepsy or those living in regions infested with malaria, would probably wear their talismans throughout the course of their lifetime and finally carry them to their graves. But if the bearer were healed from a temporary injury or an intermittent disease (we must accept that this was possible, whether by virtue of the charm or not), that person may have continued to wear the charm as a sort of protective extension. On the other hand, he or she may have discarded it or even have presented it as a sort of votive offering, in thanks for having been healed by the gods.[99]

This ambiguity about general "salvation" is even more pronounced in the Dumbarton Oaks[100] gold phylactery (third century A.D.), a charm whose text shows strong Jewish or Christian influences:

> *ὁ ἄγγελος ὁ φυλάσ⟨σ⟩ων καὶ διασῴζων τὰς ψυχὰς τῶν ἀνθρώπων, διαφύλαξον τὴν ψυχὴν Μασταριῶνος Σαλαμισίου καὶ διάσωσον αὐτὸν ἐκ παντὸς κινδίνου καὶ φεῖσε αὐτοῦ τὴν ψυχὴν. κ(ύρι)ε ἐλέ⟨ησ⟩ον αὐτὸν, κτλ.*
>
> O Angel who guards and rescues mens' "souls," protect the "soul" of Mastarion Salamisios; rescue him from all danger and spare his "soul." O Lord, have mercy on him. . . .[101]

The word ψυχή (lit., "soul") should probably be understood here as "life," that is, one's earthly life; and although at first glance one might be persuaded that the reference "spare his soul" addresses the afterlife, the charm probably has less to do with bringing the bearer to eternal salvation than with deliverance from imminent peril.[102] The two need not be mutually exclusive, especially in view of the Christian overtones of the spell. But the phylactery is obviously of the "preventive" type and not "curative." One can imagine that this Mastarion found himself in a risky employment or in some type of enterprise in which the element of danger was a constant threat; an amulet with such urgent pleas as "Spare his life, O Lord!" would, for example, prove suitable for a soldier in combat or a gladiator. This sort of "global" protection from danger is the concern of several magic phylacteries dealing with a nonspecific danger; a silver tablet from Badenweiler written in Latin with Greek letters[103] reads in part, . . . *serva Te*[. . .]*um quem peperit Leib*[. . . *mate*]*r ab omni periculo,* [*serv*]*a Chilon*(?), *serva Luciolum, serva Mercussam.*[104] These are evidently "good luck" charms to be worn on all occasions and not intended for special problems. An unpublished gold charm from Bet She'an in the New York Public Library simply says, *Felicitas* Κεμω ("Good luck, Kemo[?]").[105]

A spell in the Greek magical papyri (*PGM* LXX. 4–25) highlights the problem that one faces when presented with prayers for salvation that seem embedded in an indisputably magical context. The spell, headed *Charm of Hekate Ereschigal against fear of punishment,* was clearly designed to be used in the underworld, like the "Orphic" *lamellae* discussed above, to protect against the punishments and daemons in the underworld:

> If he comes forth, say to him: "I am Ereschigal, the one holding her thumbs, and not even one evil can befall her."
>
> If, however, he comes close to you, take hold of your right heel and recite the following: "Ereschigal, virgin, bitch, serpent, wreath, key, herald's wand, golden sandal of the Lady of Tartaros." And you will avert him.
>
> ASKEI KATASKEI ERŌN OREŌN IŌR MEGA SAMNYĒR BAUI (3 times) PHOBANTIA SEMNĒ, I have been initiated, and I went down into the [underground] chamber of the Dactyls, and I saw the other things down below, virgin, bitch, and all the rest." Say it at the crossroad, and turn around and flee, because it is at those places that she appears. Saying it late at night, about what you wish, it will reveal it in your sleep; and if you are led away to death, say it while scattering sesame seeds, and it will save you. (trans. H. D. Betz)

It is of great interest that part of the protective spell consists of hexameters that contain the ASKI KATASKI formula (i.e., the *Ephesia grammata*) and some other liturgical bits (from some lost mystery religion) that are also found on the fourth-century-B.C. lead phylactery from Phalasarna discussed above (pp. 111–12).[106] The *Charm of Hekate Ereschigal* ends by suggesting two ways in which it can be used in this world; a rite of prognostication by dreams is described briefly and followed by the claim that the same charm can save a person from death. The *historiola* in the Philinna papyrus concerns an immolated "initiate of the mysteries," and it, too, has been connected to *hieroi logoi* of some hybrid Greco-Egyptian mystery religion, which also seem to have offered some kind of eternal salvation.[107] This repeated overlap of eternal and earthly salvation, of practical "handbook magic" and mystery religion points to a more general category of protective incantation and prayer that cannot be easily separated into the two distinct categories of "magic" and "religion."

The use of the "Orphic" tablets to protect people in this life and the use of a mystery liturgy on the fourth-century B.C. lead phylacteries are not to be dismissed as local aberrations to our categories; rather they should force us to rethink them. Many scholars insist that in both cases we can see an earthly (re)application of protective incantations designed (originally) for the afterlife, either as a result of a conscious and outright "theft" of the religious material or as the result of a long period of degeneration of the "pure" religion whence it came.[108] Unfortunately our earliest examples of these allegedly secondary creations are often contemporaneous with the earliest evidence for their alleged religious "prototypes"; the use of the *Ephesia grammata* and mystery liturgies on the Phalasarna and Getty lead tablets (and in the fragment of Anaxilas) all date to the classical period, as does Zuntz's "Orphic" tablet C (the tablet that had been folded up like an amulet).[109] A similar phenomenon occurs later on in the Jewish and early Christian amulets, which employ the *σώσατε* prayer formula, where one simply cannot decide whether the aim of the prayer is earthly protection or eternal salvation.[110]

In addition to the blurring of eternal and worldly "salvation" in the texts that promise global protection, the coexistence of prayer formulae and automatic incantations on the more narrowly focused medical amulets also presents problems for those who wish to maintain a strict dichotomy between magic and religion. It is simply wrongheaded to suspect the piety of the person using an abbreviated prayer formula simply because it is found in the context of an otherwise entirely "magical" ritual. Graf's essay (chap. 7) underscores the fact that the language of prayer in magic texts indicates normative religious sentiments and values and vitiates the supposed antithetical dichotomy between "magic" and "religion" still expressed or tacitly assumed by scholars still unduly influenced by the antiquated anthropological views of Sir James Frazer. The major concern of the preserved magic tablets, a concern that goes back to the headache spell of Socrates (or even to the binding of Odysseus' wound), was the prevention or healing of specific diseases and ailments. Two strategies emerge from the extant medical amulets that cannot really be distinguished in terms of their goals or social context: inscribing "automatic" incantations on an amulet and writing down a short prayer to a powerful deity. The petitioners, like Pericles and Bion (who were mentioned at the outset of this essay), find themselves in dangerous, life-threatening situations. From a purely psychological point of view, to a person who is thus racked with pain or wasting away with fever, any and all techniques for empowering an amulet were acceptable. The prayer formula aims at persuading the god to bring about the desired result. The similarly inscribed *ἐπῳδαί* simply represent another approach to solving the same problem. These incantations, most probably accompanied by some ritual gesture, were believed to act automatically on the disease through some sympathetic process (e.g., the *historiolae* and *φεῦγε* formulae respectively), much like the two automatic strategies employed in the *defixiones* discussed by Faraone in Chapter 1. And, as in the case of the *defixiones,* it is difficult-if-not-impossible to distinguish among amulet inscriptions between the function of prayer on the one hand and that of incantation on the other.

Notes

The following oft-cited works will be referred to by the author's last name and date of publication or by the abbreviation given in square brackets:

H. D. Betz, ed. *The Greek Magical Papyri in Translation. Including the Demotic Spells,* vol. 1, *Texts* (Chicago, 1986) [*GMPT*].

C. Bonner, *Studies in Magical Amulets Chiefly Graeco-Egyptian* (Ann Arbor, 1950).

R. Heim, *Incantamenta Magica Graeca Latina,* Jahrbücher für classische Philologie Suppl. 19 (Leipzig, 1892), 463-576.

Th. Hopfner, *Griechisch-ägyptischer Offenbarungszauber,* 2 vols. (Leipzig, 1921–24; repr. Amsterdam, 1974) [*OZ*]

P. Laín Entralgo, *The Therapy of the Word in Classical Antiquity,* ed. and tr. L. J. Rather and J. M. Sharp (Yale, 1970).

C. C. McCown, "The Ephesia Grammata in Popular Belief," *TAPA* 54 (1923): 128–40.

K. Preisendanz and A. Henrichs, eds., *Papyri Graecae Magicae: Die griechischen Zauberpapyri,* 2d ed., 2 vols. (Stuttgart 1973–74) [*PGM*].

D. M. Robinson, "A Magical Text from Beroea in Macedonia," in *Classical and Mediaeval Studies in Honor of Edward Kennard Rand,* ed. L. W. Jones (New York, 1938).

G. Zuntz, *Persephone: Three Essays on Religion and Thought in Magna Graecia* (Oxford, 1971).

1. What "magic" is by definition remains a complex question that cannot be investigated here. A working definition is suggested in David E. Aune, "Magic in Early Christianity," *ANRW* 2.23.2 (1980): 1507–57: "Magic is defined as that form of religious deviance whereby individual or social goals are sought by means alternate to those normally sanctioned by the dominant religious institutions," and "goals sought within the context of religious deviance are magical when attained through the management of supernatural powers in such a way that results are virtually guaranteed" (p. 1515); however, Aune's two-pronged definition is valid only for the social-political and historical context in which such dichotomies between "dominant religious institutions" and "religious deviance" can flourish. The definition does not address the phenomenon of "magic" as a religious expression from the believer's perspective (nor of "religion" being an expression of magic, from the point of view of an unorthodox adherent). As J. E. Lowe in her concise *Magic in Greek and Latin Literature* (Oxford, 1929) states, "Many definitions of the word 'magic' have been attempted: none, perhaps, is wholly satisfactory. The word connotes so much, the boundary line between it and religion is so hazy and indefinable, that it is almost impossible to tie it down and restrict it to the narrow limits of some neat turn of phrase that will hit it off and have done with it" (p. 1). On the whole matter, see also the sobering remarks of A. F. Segal, "Hellenistic Magic: Some Questions of Definition," in *Studies in Gnosticism and Hellenistic Religions,* ed. R. van den Broek and M. J. Vermaseren, EPRO 91 (Leiden, 1981), 349–75.

2. In this essay we cannot discuss such topics as prayer, sacrifices, and Asclepian temple therapy as modes of healing. On these alternative religious forms of therapy, or "belief" medicine, see in general the observations of G. E. R. Lloyd, *Magic, Reason, and Experience* (Cambridge, 1979), 38–41 and Laín Entralgo 1970, which characterizes the situation in the classical period as follows: "In the treatment of diseases, magical cures of mantic or purificatory character become much more frequent: various enchantments, cathartic ceremonies, medical oracles, orgiastic cults, Asclepian temple sleep" (p. 41). For prayer, the essay of H. S. Versnel, "Religious Mentality in Ancient Prayer," in *Faith, Hope, and Worship: Aspects of Religious Mentality in the Ancient World,* ed. H. S. Versnel and F. T. van Straten (Leiden, 1981), 1–64 is most valuable. (On votive offerings and the cult of Asclepius, one should also

consult van Straten's essay in the same volume: "Gifts for the Gods," pp. 65–151). On the cult of Asclepius and the inscriptions of Epidaurus, the pioneering work is R. Herzog, *Wunderheilungen: Die Wunderheilungen von Epidauros,* Philologus Suppl. vol. 22, pt. 3 (Leipzig, 1931); see also E. J. Edelstein and L. Edelstein, *Asclepius: A Collection and Interpretation of the Testimonies,* 2 vols. (Baltimore, 1945). The whole matter of the use of music for healing similarly cannot be dealt with here (apart from the fact that ἐπῳδαί are indeed sung or chanted); see Lloyd (above), 42f. and n. 18 below.

3. Lloyd (see n. 2) pp. 38–39 draws attention to those social figures, who, apart from the standard "doctors" (ἰατροί), claimed to be able to heal disease. These include "rootcutters" (ῥιζοτόμοι) and "drugsellers" (φαρμακοπῶλαι), midwives, gymnastic trainers, and barbersurgeons, as well as the priests and attendants at temples of healing gods. On the rootcutters, see also J. Scarborough below (chap. 5).

4. The spells and charms contained in the late ancient magic handbooks could often be adapted to virtually any wish; see, for example, the use of the scribal formula κοινόν or κοινά ("and so forth, et cetera.") and by such expressions as ὡς θέλετε ("whatever you wish"). Otherwise rubrics, like those found in contemporary medical treatises, divide the recipes of the magical papyri into categories according to the charm's function, even though sometimes (due mostly to erroneous textual transmission or misundertandings) the title and spell's content do not match. The practice of marking such titles in red (whence the word *rubric*) goes back to Egyptian practice and occurs in the demotic sections of the bilingual magic texts; see Janet H. Johnson, "Introduction to the Demotic Magical Papyri," in *GMPT,* lv–lviii. For the classification of different types of spells, see Hopfner, *OZ,* vol. 2, sec. 41ff. and idem, "Mageia," *RE* 14.1 (1928): 378.

5. Such charms encouraging good luck and prosperity are usually differentiated as talismans. The English word *talisman* (supposedly from late Greek τέλεσμα and classical Greek τετελεσμένον, perfect past participle of the verb τελεῖν, "to consecrate," via the Arabic/Turkish term *telesma*) will not here be differentiated from *amulet* (from the Latin *amuletum, amoletum,* etc. but of unsure derivation; cf. Arabic, *hamalet*). The recipes in *PGM* generally use the term φυλακτήριον (*phylaktērion*) for amulet. For a discussion of the terminology see S. Seligmann, *Die magischen Heil- und Schutzmittel aus der unbelebten Natur mit besonderer Berücksichtigung der Mittel gegen den bösen Blick: Ein Geschichte des Amulettwesens* (Stuttgart, 1927); R. Wünsch, "Amuletum," *Glotta* 2 (1911): 219–30; F. Eckstein and J. H. Waszink, "Amulett," *RAC* I (1950), cols. 397–411; see also C. H. Ratschow, "Amulett und Talisman," *RGG* I (Tübingen, 1957), 345–47; F. X. Krause, "1: 48–51; P. Wolters, "Faden und Knoten als Amulett," *ARW* 8 (1905): 1–22; and U. Wilcken, "Amulette," *APF* 1 (1900–1901): 419–436; G. Kropatscheck, *De Amuletorum apud Antiquos Usu Capita Duo* . . . (Gryphiae, 1907), esp. 9–12. Few of these have much discussion of amulets that are inscribed; however, a substantial treatment is found in H. Leclercq, "Amulettes," in *Dictionnaire d' archéologie Chrétienne et de liturgie* 1, 2 (Paris, 1907), cols. 1784–1860 (see also the *RAC* article mentioned above); and especially L. Robert, "Amulettes grecques," *Journal des Savants* (1981): 3–44.

6. In the Greek magical papyri and elsewhere the verb περιάπτειν should be regularly translated cognately, viz., "to wear/attach/suspend a περίαπτον," or the equivalent.

7. Understandably, a personal charm would be small enough to be portable, while a phylactery for a city could presumably be much larger. For the early use of phylacteries in the form of statues or other monumental *apotropaia,* see C. A. Faraone, "Hephaestus the Magician and Near Eastern Parallels for Alcinous' Watchdogs," *GRBS* 28 (1987): 257–80.

8. To what extent the early Greeks adopted the magic practices of their neighbors has not and cannot be investigated here. Suffice to say that written charms among the Egyptians and Phoenicians (see nn. 54 and 55) were in widespread use and that the colonial Greeks, in the process of adapting the Phoenician system of writing, also borrowed their practice of writing

amulets. P. W. Schienerl (see n. 26) suggests that borrowings also came directly from contacts with the Egyptians of Naucratis at the end of the 7th century B.C.

9. G. Lanata (*Medicina magica e religione popolare in Grecia* [Roma, 1967], 46–51) discusses healing incantations in general. Elsewhere in Homer we read of Poseidon casting a spell on Alcathous (*Il.* 13.434); of Circe's potions, herbs, and charms (*Il.* 10.214, 234–36, 301–6); of the famous *moly* discussed by J. Scarborough below (chap. 5); and of Hephaestus' magic chains (*Od.* 8.272–75), as well as other magic implements of the gods and goddesses. For these and more, consult S. Eitrem, "La magie comme motif littéraire," *SO* 21 (1941): 39–83, esp. 39–42.

10. See Laín Entralgo (1970, 21), who cites H. Pfister, "Epode," *RE* Suppl. 4 (1924), cols. 323–44, 325; and I. Scheftelowitz, *Das Schlingen- und Netzmotiv im Glauben und Brauch der Völker,* in RGVV vol. 12, pt. 2 (Giessen, 1912). Cf. also Wolters (see n. 5).

11. The didactic tradition of Chiron is well established; for Chiron as a teacher of the healing arts, cf. Hom. *Il.* 4.219; 11.831–32. See also Emmet Robbins, "Jason and Cheiron: The Myth of Pindar's Fourth Pythian," *Phoenix* 29 (1975): 205–13, esp. 210 for a discussion of *Pyth.* 3.52–53. Laín Entralgo (1970, 45–47) counters Edelstein's view (*Ancient Medicine: Selected Papers of Ludwig Edelstein,* ed. O. and C. L. Temkin [Baltimore, 1967], 226) that Asclepius' "soft incantations" in Pind. *Pyth.* 3.52–53 refer to music and not to verbal charms. As far as the use of incantations is concerned, Edelstein repeatedly turns a blind eye to the evidence and tries to "filter out" the magical element in many passages like this one. Pindar's use of ἐπῳδή in *Pyth.* 4.218 and *Nem.* 8.49 shows they are the sung incantations of words.

12. This passage stands in contrast to a similar type of catalogue of medical methodology used by the "rationalistic" school of Hippocrates: "Those diseases that medicines (φάρμακα) do not cure are cured by the knife. Those that the knife does not cure are cured by fire. Those that fire does not cure must be considered incurable" (*Aph.* 7.87; trans. W. H. S. Jones). On the "three categories of the art of healing" see Laín Entralgo 1970, 16–17 with n. 26. See also his discussion (pp. 47–48) of the use of incantations in Soph. *Aj.* 581–82 ("It is not fitting for wise physicians to recite ἐπῳδαί in cases of ailments that demand the knife") and Plato *Resp.* 4.426B (healing by medicaments, cauteries, incisions, and ἐπῳδαί).

13. See Th. Bergk, "Zwei Zauberformeln bei Cato," in *Kleine Philologische Schriften,* vol. 1 (Halle, 1884), 556–70, esp. 560, with the reservations expressed by F. Skutsch in Heim 1893, 565–66 and E. Laughton, "Cato's Charm for Dislocations," *CR* 52 (1938): 52–54. W. B. McDaniel ("A Sempiternal Superstition for Dislocated Joint: A Split Green Reed and a Latin Charm," *CJ* 45 [1950]: 171–76) provides the best discussion of the difficulties involved in interpreting this rather obtuse passage.

14. On Plato's metaphorical use of ἐπῳδή see the chapter "The Platonic Rationalization of the Charm," in Laín Entralgo 1970, 108–38; see also Lanata (see n. 9), 49–50, on the φύλλον of Socrates.

15. Cf. *μετὰ ῥηματίων, ὥς φατε, καὶ γοητείας τινὸς ἐνεργεῖν καὶ τὴν ἴασιν ἐπιπέμπειν προσαρτώμενα* (*Philops.* 8). Edelstein (see n. 11) p. 245, n. 140 cites Diodorus (frag. 30, line 43 Dindorf), who mentions people who go to θύται and μάντεις to be healed by "incantations" (ἐπῳδαί) and all sorts of "amulets" (περίαπτα).

16. On gout, see the discussion at nn. 84 and 85 below. We are not told what kind of magic power the stele fragment carried or whether or not it was inscribed. In Eur. *Hec.* 1272 we read of a tombstone that is believed to be endowed with a very strong magical name. The tombstone is not used as an amulet in this case, but the power of a name carries apotropaic force. The epitaph apparently preserves an actual ἐπῳδή, for when the name is uttered it is believed that one can call up the "shade" (μορφή) of Hecuba. See Laín Entralgo 1970, 50 for discussion.

17. For some additional references to texts describing incantations with amulets, cf. the helpful study of H. Pfister, (see n. 10), 330 and 337, lines 50ff. See also A. D. Nock, "Paul

and the Magus," in *Essays on Religion and the Ancient World,* ed. Z. Stewart, vol. 1 (Oxford, 1972), 308–30 for additional references here and there to early Greek incantations.

18. See n. 20. Laín Entralgo (1970, 45) recalls the view that Orpheus' songs were magical enchantments in themselves, citing J. Combarieu, *La musique et la magie,* Etudes de philologie musicale 3 (Paris, 1909), and suggests a transition of the ἐπῳδή from song to amulet: "Magic formulas have passed through the following phases: at first they were sung; then they were recited; finally they were written upon a material object worn in some cases as an amulet."

19. The large number of inscribed lead curse tablets (*defixiones*) dating to the classical period reflects a much different ritual; they were commonly thrown into wells (and other underground bodies of waters) or buried in graves and chthonic sanctuaries (i.e., areas protected by strong taboo), and in this way they were rarely able to be reused and wait patiently for the archaeologist. On the whole matter see Faraone's discussion in this volume.

20. Bonner 1950, 4–5. An early record of the use of a written amulet is found in the account of Bion, the third-century B.C. philosopher mentioned at the outset of this article (Diog. Laert. 4.56). That the amulet tied to Bion's neck is called an ἐπῳδή suggests that the charm must have been written, if not also spoken. In addition, organic herbals supplemented the written charm.

21. This according to Pausanias, as cited by Eustathius on Homer *Od.* 19.247; see McCown 1923 for this and other passages; W. Schultz, "'Ἐφέσια und Δελφκὰ γράμματα," *Philologus* 68 (1909): 210–28; M. Siebourg, "Zu den Ephesia Grammata," *ARW* 18 (1915): 594; A. Deissmann, "Ephesia Grammata," in *Abhandlungen zur semitische Religionskunde und Sprachwissenchaft: Wolf Wilhelm Grafen von Baudissin zum 26. Sept. 1917 überreicht von Freunden und Schülern* (Giessen, 1918), 121–24; cf. M. Huvelin, "Les tablettes magiques et le droit romain," in *Annales Internationales d'Histoire: Congrès de Paris 1900 1re section* (Paris, 1901), 47ff. For a collection of magic texts containing the "Ephesian letters" and other magic incantations, see Karl Wessely, *Ephesia Grammata aus Papyrusrollen, Inschriften, Gemmen, etc.* (Vienna, 1886). We also might mention here the second- or third-century-B.C. terracotta *Hausphylakterion* portraying Artemis surrounded by a series of *Ephesiae litterae:* K. Preisendanz, "Ephesia Grammata," *RAC* 5 (1962): 518; E. Labatut, "Amuletum," *DAGR* 1.1 (1873), fig. 303. The piece is in the Syracuse Museum and was first published by L. Stephani, "Ueber ein Ephesisches Amulett," *Mélanges gréco-romains tirés des Bulletin historico-philologique de l'Académie Impériale des Sciences de St. Petersbourg* 1 (1855): 1–5 (with pl.). For further readings and discussion, see A. B. Cook, *Zeus: A Study in Ancient Religion,* vol. 2, pt. 1 (Cambridge, 1924; repr. New York, 1965), 409–10.

22. Their appearance in verses inscribed on lead, however, suggests strongly that they must have once been meaningful hexameters in their original context (discussed below).

23. Ἐφέσια τοῖς γαμοῦσιν οὗτος περιπατεῖ λέγων ἀλεξιφάρμακα according to the *Suda* (= frag. 313 Koerte; Meineke *FCG* IV.181; Kock *CAF* III.108) as cited in McCown 1923, 131, n. 17.

24. The story is preserved in Photius, the *Suda,* Eustathius, s.v.; see McCown 1923, 131.

25. *PGM* VII.186–90, 390–93, 528–39, 919–24; XII. 270–350; XXXVI.1–42; etc.; cf. Jordan (see n. 64), 162–67.

26. In addition to the tubular capsules of metal used during the Roman period as amulet cases, one should educe as a parallel the late example of a leather case containing an Ethiopian amulet; see F. T. Elworthy, *The Evil Eye* (London, 1895), 391–92. For earlier Greek examples of amulet capsules, see the excellent treatment by Peter W. Schienerl, "Der Ursprung und die Entwicklung von Amulett behältnissen in der antiken Welt," *Antike Welt* 15 (1984): 45–54, esp. 50–54. The earliest Greek examples he cites date to the fourth and third centuries B.C.

27. *I Cret.* 2.(19).7 (ed. Guarducci). It is presently in the National Museum, Athens (inv. 9355), but apparently has never been photographed or produced in facsimile. See McCown 1923 and, for additional bibliography, D. R. Jordan, *AM* 95 (1980): 228. A similar unpublished lead text from Selinus (in the J. Paul Getty Museum) contains hexameters and portions of the same mystic letters and also seems to have been used as an *apotropaion*. (See now the remarks with some preliminary readings in D. R. Jordan, "A Love Charm with Verses," *ZPE* 72 [1988] 256–58, esp. on lines 64–73). Besides containing hexameters apparently related to the mysteries, the text cites words from the hitherto "meaningless" *Ephesia grammata,* which can now be understood as perfectly good Greek hexameters.

28. The beginning of the text suggests that it provided some kind of general protection to a household (see n. 29). A fragmentary portion of the unedited Getty text also seems to describe something like a natural calamity or plague that may have affected a large estate or community (col. 2, lines 4–8):

> [c. 6]. *οὗ κατάκουε φάσιν γλυκύν*[
> [c. 5 *ἀ*]*νθρώποισιν ἐπιφθέγγεσθαι ἀν*[
> [c. 6]*ωι κἂν εὐπολέμω*[*ι*] *ναυσὶν* h*οτα*[
> [c. 5 *ἀ*]*νθρώποις θανατήφορος ἐγγυ*[
> [c. 8] *προβάτοις καὶ ἐπὶ τέχναισι βροτ*[*οῖσι*]

. . . (not?) giving ear to my sweet voice . . . to touch men . . . one skilled in war and ships . . . bringing death near to men, to cattle, and upon the handiworks of mortals.

29. Lines 1–2: *κελεύω φευγέμ⟨εν ἡμ⟩ετέρων οἴκω(ν) ἄ*[*πο* . . . (Guarducci in *I Cret.* 2.(19).7).

30. The reading of the text is at points very dubious, but the translation of McCown, (1923, 134) must suffice until a new edition is prepared. From line 14, he translates, "Happily he who knows binding magic may pass down the highway. . . . / Damnameneu, do thou tame by force the wickedly stubborn, / Whoso may harm me and those who some charm would cast o'er me to bind me; . . . / Whoso with ointments of magic would hurt me, to him be no refuge / By ways whether trodden or trackless: to Earth, the All-spoiler, I doom him." The text has many points of contact with the unpublished Getty hexameters, whose content also shows that the hexameters had some kind of apotropaic function.

31. It is also worthwhile to note that a comparable text, a hymn from Eretria also dating to the fourth century B.C. (*IG* 12.9, no. 259), addresses the Idaean Dactyls, one of whom is named Damnameneus (line 19), famous as one of the personalities of the Ephesian *grammata* (see n. 30). See B. Hemberg, "Die Idaiischen Daktylen," *Eranos* 50 (1952): 41–59. The hymn also describes *φάρμακα ἀλεξητήρια* (line 6) presumably discovered by Eurytheos (line 5). See I. U. Powell, *Collectanea Alexandrina* (Oxford, 1925), 171–72.

32. The text was reconstructed by P. Maas, "The Philinna Papyrus," *JHS* 62 (1942): 33–38, from *P. Berol.* 7504 plus *P. Amherst* 2, col. II(A) plus *P. Oxy. ined.* (= R. A. Pack *The Greek and Latin Literary Texts from Greco-Roman Egypt,* 2d ed. [Ann Arbor, Mich., 1967] no. 1872). For a current bibliography see the translation and notes in *GMPT,* 258. In the papyri, descriptions for the making of amulets are common (e.g., texts like *PGM* I–XIV or XXXVI are portions of handbooks). Examples of actual amulets fashioned out of inscribed papyrus are *PGM* XVIIc., XVIIIa–b., XXVIIIa–c, XXXIII, XLI–XLV, XLVII–XLIX, LIX, LX and, in *GMPT,* nos. LXXXIII, LXXXVII–LXXXIX, XCI, XCVI, XCVIII, XCIX, C, CIV, CVI, CXII, CXIII–CXVI, CXX, CXXI, CXXVIII, CXXX; but none of these is particularly early.

33. We have other early evidence for magic books in the papyri: see *GMPT,* nos. CXXII (first century B.C.–first-century A.D., a collection containing *ἐπῳδαί*, one with a *historiola* for headache of Egyptian origin) and CXVII (first century B.C., with fragments of a love spell). A

feature common to all three of the early papyrus spells (*PGM* XX, CXVII, CXXII) is the duplicated verse, "carry out for me this perfect spell" (or equivalent). It is also worthwhile to note that in the hexametrical *Prayer to Selene for any spell* (*PGM* IV 2785–2890) we find a verse that describes a scepter inscribed with magic words and apparently used as an amulet (trans. E. N. O'Neil):

> As everlasting / band around your temples
> You wear great Kronos' chains, unbreakable
> and unremovable, and you hold in
> Your hands a golden scepter. Letters 'round
> Your scepter / Kronos wrote himself and gave
> To you to wear that all things stay steadfast.

On this passage, note Hopfner, *OZ* 1, sec. 764.

34. The text is reconstructed as Hymn 28 at the end of the second volume of *PGM*. The fact that they are now recorded and perhaps even written down on papyrus used as amulets also suggests the transmission of spoken charm to written amulet. Again, in a similar way, the fact that the atheist Bion (see n. 20) bids an old woman tie an ἐπῳδή to his neck implies an amulet engraved with verses.

35. Cf. P. Maas (see n. 32) and L. Koenen, "Der brennende Horusknabe: Zu einem Zauberspruch des Philinna-Papyrus," *Chron. d'Egypte* 37 (1962): 167–174; Koenen identifies this child as Horus. But see the commentary ad loc. in *GMPT* 258.

36. Cf. Heim 1893, 495–507.

37. *LSJ* and the *Supplement* list only two other occurrences of this term, both from papyrus (though one should now add *PGM* LXV.4). *PGM* VII.199–201 reads, "For migraine headache: Take oil in your hands and utter the spell: / 'Zeus sowed a grape seed: it parts the soil; he does not sow it; it does not sprout' (trans. John Scarborough). Curiously, like the Carnuntum spell, a mythic *historiola* is used for a migraine spell in a magically symphathetic way, i.e., just as the seed does not sprout, the headache will also not "sprout." For another headache spell with Egyptian background, see W. Brashear, "Ein Berliner Zauberpapyrus," *ZPE* 33 (1979): 261–78 (= *GMPT,* p. 317, *PGM* CXXII.51–55); J. W. B. Barnes and H. Zilliacus, *The Antinoopolis Papyri* II (London, 1960), no. 66, pp. 47–49 (= *GMPT,* p. 305, *PGM* XCIV.39–60); R. Kotansky, "Two Amulets in the Getty Museum," *JPGetty Museum Journal* 8 (1980): 181–87; see also A. Barb, "Klassische Hexenkunst: Aus der Verwesung antiker Religionen—Ein antikes Zaubergebet gegen die Migräne und sein Fortlegen," *Jedermann Heft* 3 (1933): 335.

38. A. Barb, "Griechische Zaubertexte vom Gräberfelde westlich des Lagers," *Der Römische Limes in Österreich* 16 (1926): 54–68 (pl. 1) and idem, "Antaura the Mermaid and the Devil's Grandmother," *JWCI* 29 (1966): 1–23 (plates 1–6). The numerous parallels, mostly late and replacing Artemis with Christ, indicate how the text of the spell probably ended.

39. For headache spells, see Kotansky (see n. 37).

40. Pliny *HN* 27.75.100 = J. M. Edmonds, *Lyra Graeca* vol. 3 (Cambridge, Mass., 1959), 542–44, no. 38a. Edmonds rightly prints the early anonymous emendation ὔμμε (it already appears in the first printed editions of Pliny) for the nonsensical αἷμα (blood) (*pace* Heim 1893, ad loc.). Aside from the insurmountable problems created by retaining αἷμα ("Flee, beetles, a savage wolf pursues blood"?), all of the other nineteen examples of the φεῦγε formula cited by Heim contain some form of the second-person pronoun. As for the importance of the form of ὔμμε, I quote Edmonds' note in full: "The period to which this and the next two songs or sayings belong is doubtful, but the Aeolic [*sic*] form of the word 'you' indicates, for this, at any rate, a pre-Alexandrine date." Edmonds also gives a cure for styes

and eyesores in which the infected area is pricked with barley corns while reciting the incantation, "Flee, flee, the barleycorn pursues you" (his no. 38b from Marc. Emp. 8.192); and a charm for excessive bile inscribed on an iron ring: "Flee, flee, bile, the skylark seeks you out" (no. 38c from Alexander Trallianus, vol. 2, p. 337 [ed. Puschmann]), where the verb ζητεῖν replaces the more usual διώκειν. The citation of these formulas in verse may also indicate their greater antiquity. See n. 86 for additional discussion of this kind of formula.

41. When the question is raised whether or not incantations have any proven effect in healing, Pliny states unequivocally that prayers, and indeed the power of words in general, carry efficacious results (*HN* 28.3.10). At 28.4.20ff. we even find a fascinating history of the use of incantations: "On walls too are written prayers to avert fires. It is not easy to say whether our faith is more violently shaken by the foreign, unpronounceable words, or by the unexpected Latin ones, which our mind forces us to consider absurd, being always on the look-out for something big, something adequate to move a god, or rather to impose its will on his divinity. Homer said that by a magic formula Ulysses stayed the haemorrhage from his wounded thigh; Theophrastus that there is a formula to cure sciatica; Cato handed down one to set dislocated limbs [see n. 13], Marcus Varro one for gout. The dictator Caesar, after one serious accident to his carriage, is said always, as soon as he was seated, to have been in the habit of repeating three times a formula of prayer for a safe journey, a thing we know that most people do today" (trans. W. H. S. Jones in the Loeb series, here and throughout).

42. Another historical case is that of Licinius Mucianus (consul three times and governor of Syria, 68–69 A.D.), who enclosed a living fly in a white linen bag for the same purpose.

43. Bonner (1950) provides the best work in this area. See also his subsequent publications: "Amulets Chiefly in the British Museum: A Supplementary Article," *Hesperia* 20 (1951): 301–345 (plates 96–100); "A Miscellany of Engraved Stones," *Hesperia* 23 (1954): 138–57 (plates 34–36). His earlier works include "Liturgical Fragments on Gnostic Amulets," *HTR* 25 (1932): 362–67; "Magical Amulets," *HTR* 39 (1946): 25–34; and numerous other articles. See also A. Delatte and Ph. Derchain, *Les intailles magiques gréco-égyptiennes* (Paris, 1964); D. Wortmann, "Neue magische Gemmen," *BJ* 175 (1975): 63–82; F. M. and J. H. Schwartz, "Engraved Gems in the Collection of the American Numismatic Society," pt. 1, "Ancient Magical Amulets," *ANSMusNotes* 24 (1979): 149–95 (plates 34–40); A. A. Barb, "Gnostische Gemme," *Enciclopedia dell'arte antica classica e orientale* (Rome, 1960) 3: 971–74; idem, "Three Elusive Amulets," *JWCI* 27 (1964): 1–22; P. J. Sijpesteijn, "Four Magical Gems in the Allard Pierson Museum at Amsterdam," *BABesch* 95 (1970): 175–77; idem, "Einige Bemerkungen zu einigen magischen Gemmen," *Aegyptus* 60 (1980): 153–160; idem, "Zu einigen Kölner Gemmen," *ZPE* 51 (1983): 115–16 (for some comments and corrections on some of the published European gemstone collections). The list of articles on the subject could be continued quite extensively; but the single most important recent treatise to appear in years is that of H. Philipp, *Mira et Magica* (Mainz, 1986).

44. Occasionally inscribed bronze and copper phylacteries are also unearthed. Nearly two-hundred magic *lamellae* are published or in private hands. Space does not allow a complete citation of the secondary literature on these. The classic treatment of the magic *lamellae* is that of M. Siebourg, "Ein gnostisches Goldamulett aus Gellep," *BJ* 103 (1898): 123–53, though it is now badly outdated. In the article he cites at least seven examples of such phylacteries known to him in addition to the one he publishes; he also includes about fourteen literary references. Jordan (see n. 64) also gives a useful survey of the material.

45. The study of these gold and silver phylacteries, including a corpus of all known examples, is the subject of my Ph.D. dissertation (University of Chicago, 1988). For a previously unpublished example not discussed in this paper, see C. A. Faraone and R. Kotansky, "A Gold Phylactery in Stamford, Connecticut," *ZPE* 75 (1988): 257–66.

46. The terms *lamella* and *lamina* are used to describe these amulets in ancient literary

sources. In addition to the generic "phylactery," Greek writers employ a number of other terms, often interchangeably: λᾶμνα (*PGM* II.15, 297, 299; IV.2153–54, 2166, 2177, 2208, 2226, 2238; VII.398, 459, 462; IX.8; X.26, 36; XXXVI.1, 37–38, 231, 234; LVIII.6); πέταλον (*PGM* III.[58], 66; IV.330, 1218, 1255, 1813, 1824, 1847, 2705; V.306, 359; VII.216, 382, 417, 487, 581, 743; IX.14; X.[36], 39; XII.197–98, 199; XIII.889, 898, 903, 1008, 1052); λεπίς (*PGM* III.410–11, 417; IV.258, 1828, 2160–61, 2216, 2228; VII.271, 919, 925; XIII.1001; XXXVI.278; LXXVIII.3); πλάξ (*PGM* IV.2187, 2194, 2212; VII.432); πλάτυμμα (*PGM* IV.329, 407; VII.438); πτύχιον (*PGM* VII.740–41; cf. also *P. Vars*. 4, 2f.; *P. Ant*. vol. III, no. 66, line 37).

47. See primarily K. Parlasca, *Mumienporträts und verwandte Denkmäler* (Wiesbaden, 1966), Taf. 50, Nr. 1–2; Taf. 17, Nr. 1; pl. 33, Nr. 1(?); idem, *Repertorio d'Arte dell'Egitto Greco-Romano* series B, vols. 1–3 (Palermo, 1969–80) with the following examples: vol. 1, nos. 35, 36, 62, 96; vol. 2, nos. 250, 257, 321(?); vol. 3, nos. 527, 533, 575(?), 621, 654, 656–57, 659–60, 663, 665, 669–70, 672, 674.

48. K. Wessely (*WS* 8 [1886]: 178–79) and F. H. Marshall (*Catalogue of the Jewellery, Greek, Roman, Etruscan in the British Museum* [London, 1911], xlvii) stressed the connection between the "Orphic" *lamellae* and the latter magical ones, but the protests of Günther Zuntz (1971, 277–86; discussed in n. 49), seem generally to have prevailed. Apart from the work of West and Janko (see n. 51) the literature on the "Orphic" *lamellae* has ignored the question entirely; see Domenico Comparetti, *Laminette Orfiche* (Firenze, 1910); Alexander Olivieri, *Lamellae Aureae Orphicae* (Bonn, 1915); M. Markovich, "The Gold Leaf from Hipponion," *ZPE* 23 (1976): 221–24; R. Merkelbach, "Ein neues 'orphisches' Goldblättchen," *ZPE* 25 (1977): 276; M. Guarducci, "Laminette auree orfiche: alcuni problemi," *Epigraphica* 36 (1974): 7–31 (= *Scritti scelti sulla religione greca e romana e sul cristianesimo* [Leiden, 1983], 71–86); B. Feyerabend, "Zur Wegmetaphorik beim Goldblättchen aus Hipponion und dem Proömium des Parmenides," *RhM* 127 (1984): 1–22.

49. The motif in the tablets may be Egyptian, as the so-called Coffin Texts deal with such protection in the afterlife; see Zuntz 1971, 370–76. Zuntz also emphasizes the differences between the Orphic tablets and late magic tablets (pp. 282–84): he argues that even though both types accompanied the dead to their graves, "the one is designed for the living, the other for the dead" (p. 282). Furthermore, the two types of tablets are chronologically separated, on the whole: "The long gap of time separating the 'Orphic' from the 'magic' gold leaves thus remains unbridged; and while the former have at least one successor in the tablet of Caecilia Secundina (which is contemporary with the first crop of 'magic' gold amulets and adapted to them) there are no pre-Roman 'magic' gold leaves, let alone any that are contemporary with, or earlier than, the 'Orphic' ones" (pp. 283–84). But with such a tiny statistical sampling, even the presence of a single exception should caution against overstressing the chronological hiatus. The fact that both types were found in tombs does not argue in the least for their different functions; both were buried with the owner after death, a typical practice. But that does not exclude the possibility that the Orphic tablets (despite the content of their texts) could have been worn as amulets, like the Petelia (B1; see n. 51) and Rome pieces. Gold leaves with verses of Homer (see n. 61), for example, give no explicit indication of their apotropaic value in the inscribed text itself but were nonetheless worn as talismans. And to argue that because we have found no gold amulets contemporary with the Orphic ones they did not exist is faulty reasoning. See, for example, Zuntz (1971, 284, n. 1), who supports Perdrizet's remark that Theophrastus would have presented his "character" with a gold amulet if one had been in use at the time! Indeed, we have such early texts on lead in hexameters (see n. 27); someday the archaeologist's spade may turn up a gold example.

50. On the whole matter of the amuletic function of the leaves, Zuntz (1971, 353) says of this one, "One cannot but conclude that these lamellae were articles of a local mass-

production, objects of a beadles' trade like the pictures of the Madonna and of saints sold at Roman Catholic churches. *This fact is significant enough, for it implies that they came to be appreciated as material objects rather than as carriers of the words inscribed on them.* We have previously protested against the designation of the lamellae as 'amulets,' and shall continue to urge their essentially different character; *but have here to admit that the facts just noted are evidence of the Gold Leaves gradually being accepted for just this*" (italics mine). Elsewhere, when discussing the Hipponion tablet, Zuntz ("Die Goldlamelle von Hipponion," *WS* 10 [1976]: 129–51, esp. 135) implies that the text was indeed a "talisman."

51. For the Petelia tablet, see also C. Smith and D. Comparetti, "The Petelia Gold Tablet," *JHS* 3 (1882): 111–18. M. L. West, "Zum neuen Goldplättchen aus Hipponion," *ZPE* 18 (1975): 229–36, implicitly supports the view that the Petelia tablet could have been worn as an amulet when he successfully uses parallels from the later magical *lamellae* to restore line 13: [ἐν πίνακι χρυσέῳ] τόδε γρα[ψάτω ἠδὲ φορείτω] ("[Let him] write this [on a gold tablet] and carry it". He is followed by R. Janko ("Forgetfulness in the Golden Tablets of Memory," *CQ* 34 [1984]: 89–100, esp. 92 and 99) who in his restoration of the archetype Ω places the verse at the beginning as an instruction given to those who are about to perish.

52. Cf. Marshall (see n. 48), 381. Guarducci (see n. 48) p. 75 rejects the possibility that the amulet capsule ever housed the tablet. Details regarding the tablet's discovery are not known, but the tablet had apparently been folded tightly at least two or three times from top to bottom and thrice from left to right.

53. Marshall (see n. 48), xlvii. The Caecilia Secundina piece, found at San Paolo fuori le Mura near Rome and acquired by the British Museum in 1899, was apparently worn in a capsule, judging from the way it had been folded. Personal names are never found on the Orphic *lamellae* of an earlier date, and they are not (with the exception of the Petelia tablet) enclosed in capsules but simply laid by or on the body of the deceased.

54. The best recent survey of the capsules is that of Jean Leclant, "A propos des étuis porte-amulettes égyptiens et puniques," in *Oriental Studies Presented to Benedikt S. J. Isserlin, by Friends and Colleagues on the Occasion of his Sixtieth Birthday 25 February 1976*, ed. R. Y. Ebied and M. J. L. Young (Leiden, 1980), 100–107.

55. For the inscribed foil amulets found within the capsules, see P. Gauckler, "Note sur des étuis à lamelles gravées, en métal précieux," *CRAI* 1 (1900): 176–204 (and addendum with Ph. Berger, 205–207); Jean Vercoutter, *Les Objets Égyptiens et Égyptisants du mobilier funéraire Carthaginois* (Paris, 1945), 311–37 and 343–44; and P. Cintas, *Amulettes Puniques*, Publ. de l'Institut des hautes études de Tunis 1 (Tunis, 1946), 66–72. Note also A. A. Barb, ("Mystery, Myth, and Magic," in *The Legacy of Egypt,* 2d ed., ed. J. R. Harris [Oxford, 1971], 149–151), who rightly connects the Punic-Egyptian gold tablets with the Orphic. For the best and most recent discussion of the inscribed texts found within the capsules, see G. Hölbl, *Ägyptisches Kulturgut im phönikischen und punischen Sardinien,* EPRO 102 (Leiden, 1986), 338–53, which provides good facsimiles of all the tablets.

56. For example, see M. Siebourg, "Zwei griechische Goldtänien aus der Sammlung C. A. Niessen," *ARW* 8 (1905): 390–410; idem, "Neue Goldblättchen mit griechischen Aufschriften," *ARW* 10 (1907): 393–96; P. Thompsen, *Die lateinischen und griechischen Inschriften der Stadt Jerusalem* (Leipzig, 1922), 113, no. 208; P. Benoit, "Nouvelles 'brattées' trouvées en Palestine," *Revue Biblique* 59 (1952): 253–58 (plate IX.2); E. Michon, "A propos d'un bandeau d'or palestinien," *Syria* 3 (1922): 214–18; W. Deonna, "Monuments orientaux du Musée de Genève," *Syria* 4 (1923): 224–33.

57. On the phrase, see M. Simon, "Θάρσει οὐδεὶς ἀθάνατος," *RHR* 113 (1938): 188–206; T. B. Mitford (*The Inscriptions of Kourion* [Philadelphia, 1971], 300, no. 156) says on this, "The formula . . . is particularly common on Christian and Jewish tombstones," and

adds that F. Cumont argues (*Les religions orientales,* p. 350) that it is of Egyptian origin. Note also I. Bilkei, *Alba Regia* 17 (1979): 28, no. 11.

58. See R. Mouterde, *Le Glaive de Dardanos: Objects et Inscriptions Magiques de Syrie,* Mélanges de l'Université Saint-Joseph, vol. 15, pt. 3 (Beirut, 1930), 105–6, no. 33 (plate 11.5 and fig. 33). This city is probably modern Afiq in the Golan.

59. Mouterde (see n. 58) only provides drawings of two of the inscriptions. The photograph is too inadequate to make reliable readings, but it is not entirely impossible that some of the letters are Greek. The piece may be compared to similar gold bands in the Niessen collection, which also have holes for tying the bands on as crowns; these, however, are inscribed with the formula, "Take courage, NN, no one's immortal." See S. Loeschcke, C. A. Niessen, and H. Willers, *Beschreibung römischer Altertümer gesammelt von C. A. Niessen* (Köln, 1911), 238, nos. 4471–2 (pl. CXXXIII) (and see n. 56). For other funerary bands, see Mouterde, ibid., 105, n. 5, and for a general discussion, see M. Blech, *Studien zum Kranz bei den Griechen,* RGVV 38 (Berlin, 1982), 89–92.

60. The long strings of magical words that occur repeatedly in different magical texts are often termed *logoi* or "formulas" and may be transmissions of foriegn languages like Egyptian. Similarly, the vowels, long held to correspond to the planets and thus to "the music of the spheres," may have been especially intoned (or to have represented such intonation) on written amulets, cf. *PGM* XIII.556–57, 206, 627–28, etc. See Hopfner, *OZ* 1, sec. 150–51, 731–87, etc.

61. See discussion below. In the magical papyri the use of Homer is frequent: *PGM* IV.467–74, 821–24, 830–34, 2145–2240; VII.1–148 (*Homeromanteia*); XXIII.1–70 (consult *GMPT,* pp. 262–64); *IG* 14.2580.2. Several short recipes in *PGM* XXIIa also mention making amulets from verses of Homer. One of these (XXIIa.2–9 = *GMPT,* p. 260), written for hemorrhage, is of interest for comparative study because it mentions *speaking* a verse (*Il.* 1. 75) to the blood to stop the flow; but if the patient is ungrateful, another verse (*Il.* 1.96) is to be *written.* So also in lines 11–14 a contraceptive spell is to be inscribed on a papyrus amulet or (it says) even spoken. Examples and discussion are also given in Heim 1893, 514–20.

62. See Heim 1893, 544–50.

63. The use of the genitive where one might expect rather *dativus commodi* is rare, but the sense and meaning are inescapable. Cf. H. W. Smyth, *Greek Grammar* (Cambridge, 1956), sec. 1408 for the genitive of purpose.

64. Bonner (1950, 51–95) discusses the diseases mentioned in the texts of the gemstones. A thorough study of the naming and description of diseases from medical writers compared with that of the magical texts would yield a valuable monograph. For a useful (albeit outdated) attempt at such a study, E. Tavenner (*Studies in Magic from Latin Literature* [New York, 1916], 76–112) treats the texts discussed below as well as many others. Cf. also D. R. Jordan, "The Inscribed Gold Tablet from Vigna Codini," *AJA* 89 (1985): 164, n. 9.

65. The belief that daemons were responsible for a disease is found already in Hes. *Op.* 100; cf. Soph. *Phil.* 757; Aristoph. *Vesp.* 1037. For a full bibliography, see J. Z. Smith, "Towards Interpreting Demonic Powers in Hellenistic and Roman Antiquity," in *ANRW* 2.16.1 (Berlin, 1978), 425–39.

66. For example, a silver tablet from Antiochia Caesarea in Pisidia is written "for spirits." Unfortunately, the reading of the text is very uncertain; see David M. Robinson, "A Magical Inscription from Pisidian Antioch," *Hesperia* 22 (1953): 172–74. Another piece, a gold one in the British Museum, is written to protect Phaeinos son of Paramona from every male and female daemon, a description found often in the Aramaic incantation bowls. For a reedition of the piece, see R. W. Daniel, "A Phylactery from Amphipolis," *ZPE* 41 (1981): 275–76.

67. F. Fiorelli, *NotSc* ser. 4, vol. 3 (1887): 157; Zuntz 1971, 281; cf. Siebourg, *BJ* 103 (1898): 135, no. 3; Zuretti, *Rivista di Filologia* 20 (1891), fasc. 1–3; Heim 1893, no. 234.

68. M. Siebourg, "Ein griechisch-christliches Goldamulett gegen Augenkrankenheiten," *BJ* 118 (1909): 158–75. See below, n. 94. On the basis of the Christian formula Siebourg dates the tablet to the first century A.D. This, however, cannot be accepted as reliable since no facsimile or photograph of the piece is provided.

69. For the text see M. Niedermann, *Marcelli "De Medicamentis" Liber,* Corpus Medicorum Latinorum 5 (Leipzig, 1916). At *De Med.* 8.59, we read, *In lamella aurea acu cuprea scribes ορυω ουρωδη et dabis vel suspendes ex licio collo gestandum praeligamen ei, qui lippiet. Quod potenter et diu valebit, si observata castitate die Lunae illud facias et ponas.* "On a small plaque of gold you should inscribe with a copper stylus ORYO OURODE; then you must hang by a thread an amulet from the neck of the wearer who suffers from the inflammation of the eyes. This proves effective and long-lasting, if you perform and carry out the rituals while (sexually) pure on a Monday."

70. Although the magical papyri do describe amulets for eye inflammation, none recommends the making of gold or silver phylacteries; see *PGM* XCVII.1–6 (= *GMPT,* p. 306; Wortmann, "Neue magische Texte," *BJ* 168 [1968]: 109–11); *PGM* XCIV.4–6; cf. Heim 1893, nos. 31, 32, 35; etc. Cf. also Tavenner (see n. 64), 84–87; and L. C. Youtie, "A Medical Prescription for Eye-salve," *ZPE* 23 (1976): 121–29. The orator Libanius, as well as his brother, resorted to magicians for their eye ailments; see C. Bonner, "Witchcraft in the Lecture Room of Libanius," *TAPA* 63 (1932): 34–44. Bonner (1950, 69–71) also discusses ailments of the eyes, but the inscribed gemstones do not mention such diseases explicitly.

71. See M. K. Kubinya, *ArchÉrtes* 3rd ser., vol. 8–9 (1946–48): 276–79; I. Bilkei, *Alba Regia* 17 (1979): 33, no. 31, pl. II.4; and A. Sz. Burger, "A Late Roman Cemetery at Sagvar," *Acta Archaeologica* 18 (1966): 110, fig. 86.

72. Jordan (see n. 64) pp. 165–66 has identified from the printed plate that the spell is written "to avert swelling."

73. Th. Homolle, "Inscriptions d'Amorgos," *BCH* 25 (1901): 412–56: "II. Exorcisme contre la tumeur maligne."

74. See D. R. Jordan, "Two Inscribed Lead Tablets from a Well in the Athenian Kerameikos," *AM* 95 (1980): 228, n. 12 (cf. also app. I): "Th. Homolle . . . has published, unseen, from two transcriptions made by D. Prasinos of Arkesine, the text of a Jewish or Christian invocation of the Archangels to banish an evil tumor; the tablet is apparently lost, and it is accordingly impossible for me to verify Prasinos' description of the metal as lead." (Jordan then goes on to describe three tablets published as lead that he has subsequently proved to be silver).

75. Cf. the Greek text of *PGM* XCV.14 (= *GMPT,* p. 306), which is unfortunately too fragmentary to yield a completely secure reading. *PGM* CXIV (= *GMPT,* p. 313) is a *φυλακτήριον* for epileptic seizures caused by daemonic attacks. For epilepsy in general, see O. Temkin, *The Falling Sickness,* 2d ed. (Baltimore, 1971); E. Lesky and J. H. Waszink, "Epilepsie," *RAC* 5 (1965), cols. 819–31; note also A. Delatte, *Anecdota Atheniensia* (Paris, 1927), p. 487, line 6.

76. The piece (5.6 by 3.0 centimeters) is located in the Cabinet des Médailles. See P. Perdrizet, "Amulette grecque trouvée en Syrie," *REG* 41 (1928): 73–82 for the editio princeps and Robinson 1938, 252 for the restored reading *κεφαλαλ]γίαν*, which was also suggested in L. Robert, *BE* (1971): 406, no. 68. Where Robinson reads [*παιδί*]*ον*, one should supply instead a personal name. See also, B. Lifschitz, "Notes d'épigraphie grecque," *Revue Biblique* 77 (1970): 81–82;.

77. Kotansky (see n. 37).

78. Including the new listings in *GMPT,* the following represent actual amulets for fever (as opposed to descriptions from formularies): *PGM* XVIIIb, XXXIII, XLIII, XLVII, LXXXIII, LXXXVIII–XCI, CIV, CVI, CXV, CXXVIII, CXXX. For fever amulets, see also Kotansky (see n. 37), 187–88; D. Wortmann, "Der weisse Wolf," *Philologus* 107 (1963):

157–161; P. J. Sijpesteijn, "Amulet against Fever," *Chron. d'Égypte* 57 (1982): 377–81; and Bonner, 1950, 67–68.

79. See above p. 119 with n. 87 for the phylactery formerly in the Louvre; for the Acrae piece see A. Vogliano and K. Preisendanz, "Laminetta Magica Siciliana," *Acme* 1 (1953): 172–78; L. Bernabò Brea, *Akrai* (Catania, 1956) 170–71, no. *52 (pl. 39 [52]); and Robert (see n. 5), 14, n. 23.

80. See n. 60.

81. The editio princeps is that of D. Comparetti, *Notizie degli Scavi* (1923): 207–8 (who incorrectly dates the piece to the fourth century B.C.). Subsequent scholars note that it is a late amulet; see A. Olivieri, "Laminetta d'oro iscritta di Brindisi," *RIGI* 7 (1923): 53–54; for discussion of the earlier bibliography see Zuntz 1971, 283.

82. It is the title of a Byzantine collection of late Roman veterinary writers; the remedies, for the most part, were written for horses; see E. Oder and C. Hoppe, *Corpus Hippiatricorum Graecorum,* vol. I, *Hippiatrica Berolinensia* (Leipzig, 1924), vol. II, *Hippiatrica Parisina Cantabrigiensia Londinensia Lugdunensia* (Leipzig, 1927). G. Björck, *Zum Corpus Hippiatricorum Graecorum* (Uppsala, 1932); idem, *Apsyrtus, Julius Africanus, et l'hippiatrique grecque* (Uppsala, 1944); A. M. Doyan, "Les textes d'hippiatrie grecque," *L'AntClass* 50 (1981): 258–73; note also Klaus-Dietrich Fischer, "Two Notes on the *Hippiatrica,*" *GRBS* 20 (1979): 371–79, esp. 372, no. 6, with suggestions for a new edition of the text of Oder and Hoppe; consult also Fischer's recent edition of Pelagonius in the Teubner Library, *Pelagonii Ars Veterinaria* (Leipzig, 1980) and his article, "The First Latin Treatise on Horse Medicine and Its Author Pelagonius Saloninus," *Medezin historisches Journal* 16 (1981): 215–26 for a general discussion of veterinary medicine in antiquity.

83. The piece is of unknown provenance and currently resides at the Dumbarton Oaks Center for Byzantine Studies, Washington D.C. (Acq. no. 53.12.52; acquired in 1953; 4.0 by 1.5 centimeters). M. C. Ross (*Catalogue of the Byzantine and Early Mediaeval Antiquities in the Dumbarton Oaks Collection,* vol. 2, *Jewelry, Enamels, and Arts of the Migration Period* [Washington, D.C., 1965] no. 29 [plate XXV]) reports that the piece had been rolled up, presumably for insertion into a tubular case. I give the text in Heim 1893, no. 152, which has *ἐστοναχίζετο* in lieu of *στεναχίζετο*. For the use of verses from Homer in magic, see n. 61.

84. The text is in F. L. Griffith and H. Thompson, eds. *The Demotic Magical Papyrus of London and Leiden,* 3 vols. (London, 1904), verso, col. 10 lines 1–12 with translation (and transliterated text). An abridged version has been published by Dover Press: *The Leyden Papyrus: An Egyptian Magical Book,* eds. F. L. Griffith and H. Thompson, (New York, 1974). The text has been retranslated with new readings from the original manuscript by J. H. Johnson in *GMPT,* p. 244. PDM xiv. 985–1025 contains several short recipes for gout and (arthritic) stiffness. One recipe (PDM xiv. 1003–14) reads as follows: "Another amulet for the foot of the gouty man: You should write these names on a strip of silver or tin. You should put it on a deerskin and bind it to the foot of the gouty man named, on his two feet: 'THEMBARATHEM OUREMBRENOUTIPE AIOXTHOU SEMMARATHEMMOU NAIOOU, let NN, whom NN bore, recover from every pain which is in his knees and two feet.' You should do it when the moon is [in the constellation] Leo." For the treatment of gout, see now D. Gourevitch, *Le triangle hippocratique dans le monde gréco-romain: Le malade, sa maladie, et son médecin,* BEFRA 251 (Rome, 1984), 517–16 for additional bibliography.

85. For the Coptic spells, see W. H. Worrell, "Coptic Magical and Medical Texts," *Orientalia* n.s. 4 (1935): 1–37 and 184–94. The relevant portion of the text reads (following Worrell's translation, p. 18, lines 10ff.): "*For gout:* a proven (remedy). (Magical signs and letters). Write upon a piece of silver when the moon is waning; then, pouring warm (water) of the sea, read the name. Labor diligently. Do this for forty-four days (saying): 'I invoke thee,

great Isis, ruling in the absolute blackness, Mistress of the gods of heaven by birth." The text then breaks into a series of magic words followed by a prayer to rectify a dislodged uterus. Evidently we have here a new spell altogether, or the text has become conflated in transmission. Other amulets for specific maladies within the same manuscript are for "the wandering of the uterus," (p. 29, lines 25ff.); spleen disorders (lines 35ff.); disorders of the eyelids (lines 39ff.); chills (lines 41ff.); inflammation (p. 30, lines 56ff.); child bearing (pp. 30f. lines 60ff.); stomach- and headache (p. 33, lines 125ff.); fire(?) disease (lines 134ff.); uterine pains (p. 34, lines 162ff.); mental illness (lines 169ff.); pustule (p. 35, lines 184ff.); as well as other minor ailments.

86. See Bonner 1950, 76–77; note also O. Neverov, *Antique Gems in the Hermitage Collection* (Leningrad, 1976), nos. 143 and 143a and G. Schlumberger, *REG* 5 (1892): 88. Additional references are also given in Eckstein and Waszink (see n. 5), 398. I also note that like the *φεῦγε* formula preserved in Pliny (see n. 40), this inscription seems to be fragments of a longer hexametric line and may be much older than the stone on which it is incised. Bonner, ibid. also discusses an inscription on a bronze prism published by H. Seyrig, "Notes archéologiques," *Berytus* 2 (1935): 483, which also may be for gouty feet (though Seyrig suggests—incorrectly I believe—that it was for horses in the circus). See L. Jalabert and R. Mouterde, *Inscriptions grécques et latines de la Syrie* vol. 3, pt. 2 (Paris, 1953), no. 1083. For the use of the *φεῦγε* formula see n. 40 and Heim 1893, 479–84, where about twenty examples are given (including the Perseus gem [= no. 50], cited above); see also L. Robert, "Appendice 5. Échec au Mal," *Hellenica* 13 (1965): 265–71; Bonner, *HTR* 35 (1942): 89 and esp. Björck, *Apsyrtus* (see n. 82), 61–62.

87. The tablet was formerly in the Louvre (inv. Bj 87) but has been long lost; for the text, see W. Froehner, *Bulletin de la Société des Antiquaires de Normandie* 4 (1866–67): 217ff; F. X. Kraus, *Annalen des Vereins für Nassauische Altertumskunde* 9 (1968): 123ff. and G. Pelliccioni, *Atti e memorie dell RR. deputazione de storia patria per la provincia dell' Emilia* n.s. 5.2 (1880): 177–201.

88. All editions read *φύλαττε*, but the reproduction of the tablet clearly shows the prefix *δια-* at the end of the preceding line. The literature on this piece is simply too extensive to cite in its entirety: see, principally, G. C. Boon, "Excavations and Discoveries," pt. 2, "Roman," *Bulletin of the Board of Celtic Studies* 21 (1964): 96–99 (plates Ib, II); E. L. Barnwell, "The Carnarvon Talisman," *Archaeologia Cambrensis* ser. 4, vol. 10 (1879): 99–108. The piece was found in 1827 at the site of an ancient gravefield outside the Roman fort at Segontium during the excavations of a house (of later date) called Cefn Hendre; on the site see F. Haverfield, "Military Aspects of Roman Wales," *Cymmrodorian Society Transactions* (1908–1909) [1910]:85–86 (fig. 8). See, further, n. 103.

89. D. M. Robinson (see n. 66).

90. Zuntz 1971, 281; Bonner 1950, 96; E. Hübner, *Monatsberichte der königlichen Preussischen Akademie der Wissenschaften zu Berlin 1861* (1862): 533; *IG* 14, 2413, line 13.

91. See n. 37 and 76.

92. See n. 76.

93. The verb also occurs in a spell, written on a silver tablet and found in a grave at Amisos (Pontus). The spell's purpose, however, seems far from certain; lines 10–13 read, "Let no evil (*κακόν*) appear. Drive away, drive away *ὑπόθεσιν* from Rufina." The meaning of *ὑπόθεσις* in this context is open to numerous interpretations, which cannot be addressed here. See S. Pétridès, "Amulettes Judéo-Grecque," *Échos d'Orient* 8 (1905): 88–90 and R. Wünsch, "Deisidaimoniaka," *ARW* 12 (1909): 24–32, no. 4.

94. See n. 68. The verb seems to have this "medical" sense in later Koine Greek; cf. Bauer-Arndt-Gingrich, s.v.; cf. also *LSJ* s.v. *ἐπιφορά*, 5b. The use of *ἐμβολή* speaks of preventing any further "setting in" of the disease.

95. See the editio princeps of S. Sciacca, *Kokalos* 26–27 (1980–81) [1982]: 459–63, lines 13f., 20ff. Such references to "the bearer of a charm" occur regularly on amulets, but here the bearer carries, it seems, the Law of God itself, perhaps understood in the traditional sense of the Jewish phylacteries, the *tefillin* (parchment on which a text of the Torah was inscribed). One cannot be certain whether a lost portion of this tablet actually cited a verse from the Torah or the Old Testament, though such verses do occur on the Jewish-Aramaic pieces.

96. Robinson 1938, 245–53 with plate I. I have not been able to locate the present whereabouts of the piece.

97. Siebourg, "Neue Goldblättchen" (see n. 56), 398–99. See note 98 for additional references.

98. See also *IG* 14.2413, line 3; Jordan (see n. 64), 164, n.13. The piece is actually quadrangular, engraved on four sides, one of which reads "O Publicianus." Siebourg ("Neue Goldblättchen" [see n. 56], 398–99) rightly interprets this as a phylactery for the deceased Publicianus, upon whom Zeus Serapis is to show his mercy.

99. Indeed, this seems to be the case with a silver phylactery from El Jem in Tunisia that was found among the steles dedicated to Saturn; see R. Cagnat, *BCTH (*1928–29 [1930]: 54, no. 51). On dedicating materials associated with personal cases of healing, cf. esp. F. T. van Straten (see n. 2), who includes an appendix entitled "Votive offerings representing parts of the human body."

100. Provenance unknown, acq. no. 48.3 (D.O.H. no. 162): Ross (see n. 83), no. 28.

101. The last line of the text is corrupt and cannot be read.

102. See Bauer-Arndt-Gingrich, s.v. The translation given by Ross may be misleading.

103. The use of Latin written with Greek characters may have served a magic purpose; however, for such conventions, see the examples (and notes) in P. Collart, "Inscription de Sélian-Mésoréma," *BCH* 54 (1930): 378. The phylactery from Segontium (see n. 88) has some Hebrew text written in Greek, reading "Lord God of Hosts" (lines 1–3), "Elyon the terrible" (lines 8–10; cf. Deut. 10:17; Neh. 9:32), and "Blessed Thou and blessed Thy glory for ever, ever (in Greek here: *ἀεί*), for ever" (lines 11–15); see R. G. Collingwood and R. P. Wright, *RIB,* vol. 1 (Oxford, 1965), 144.

104. *CIL* 13.5338; G. Grimm, *Die Zeugnisse Ägyptischer Religion und Kunstelemente im Römischen Deutschland* (Leiden, 1969), 212–13, no. 128 (with additional bibliography). The third-century A.D. tablet was found not in a private burial but at the public baths of Badenweiler in 1784 (Badisches Landesmuseum Karlsruhe, inv. no. C625); accordingly, it can be argued from this and more that since more than one person is named on the tablet (I use the reading of *CIL,* with the notes of the apparatus criticus) and since the piece was *not* found rolled up in a tube (and thus probably was not carried by a single person), the piece was not a protective amulet of the usual sort. Since the word *periculum* (i.e., the situation from which the group of suppliants seek deliverance) can refer to a court trial or even the sentencing resulting from such a trial, perhaps the charm was meant to deliver the group as litigants in an impending trial.

105. New York Public Library, cat. no. 1 (5.7 by 1.7 centimeters). I am preparing a full reading of this text for publication; note R. Gottheil, *Journal Asiatique* ser. 10, vol. 9 (1907): 150–52. A similar Greek expression, *ἐπ' ἀγαθῷ* (found often on Greek inscriptions) occurs on a gold plaque apparently worn open to view and not rolled up (judging from the single suspension ring attached). The plaque of gold sheet shows two repoussé figures facing, wearing Egyptian garb, carrying ankhs and with solar disks atop their heads; their two children stand between with the inscription below; see L. Habachi, *Tell Basta, Suppl. aux Annales du Service des Antiquités de l'Egypte* (1957), plate 25b. For this common expression found on gemstones, see E. LeBlant, *750 Inscriptions de pierres gravées inédites ou peu connues,* MAI

36 (Paris, 1898), no. 89 (with references). Note also the expression on an Egyptian stele: R. Noll, "Römerzeitliches Sphinxrelief mit griechischer Weihinschrift aus Ägypten," *JOAI* 42 (1955): 67–74. For other examples of deities represented on plaques that are not necessarily "magical," see Ch. Clermont-Ganneau, "Plaque d'or représentant Esculape, Hygie et Télesphore," *Recueil d'Archéologie Orientale* 5 (1903): 54–55, plate IIIC.

106. H. D. Betz ("Fragments from a Catabasis Ritual in a Greek Magical Papyrus," *HR* 19 [1980]: 287–95) discusses the spell and its mystery liturgy in detail, but he failed to point out that the *aski kataski* formula preserves a hexametric verse (already recognized by Bonner in his editio princeps; see the reference in Preisendanz, *PGM,* vol. II, p. 202; cf. also W. Roscher, *Philologus* 60 [1901]: 89). These same words (*καταςκει ερων ορεον μεγα*) appear in slightly altered form as dactylic hexameters in the unpublished Getty lead tablet: *κατὰ σκιαρῶν ὀρέων μελαναυγεῖ χώρωι* (see n. 27).

107. Cf. Koenen (see n. 35).

108. This is Betz's interpretation (see n. 106) of *PGM* LXX and Zuntz's interpretation (see n. 48) of the Caecilia Secundina tablet. For the broader theoretical argument that most magic is "degenerate religion," see A. A. Barb, "The Survival of the Magic Arts," in *The Conflict between Paganism and Christianity in the Fourth Century,* ed. A. Momigliano (Oxford, 1963), 100–25.

109. See also the fragmentary third-century B.C. papyrus from Gurob (= Kern *Orph. Frag.* no. 31), which invites comparison at many points with the texts under discussion here. In addition to the apparent allusions it makes to the Orphic gold leaves (see M. L. West, *The Orphic Poems* [Oxford, 1983], 170–71 and 205), it also mentions the same tokens or *symbola* (i.e. "virgin, bitch, serpent, wreath, key, etc") that appear in the Michigan papyrus (quoted above) and the lead phylactery from Phalasarna (see n. 27); M. G. Tortorelli, *PP* 164 [1975]: 365–70. Furthermore, in the incantatory sections several deities associated with the *teletai* are addressed and urged, "save me!" (*σῶισόμ με*). The context, moreover, suggests salvation from death is to be understood here (cf. lines 18–19: *κἂν ἐπὶ θάνατον ἀπάγῃ . . . καὶ σώσει σε*).

110. For the use of *σῴζειν* in a context that blurs the theological sense with the more practical aim of saving one from earthly danger, see *PGM* I. 195–222 (containing a hymn called a *ῥυστική* [rescue prayer] that summons the "eternal god" to save the believer from "fate" and concluding with the plea, "Rescue me in my hour of need"); *PGM* IV.1167–1226 ("It even delivers one from death"); and *PGM* V.96–172. For a limited discussion of these texts, see M. J. Vermaseren, "La sotériologie dans les papyri graecae magicae," in *La soteriologia dei culti orientali nell' impero romano,* EPRO 93, ed. U. Bianchi and M. J. Vermaseren (Leiden, 1982), 17–30 and M. Smith, "Salvation in the Gospels, Paul, and the Magical Papyri," *Helios* 13 (1986): 63–74.

5

The Pharmacology of Sacred Plants, Herbs, and Roots

John Scarborough

Greek botany produced the landmark *Inquiry into Plants* by Theophrastus of Eresus (c. 370–288 B.C.),[1] and this work stands as one of the most remarkable documents of the Peripatetic school.[2] Along with his *Causes of Plants,*[3] Theophrastus' *Inquiry* set a pattern followed in Western botany until the invention of a sexual parts nomenclature by Linnaeus in the eighteenth century.[4] Yet, although morphology dominates Theophrastus' *Inquiry,* one finds at the end of the tract an account of herbal lore, the first such work on herbal pharmacy to survive in Greek.[5] The *Inquiry* was set down by about 300 B.C. and reflects the standards of Aristotle and his general ἱστορίαι, in which data were gathered and organized into particular subjects of knowledge, for instance, a history of medicine by Meno,[6] numerous histories of city constitutions (of which only one survives, the famous *Constitution of the Athenians*),[7] and so on.

Book IX of Theophrastus' *Inquiry* from time to time employs "facts" that come from folk medicine sources, generally labeled the ῥιζοτόμοι;[8] these were a semiprofessional class of "rootcutters" who had their own standards of knowledge and whose folklore about various roots and herbs mirror the deepest traditions of Greek "inquiry" on several, simultaneously applied levels from pure "magic" to utter rationalism. This mixture of assumptions about plants observed in book IX of the *Inquiry into Plants* has caused some modern scholars great discomfort as they attempt to "explain" the irrational elements, while preserving the basic format of Theophrastus' presumably "scientific" gathering of data about herbs and drugs made from plants.[9]

In some respects, this medley of approaches need not be dissected, since the multicombination of views—whether described as magical or religious or rational—had been characteristic of Greek thought since the days of Homer.[10] The amalgamation of the rational with the irrational regarding herbs and drugs received an initial form in the *Iliad* and *Odyssey,* a form that would remain rather typical of Greek thinking on drug lore from Homeric Greece through the later centuries when the *Papyri Graecae Magicae* were composed. Quite seriously, Herophilus could say that "drugs are the hands of the gods,"[11] a significant quotation when one considers that the man who reputedly made this statement was one of the famous medical researchers at the Ptolemaic Museum in the 270s and 260s B.C.;[12] moreover it is Galen of Pergamon (129–after 210 A.D.) who embeds Herophilus' phrase in a consideration of drugs, apparently with tacit approval. Thus the blend of the rational with the divine would also be characteristic of the so-called scientific levels of Greco-Roman pharmacy, while even such as Dioscorides (fl. c. 65 A.D.) and Galen sought to "explain" drug action within intellectual patterns of natural theory that evolved gradually from pre-Socratic philosophy.

THE INITIAL INTELLECTUAL CONTEXT: HOMER'S "ILIAD" AND "ODYSSEY"

Even though there is some evidence to suggest a carryover into Homeric times of Mycenaean practices in drug lore,[13] it is in the Homeric epics that one first discerns a clear idea of a "drug" among the Greeks, represented in its broadest meaning by *φάρμακον*. Standing alone, the word means "magic" or "charm" or "enchantment," but eight adjectives used by Homer divide *φάρμακον* into variations of effects, ranging from the extremely harmful (a poison) to very beneficial (a remedy).[14] The contrast in effects of *φάρμακα* is well illustrated in a single episode from the *Odyssey*.[15] Circe has agreed to change Odysseus' men back into their proper forms, and she administers *φάρμακον ἄλλο* for this purpose,[16] as contrasted to the *φάρμακον οὐλόμενον* that had changed them into swine.[17] The "other drug" (or "spell," here) is for benefit, whereas the "evil drug" previously given has turned men into pigs. Homer, of course, is not interested in speaking about the actual substances that might be part of either *φάρμακον* but only about their effects. Circe herself is one of the first figures in Western literature to represent a skilled sorceress, and her talents include manipulation of the poisons and remedies, known apparently from mythic and folk traditions.[18] Yet her craft is powerless against a *φάρμακον* that Homer calls *μῶλυ*,[19] described as having a black root and a "flower like milk,"[20] that protects the hero against Circe's wiles as she "prepared . . . a potion (*κυκεών*) in a cup of gold . . . and put in it a drug (*φάρμακον*) with evil intent in her heart."[21] Μῶλυ is a gift to Odysseus from Hermes, and the poet notes that this is what the gods call it,[22] but that the root is difficult for mortals to dig up.[23] Homer does not say *how* the herb (if that is what *μῶλυ* is)[24] protects Odysseus, and one learns neither whether the hero drank it in a countermeasure before his meeting with the sorceress or chewed the root nor anything substantive that might show how Odysseus' shield against Circe's *φάρμακον* in the *κυκεών* was achieved.

Although Homer does not specify the particulars of his drug lore, it would not be completely accurate to characterize these veiled lines about *φάρμακα* as simply magical.[25] Lloyd suggests that there is a combination of divine and nondivine assumption in the *φύσις* of the *μῶλυ* that Hermes showed Odysseus,[26] and another interpretation of this passage argues that this early use of *φύσις* means a thing's "appearance."[27] A further level of meaning may be gleaned if one also assumes a primary sense of *growth*, "the natural form being thought of as the result of growth."[28] These may be more sophisticated analyses than the poet ever intended, but coupled with the vague generalities about healing and harmful substances are clear indications of an ongoing inquiry into such matters, albeit rather muffled. In the *Odyssey* one reads of a drug that "quiets all pains and quarrels," a beneficial drug that comes from Egypt, where the land brings forth many φάρμακα, both good and ill "when they are mixed."[29] The poet adds that Egypt is a country where every man is a physician, "learned above the rest of mankind."[30] The beneficial drug from Egypt is probably the opium poppy, mentioned in the *Iliad* as a plant that "bows its head to one side [and] in a garden is heavy with its fruit and the spring rains."[31] *Fruit* appears to be poetic license for the "poppy juice" as it drips from the head when harvested by slitting,[32] and the ripe opium poppies are indeed slit for their valuable exudations in the spring.[33] This is solid information interwoven by the poet into his episode about Gorgythion's death from an arrow shot by Teucer, and it can be argued that those who were listening to the poet's song[34] would have known what the pain-killing drug from Egypt actually might be. Poets and playwrights certainly must make allusions to substances with which their audiences would be familiar, and it seems reasonable to suppose that Greek listeners of the eighth or seventh centuries B.C. were acquainted with the powers and properties of the opium poppy (*Papaver somniferum* L.), which had already enjoyed a long history in ancient drug lore[35] and would continue to do so.

Homer couches his account of mixing opium with wine in a context of god-delivered and god-derived powers and knowledge,[36] and it is again a woman (Helen, daughter of Zeus) who possesses this specialized skill, making a link with the drug and sexual sorcery recorded of Circe.[37] Yet even though Homer emphasizes a mythological setting of treacherous females who "know" the plants and drugs, modern scholars are ill advised in presuming that the lore of drugs and poisons is used exclusively by women, in spite of the perpetuation by males of this quasi-mythology in texts ranging as widely as Sophocles, Seneca, and Petronius: anonymous gathering of pharmacological data, especially those of magical importance, has both male and female antecedents.[38] Moreover, Homer specifies a geographic origin for his drug, indicating something more than purely mythical explanation or a complete dependence upon the well-known tales he fused into the *Iliad* and *Odyssey*. Some modern scholars may be right as they connect the myths of Greek antiquity with a more generalized and universal mythology as revealed by sexual themes and similar motifs,[39] but Homer designates too many specific plants for one to assume a totally mythical context. Curiously enough, *Papaver somniferum* originated in Asia Minor,[40] so that the poet would be speaking of a "local" plant, even though the drug allegedly came from Egypt. Importantly, the episode of Hermes's gift of a φάρμακον to Odysseus is one of the rare examples of the appearance of magic in the Homeric poems.

In its broader setting, the "profession" of medicine is mentioned with some respect by Homer, and it is significant that those who are knowledgeable of matters medical and herbal were among the few traveling, skilled craftsmen (δημιουργοί) made welcome in the settlements of Homer's world, perhaps a reflection of a continuous and gradual infiltration of medical and herbal lore from the Near East and Egypt, as argued by Burkert.[41] If correct, Burkert's hypothesis may indicate why later Greek medicine—especially among the Hippocratic physicians—remained tied to a long-standing custom of being a "family trade," in which specifics of the medical craft descended through the generations by means of elders imparting skills to the youths of the family. Δημιουργός is the term for "physician" as late as Plato and the Hippocratic *Ancient Medicine*,[42] but δημιουργός had become the word applied to a less-than-expert medical man if Aristotle's famous distinction between δημιουργοί and ἀρχιτεκτονικοί[43] records an aspect of this hoary tradition going back into Homeric Greece. The semilegendary Epimenides, a ῥιζοτόμος ("rootcutter") and traveling purifier a century before Socrates and Plato[44] is probably typical of the δημιουργοί in the sixth and fifth centuries B.C., and the common presence of such practitioners may have induced Plato to observe how Homeric medicine failed to consider internal diseases or ailments,[45] not to speak of the medical theories assumed for regimen and specific treatments. Later writers provide similar observations.[46] Celsus (*Med.*, proem. 3) gives a typical description of just what Homeric medicine and pharmacy was: "They only treated wounds with the knife and with drugs."[47] *Medicamenta* were part of this earliest form of medical care, as far as later Greek and Roman writers were concerned, but the aspect that Celsus terms (from his Greek sources) διαιτητική came only "after greater attention was paid to literary discipline,"[48] by which (he explains) he means the rise of Greek philosophy and philosophers. And since *vulneribus ferro mederi* was common in Homeric times, "the knowledge of anatomy evidenced in the poems is almost as sophisticated as that in the Hippocratic writings, and indeed anywhere before the serious study of anatomy in the Hellenistic period."[49]

One can conclude that Homer's warriors and the itinerant δημιουργοί might have reasonable knowledge of anatomy as it would be appropriate for the treatment of wounds, but there is something faulty in the "doxographical" approach to the Homeric poems. Searching for verbal carryovers that can be specified in later Greek literature falsifies the living quality of the Greek in Homer's poems, although the language of epic generally is conservative and archaic and tends to be dignified. Onians has aptly observed that early Greek was marvelously fluid in description of thought, intelligence, and consciousness,[50] and it becomes somewhat futile to chase identities of plants in the *Iliad* and *Odyssey,* unless there is particular evidence (internal or external) supporting special nomenclature. Reading, for example, "onion" or "garlic" or "snowdrop"[51] for Homer's μῶλυ is founded simply on the presumption that later Greek sources record an earlier meaning or that specific antidotes can be matched with specific poisons (snowdrop to counteract stramonium, for example) when there is no evidence in the texts that such identities were assumed. In fact, what Plato, Celsus, and Galen say about early medical and pharmacological thinking, as revealed in the Homeric epics, suggests a lack of concern regarding "naming" things except in association with deities, events, or repeated epithets. In later Greek botany and pharmacy, the characteristic use of particular nomenclatures would find full expression after the rise of philosophy and philosophers—to paraphrase Celsus.

All cultures have basic assumptions about health and disease,[52] so that one can presume Homer's Greek world had appropriate concepts that included "how drugs worked." In contrast to the exorcism displayed in many worldwide systems of treatment,[53] Homeric medicine and drug lore do not exhibit an "expelling" function, although occasional chants might be added after soothing medicines were applied to wounds.[54] One can argue that since the "causes" of the wounds and fractures mentioned by Homer are obvious, there would be little need to assume divine or pseudodivine forces to explain the nature of these manifest injuries; but commonly occurring fevers, headaches, and like "diseases" ("symptoms" in modern diagnostics) probably required religious expertise to ascertain their causes and thereby to prescribe treatments. If Homeric medicines did not perform an "expelling" role, there must have been other fundamental actions assumed for drugs, explanations for "how drugs work." If Smith is correct, early Greek medicine and pharmacy combine theurgy (in its widest sense of supernatural or divine agencies in both diseases and their treatments) with the practical application of drugs,[55] foreshadowing later abstractions so common in medicophilosophical thought in later Greek medicine. Theurgy remained fundamental throughout Greek history even after the accession of Christianity,[56] and theurgy continued to exist side by side with other "medical intervention" systems. Early Greek theurgy and theurgic medicine and drug lore had no specific concepts about what a disease would be "in advance of an attack," unlike other theurgic approaches that supposed invading spirits (among a number of agents) against which theurgic medicine and pharmacy might be directed.[57] Homer's poems depict gods and goddesses acting on and through "natural" forces. Thus, on Smith's hypothesis, Homeric theurgic medicine and pharmacology would presume "natural causes" for illnesses but would also use supernatural powers as a portion of the "explanation" of how and why the disease came into being. Homeric similes give only the top layer of assumptions about how man is related to the life of other things generally, and Homer's comparisons between men and stalks of wheat, leaves, and animals are familiar. Underneath such similes is a series of primary concepts about man and his growth, maturity, and withering, processes analogous to those observed among plants and in an everyday manner in agriculture.

Homer's *Iliad* and *Odyssey* supply plain evidence, in early Greek intellectual contexts, of concepts about the workings of drugs; and by employing certain adjectives to focus these concepts[58] Homer "explains" how "good drugs" or "bad drugs" might function. Also there *is* an ideal environment, a place where good health is normal and where *φάρμακα* apparently would be neither feared nor used for cures: this is the Elysian Plain, where Menelaus will go instead of dying, and in this perfect setting there are no harsh weather patterns. To the Elysian Plain continually blows Zephyros' breeze, always "to revive men."[59] Weather magic and medical skills seem intimately joined by Homer, a connection also observed in the Homeric *Hymn to Pythian Apollo* (lines 189–94) and especially in Empedocles (Frag. 101 Wright = Frag. 111 DK): "You will learn drugs [*φάρμακα*] for ailments and for help against old age. . . . You will check the force of tireless winds, which sweep over the land destroying fields with their blasts, . . . you will bring back restorative breezes, . . . you will bring out of Hades the life-force of a dead man."[60] Homeric imagery (certainly echoed by Empedocles) suggests how drugs are compounded from elements of divinity—as is revealed in commonly known myths—and then are fused by means of an ever-present lore of agriculture, providing an essential context for hearers of the poems.

FROM HOMER TO THEOPHRASTUS

Even as one leaves the era of the Homeric epic, one notes that there is much disagreement about what *should* be accepted as knowledge of plants and herbs, particularly in regard to basic farm lore itself. Hesiod most assuredly knows practical botany, but except for some carefully sarcastic lines about asphodel and mallow as foods available to the poorest of peasants,[61] the poet from Ascra (fl. c. 700 B.C.) has little to say about presumed properties of plants, being apparently far more concerned with customs and superstitions associated with sex and general purification.[62] Lloyd notices the diverse opinions held by experienced farmers,[63] and Hesiod grumbles about these contradictory notions as they apply to the planting season, when "one man praises one day, another another,"[64] in spite of the presumably objective evidence signaled by the rising and setting of various stars and constellations.

Murky as is the evidence for pre-Theophrastean concern about plants and their properties, some of the various bits of texts and traditions that do survive firmly indicate not only an interest but also a continual debate on several simultaneous and intertwined levels, which incorporated the wide swath of Greek thinking on magic and agriculture, plants and herbal remedies, religious customs and the world of nature. Argument began early about the "nature" of plants in relationship to other forms of life, and when Aristotle (*De An.* 410b22) mentions the faulty theories of the Orphics on how plants breathe ("Orpheus," Frag. 11 DK), he shows how very old this debate was. Speculation began quite early about the *φάρμακον* used by Medea to lull the guardian reptile of the golden fleece, and the lost poems of the semi-legendary, pre-Homeric Musaeus may have stated that Medea used a *φάρμακον* with *ἄρκευθος* (*Juniperus oxycedrus* L., the prickly juniper),[65] a tradition faithfully reproduced many centuries later in the *Argonautica* by Apollonius of Rhodes.[66] There are hints that Musaeus composed poems specifically on the healing properties of plants, suggested by Theophrastus' *Inquiry into Plants* (IX.19.2), where Musaeus[67] and Hesiod[68] are cited as authorities on the magical properties of *τριπόλιον* (probably *Aster tripolium* L., the sea aster or sea starwort):[69] "[It is] useful for every good treatment, [and] they dig it up at night, pitching a tent there." Pindar (*Pyth.* III.51–53) says that the traditional medical treatments of Asclepius consisted of incantations (*ἐπῳδαί*), surgery, soothing drinks or potions, and amuletic drugs (*φάρμακα περιάπτειν*), perhaps in reflection of medicine's venerated and dual methods (drugs and surgery), with the addition of magical herbs that could be hung or worn appropriately. The combination of incantations, potions, amulets of herbs, and surgery as the assumed ideal of medical practice in Pindar has no trace of discomfort or conflicting feelings; one can presume that all four approaches to medicine and the lore of herbs existed in the Greek world of Pindar's time (518–438 B.C.) without too much notice, even though there would be certain practioners who would argue against supernatural explanations for disease (as in the Hippocratic *Sacred Disease*) or against the excessive influence of philosophical theory in medicine (as in the Hippocratic *Ancient Medicine*).

Athenian playwrights have unwittingly left clear indications that there was an ordinarily accepted "common knowledge" of drugs and herbal remedies in the fifth century B.C. and that such pharmaceutical lore was generally accepted simultaneously in both magicolegendary and empirical-practical ways. Playwrights, whether of tragedy or comedy, must use allusions readily comprehended by their audiences, so that mentions of herbs or drugs in the plays can be presumed to be understood by Athenians who sat through the productions, staged in honor of Dionysus. Two examples will be illustrative: a fragment from the lost *Rhizotomoi* (*Rootcutters*) by Sophocles, and puns and allusions made on particular herbal contraceptives by Aristophanes in his *Peace* and *Lysistrata*.

Among the many proofs adduced by Macrobius (fl. 430 A.D.)[70] in his *Saturnalia* to demonstrate Vergil as the complete rhetorician is Vergil's command of the details of Greek literature and the underpinnings of beliefs that produced Greek tragedy. Macrobius argues that Vergil's Dido is modeled after Sophocles' Medea in the *Rhizotomoi* (*Rootcutters*), and Macrobius quotes *Aeneid,* IV.513 to introduce his evidence for Vergil's borrowing of Sophoclean imagery:

> Herbs she had gathered, cut by moonlight with a bronze knife
> Poisonous herbs all rank with juices of black venom.[71]

Macrobius then writes that Medea in Sophocles' *Rhizotomoi* is presented "cutting poisonous herbs with her face turned away lest she perish from the strength of their noxious aromas, then pouring the herbs' juice into bronze jars, the herbs themselves being cut with bronze sickles."[72] Macrobius next quotes two passages from Sophocles' lost *Rhizotomoi:*

> [Medea] receives the juice whitely clouded, oozing from the cutting, while she averts her eyes from her hand; she receives the juice in bronze jars.

> These bark baskets shield and hide the ends of the roots that [Medea] cut with bronze sickles while she was naked, shrieking and wild-eyed.[73]

Apart from the ceremonial use of bronze implements and the fearful caution displayed in the gathering of this mysterious root sap, these opaque passages relate some important details: the sap is a milky or cloudy white, and the herb is particularly valued for its root. It is probable that these two quotations by Macrobius from Sophocles' *Rhizotomoi* are from the *parodos* of the play, introducing Medea in her "professional function" of harvesting herbs for magical purposes. Yet there is just enough information in these lines to suggest that the Athenian audience might be familiar with the plant and its poisonous odors, produced when its roots were severed. Local drugsellers (*φαρμακοπῶλαι*) and rootcutters (*ῥιζοτόμοι*) who plied their trades in Attica a century after Sophocles would relate *their* cautions when they cut the roots of *θαψία* (*Thapsia garganica* L., the so-called deadly carrot): one should not stand to the windward while cutting *θαψία* roots, and as a precaution one should anoint oneself with oil to prevent swelling and blistering from the *θαψία* odors and exudations.[74] *Θαψία*, with its black-barked root and white interior (the sap is milky),[75] was put in the "dangerous class" of roots,[76] alongside hemlock (*Conium maculatum* L.), as a matter of ordinary knowledge simply because it was so very common as a wild plant in Attica and particularly dangerous to cattle brought in from other disticts of Greece.[77] Perhaps Athenians sitting through the performance of Sophocles' *Rhizotomoi* would have nodded in recognition at the chorus of rootcutters and also sensed personal connections with Medea's roots and magical implements, connections in their turn interlocked with the legends of Circe and Medea as well as the farmers' lore about the "deadly carrot."

In the Homeric *Hymn to Demeter* (line 209), one reads that the goddess willingly drinks a potion (*κυκεών*) made by mixing meal, water, and pennyroyal (*γλήχων* [also βλήχων]-*Mentha pulegium* L.). Pennyroyal apparently retained its quasi-mythical associations with the functions of birthing and nursing the newborn for many centuries, as attested by Soranus' recommendation of its strong, sweetish odor (somewhere between peppermint and camphor) in the delivery chamber.[78] Yet this tight link between pennyroyal and "female functions" has another aspect than the simply magical or legendary, neatly given precision by puns in Aristophanes' *Lysistrata* (line 89) and *Peace* (line 712). In *Lysistrata,* one chuckles at the nicely obscene pun on βλήχων,[79] but the *Peace* line provides a vivid allusion to a woman's sexual attractiveness, with the suggestion of *κυκεῶνα βληχωνίαν* ("pennyroyal potion") as a remedy for too "much fruit."[80] Aristophanes is not emphasizing pennyroyal's venerated associations with Demeter (although Athenians would certainly appreciate those traditions) but rather the well-known fact that pennyroyal quaffed in solution prevented pregnancies—a most useful detail in the folk traditions of the day, since a prostitute, or "flute girl," stayed in business only as long as she did not become pregnant.[81] The playwright assumes that everyone in the audience (both men and women) would know pennyroyal by sight in the fields and also would know the "ordinary use" for the "pennyroyal potion" by women—or the allusions and puns on the name of the herb would not have gained laughter from the mixed audience.

Pennyroyal's reputation as a female contraceptive and abortifacient is verified in the Hippocratic writers,[82] Dioscorides,[83] and Galen.[84] One may presume that the pharmaceutical details on the effects of pennyroyal solutions or pessaries, as they are listed in Hippocratic writings, are extracts gathered from midwives' oral traditions or (perhaps more relevantly) from an ever-present prostitutes' lore. Aside from folkloristic connections with aphrodisiacs generally,[85] pennyroyal's pharmaceutical properties (as understood by modern pharmacognosy) confirm antiquity's basic empirical observations.[86] Until quite recently pennyroyal and its extract, the ketone pulegone, was commonly employed as an emmenagogue and as an abortifacient. Generally called "pennyroyal oil" or "*pulegium* oil," this extract acts as a mild irritant to the kidneys and bladder in excretion and reflexly stimulates uterine contractions. Contrasted to the "pennyroyal potion" of Greek antiquity, with its aparently nontoxic action, the modern extract of pennyroyal oil exhibits very poisonous effects. Convulsions result from as little as four milliliters.[87] Pennyroyal oil contains not less than 85 percent pulegone, while the natural extract gained from the dried leaves and flowering tops of *Mentha pulegium, Mentha longifolia* (L.) and *Hedeoma pulegioides* (L.) Pers. (US sp.) has about 1 percent of the volatile oil, along with tannin and bitter principles.[88] The ancient preparation apparently served well in its intended use, and its urinary effects were observed in conjunction with the action of *κανθαρίς* (the aphrodisiac today called "Spanish fly").[89] Aristophanes' puns certainly reflect a "common knowledge" of pennyroyal among the citizens of fifth-century-B.C. Athens, and one also receives a brief glimpse of the technical expertise of midwives and call girls as they plied their trades in the same era. That men as well as women would be presumed aware of what the "pennyroyal potion" was supposed to do suggests how common this particular herb was and also that the use of "sexual drugs" (whether magical or not) was not the particular knowledge of women alone.

THEOPHRASTUS AND BOTANICAL FOLKLORE

Homer's *μῶλυ* was to later Greeks an averter of drugs and witchcraft,[90] something that had the power to ward off noxious effects, a substance termed an *ἀλεξιφάρμακον*. In his *Inquiry into Plants,* Theophrastus offers an opinion as to why squill (*Urginea maritima* Baker) should be a good *ἀλεξητήριον* (averter): he suggests that because plants that grow from bulbs are very long-lived (his phrase can be rendered as something like "tenacious of life"), they are thus suited to be planted in front of the entrance to a house where the squill "fends off trouble (*δήλησις*) that threatens the dwelling."[91] Theophrastus is *not* saying squill actually does what "it is said" (*λέγεται*) to do (ward off spells from a house), but he is attempting to explain why many in his culture might believe squill has such powers and properties, carefully linked by Theophrastus to its property in being able (*δύναται*) to aid the storage of other fruits and vegetables: "If the stalk of the fruit of the pomegranate [*ῥόα-Punica granatum* L.] is set in squill," it will keep for a long time; and even when hung, squill bulbs 'live' for a long period, as do many bulbous plants. Elsewhere, Theophrastus records instances in which squill becomes part of a purification ritual. A superstitious man has chanced to see a statue of Hecate wreathed in garlic at one of the countryside altars set at forks in the road, and he rushes home to hire priestesses, who carry squill (bulbs?) around to cleanse him.[92] Parker believes this is part of an ancient tradition of "blood purification,"[93] and one notes how the stereotypical superstitious person employs several plants in an ordinary day, using a leaf of the *δάφνη* (the sweet bay, *Laurus nobilis* L.) carefully chewed in the early morning (a "sacred bay"),[94] and how he will purchase myrtle (*μύρσινη-Myrtus communis* L.) and frankincense (*λιβανωτός-Boswellia* spp.) every fourth and seventh day of the month for sacrifices to the Hermaphrodites.[95] Again, Theophrastus carefully depicts folkloristic practices without necessarily condemning them (although the stereotypes in his *Characters* generally bring laughter at the extreme antics portrayed), and the underlying questions posed by the skilled botanist are significant: Why are *these* particular plants associated with religious or magical practices and how can they be related to known properties (*δυνάμεις*) of herbs and herbal drugs, especially as understood in agricultural lore?

Squill is a particularly apt example of the mixture in the Greek mind of practical botany, magicoreligious rituals of great antiquity, and precise knowledge of pharmacological and medical utility.[96] The first-known mention of squill is in a fragment of an elegiac poem by Theognis (fl. c. 544 B.C.), and this small bit establishes that squill was widely understood in its botanical and agricultural context for its pungent properties: "Neither a rose nor a hyacinth grows on a squill."[97] A close second, in terms of earliest mention of squill in Greek, is in one of the scrappy remnants of the poems of Hipponax (fl. 540–537 B.C.), and the two lines show immediately a close link between the use of squill and the religious practice of expelling a scapegoat (*φαρμακός*) to cleanse a community of perceived impurities or pollution: "Pelting him in the meadow and beating / With twigs and squills like unto a scapegoat."[98] The brutality of this use of a *φαρμακός* to cleanse a diseased city is summarized by a late Byzantine polymath, John Tzetzes (fl. c. 1130), who describes how a *φαρ-*

μακός was selected and sacrificed in "ancient times."[99] Significantly, Tzetzes cites Hipponax as "describing the custom best,"[100] although most of the data in Tzetzes' account are unattributed: "After hitting [the *φαρμακός*] seven times on the penis with squills and [branches] of wild figs and other uncultivated [plants], in the end they [sacrificed the *φαρμακός*] in a fire burning wood of wild trees."[101]

Squill and certain other herbs (for example, the so-called chaste tree [*Vitex agnus castus* L.])[102] are given special significance in the Greek custom of choosing a *φαρμακός* to rid a community of plague or pestilence or a household of its sense of hunger (perhaps a carryover from the hoary fear of famine as suggested by the term *βούλιμος*). Tzetzes labels his account as one of *κάθαρμα* (ritual cleansing); but squill as an acrid cultivar and widely known herbal remedy for coughs, asthma, and *ἀλεξιφάρμακον* when "hung whole in front of entrances [to houses]"[103] is closely intertwined with the concept of the *φαρμακός*, an obvious analogue of *φάρμακον*, first noted as a "poison" or "beneficial drug" or "spell" in the Homeric epics. It is, however, not until hellenistic Greek, that the ancient *φαρμακός* "scapegoat," (accented on the final syllable) was wedded to a second meaning, namely, "poisoner," "sorceror," or "magician" (accented on the initial syllable),[104] quite probably a revival of the Homeric sense, much as Theophrastus wrestles with defining plants with "medicinal properties,"[105] as contrasted to ordinary plants.[106] Squill, as Theophrastus acknowledges, has peculiar properties (*δυνάμεις*) that span the range from the purely pharmacologic to venerated folkloristic practices, mirrored partially in his spoof of the superstitious man in *The Characters*. As Stannard has noted, squill attained such widespread use that there was a variety called "squill of Epimenides,"[107] identified as French sparrow grass (*Ornithogalum pyrenaicum* L.) by Hort and his sources.[108] If the tradition preserved in Apuleius reflects historical fact,[109] Epimenides was a magician or student of magic (a magus), and according to Theophrastus' normal custom of not naming living authorities,[110] Epimenides probably lived and "gathered roots and herbs"[111] quite some time before Theophrastus' century.[112] Epimenides' "cleansing" activities were intimately associated with "squill," although several herbs were called by that name in Greek.[113] One may also note the reappearance of the very ancient and very muted Near Eastern ties with the magicoreligious association of squill, as Lucian pokes fun at the "Chaldean" practice of "cleansing with torches and squills."[114] One ascertains the survival of various forms of purification rites employing squill at least as late as the second century A.D. in the eastern Roman Empire, and the unattributed sources among John Tzetzes' "the ancients" may include texts from Roman and Byzantine times,[115] as well as his cited names and quoted lines from Hipponax and Lycophron of much earlier Greek centuries. The properties of squill were highly esteemed for their powers of *κάθαρμα* throughout a millennium among Greeks and Romans. "It is tempting," writes Parker, ". . . to see [squill] as the vegetable equivalent of an animal, the impure puppy, a dishonourable plant appropriately used in a ritual applied to polluted persons."[116] Parker, however, admits difficulty with such an interpretation of the symbolism of a plant like squill, because there were so many possible uses,[117] from pure rituals in magicoreligious observances to straightforward herbal lore and pharmacology.

Theophrastus is well aware of the quandary in describing a plant that had both traditional and sacred uses and also a history of employment as a drug among herbalists and physicians. His morphological botany could be secured by close and patient observation,[118] but some other method than either pharmacology or botany was necessary to depict squill—and similar plants—as they were comprehended by the varying practitioners in the Greek world of c. 300 B.C. Not surprisingly, Theophrastus meets this problem within the structure and format of book IX of his *Inquiry into Plants,* the part of the treatise that considers plants of "medical utility," as well as important questions as to sources of information about such plants.[119]

Theophrastus begins his *Inquiry* (IX.18) by writing, "As has been noted, there are roots and shrubs that have many powers (*δυνάμεις*) affecting not only living bodies but also bodies without life (*τὰ ἄψυχα*)," and his first example is an *ἄκανθα* (probably gum arabic), which, "as they say" (λέγουσι), thickens water when put into it. His second example is the root of marsh mallow (*ἄλθαια-Althaea officinalis* L.), which has the same property if it is shredded into a container of water and allowed to stand in open air; but one identifies marsh mallow by its similarity with mallow (*μαλάχη*-either *Malva rotundifolia* L., or *Malva sylvestris* L.), and Theophrastus describes differences between the two species in terms of their leaves, stalks, flowers, seeds, and roots. Then follows the medical employment of *ἄλθαια*: "They employ (*χρῶνται*) marsh mallow for fractures, and in sweet wine for coughs, and in olive oil for open ulcers [or wounds]." The pharmacology implied by Theophrastus is the property of marsh mallow to make a glue when mixed with water (similar to the gum arabic), a property acting on a "lifeless body," in turn rendering it useful in the treatment of fractures.[120] The choice of marsh mallow seems deliberate, rather than simply one gum-producing plant selected from many examples: *ἀλθαίνω* is a common verb (heal),[121] and the noun *ἄλθεξις* as "healing" or "cure" appears in both the Hippocratic *Fractures* and Nicander's coinage as *ἀλθεστήρια* (healing remedies).[122] Theophrastus' deft connection of medicinally useful plants with their effects on nonliving things has thus been precisely laid down, beginning with roots and shrubs that have various properties, spanning a range from straightforward drugs that act in living patients to the plants that have powers to affect nonliving matter. By this simple technique of association Theophrastus is able to proceed from herbal remedies to plants that have other effects and uses, including those of magicoreligious and sacred importance. His manner of consideration, however, is carefully paced step by step as he advances from plants affecting nonliving matter through plants that act in or on animals other than human beings[123] to those that affect both "body and intellect [or soul]."[124] Parenthetically, Theophrastus precedes his "body and intellect" plants with, "And legends are concocted not without reason."

Inquiry into Plants IX.18.3–11, takes up the subject of fertility and antifertility herbs, the curiosa of aphrodisiacs and anaphrodisiacs. Most importantly, however, Theophrastus notes that his first instance of a plant that has *both* aphrodisiac properties (*ὄρχις*, lit. "testicle," probably early purple orchid, *Orchis mascula* [L.] L., from which a drink called salep is made from the macerated tubers[125]) has "leaves like a squill,"[126] indicating immediately how the botanist is investigating such plants, which are firmly linked to sacred ceremonies well known to his readers. Ὄρχις is also an excellent example of the intermeshing of a common "doctrine of sympathy" or—more loosely—the "doctrine of signatures,"[127] which Greek and Roman herbal lore continually interwove with more empirical pharmacological observations. Much as Theophrastus employs the ordinary name *testicle* for the double-propertied *Orchis mascula* (and he does so without any hint of self-contradiction or of the insertion of dissonant facts), so also one recalls Theophrastus' connections between the quasi-religious practice of planting squill to avert spells on a house and his concept of long-lived squill bulbs (VII.13.4), and one can thereby sense also the numerous but unstated folkloristic and magicoreligious customs that hover in the shadows of *Inquiry* IX.18.3–11. Preus argues that any effect such plants would have regarding conception or contraception would be expressed by the modern term *psychosomatic,*[128] but the application of any modernisms to Theophrastus' delicately skilled arguments and attempts to document pharmacological properties of sacred plants and herbs does the injustice of inserting presentism. Moreover, except for a direct reference to an otherwise unknown Aristophilus of Plataea (called a "drugseller," φαρμακοπώλης), who said he "had some [drugs] that could engender greater potency as well as eliminating it completely,"[129] almost all of the data in *Inquiry* IX.18.3–11 are qualified by "it is claimed," "are said," "it is said," "if this account is true," and "they say," phrases that delineate Theophrastus' manner of simply recording what he has heard, without necessarily approving or denying the purported facts. He does state his basic acceptance, however, that such things and effects are quite possible, with "It is paradoxical, as I have said, that opposite effects result from a single nature [φύσις]; but it is not paradoxical that such properties [δυνάμεις] exist."[130] Implicitly he has admitted the possibility of traditional claims for sacred plants like squill, but Theophrastus firmly rejects generally (ὅλως) the assertions made for ἀλεξιφάρμακα that might be worn as charms or attached to a house.[131] Exceptions could perhaps be allowed for sacred herbs; but if they are unsanctioned by hallowed magicoreligious traditions of either a public or private nature, Theophrastus is unwilling to entertain possible effects from herbs simply said to be powerful.[132]

The majority of *Inquiry* IX is taken up not with considerations of claims for plants that are magical, semioccult, or plants that have religious connections, but with a careful analysis of medicinal properties of somewhat less than sixty major herbs and herbal remedies.[133] Yet Theophrastus' main source of information for such plants are the ῥιζοτόμοι, a professional group of herbalists who collected medicinal roots and herbs, selling them at country fairs, hawking their virtues for pains and ailments of many kinds; added to the ῥιζοτόμοι as sources of data on herbs are the φαρμακοπῶλαι (drug vendors), who also touted their products in the venerated manner of folk medicine to country and city dwellers alike.[134] *Inquiry* IX.8.5–8 shows rather vividly these sources of Theophrastus' data and how he sorted out useful facts from the merely mythical: the drug vendors and rootcutters suggest that one should cut roots only while standing to the windward, especially in the case of θαψία (the deadly carrot, *Thapsia garganica* L.),[135] and that one should coat one's body with oil before trying to dig up or cut the roots. Furthermore, the herbalist has to exercise caution while gathering the fruit of the wild rose (probably the rose hips): these must be collected while the individual stands to the windward, because picking the rose hips could harm one's eyes. Theophrastus does not dismiss these assertions as "old wives' tales," but comments that there is some credence to be placed in them because of the properties (δυνάμεις) of these plants, which tend to be harmful, seizing "as they say" like a fire and burning. Digging hellebore[136] for a long time causes dizziness, and the ῥιζοτόμοι advise one to eat garlic beforehand and to quaff some undiluted wine. Theophrastus honors such reports from the rootcutters, but he derides other procedures espoused by the rootcutters as "far-fetched and irrelevant," including the practice of digging up peonies at night to avoid being seen by woodpeckers (this may cause the loss of an herbalist's eyesight) or the stipulation that one should beware of the buzzard (*Buteo* spp.) while cutting a κενταυρίς (probably *Centarium umbellatum* Gilib., the feverwort or centaury) or that while cutting the All-Heal (πάνακες) of Asclepius, one should make an offering "in its place" of fruits and baked meal. Theophrastus writes that "praying while cutting is perhaps reasonable, but the additions to this caution are ridiculous," showing he is carefully separating the patently superstititious from the empirically reasonable or from particular customs that have long-standing magicoreligious associations. One should not, however, be misled into thinking that Theophrastus advocates a medical botany bereft of magic and religion or an herbalism forsaking its ancient agricultural and sacred heritages: he says to accept rootcutters' and farmers' tales at face value and then to test them both with the logic of empiricism as well as field collection of plants, ascertaining those practices that, in effect, "make sense."

Theophrastus' grudging ambivalence about hoary rituals and his apparently reluctant piety suggest how murky are the differences in Greek antiquity between magic and religion as well as between the presumably "objective" observations gathered by farmers and those customary deferences of homage to the venerated powers of certain plants. Edelstein rightly recognizes the "specific problem of pharmacology in connection with the efficacy of plants"[137] but then proceeds to argue that Greek and Roman views are neatly categorized into a tripartite division. First the properties of medicinal plants (and by logical extension, the causes of diseases) are marked by purely "scientific" and empirical observations, seeing "in plants nothing but natural powers."[138] Second, since Nature is divine, plants and drugs derived from them are divine, which can be fitted within a class of traditional religion of a deistic character, an "interpretation which accepts the divinity of the plants because of the divinity of the intellect in the human being who applies them."[139] Third, Edelstein scornfully derides "magical belief . . . sorcery and such nonsense. . . . All the great pharmacologists rejected those things. Andreas and Pamphylus [*sic*] and [those who accepted magic] constituted a small minority; they were scholars rather than physicians; they were antiquaries. These men were isolated as were those who believed in the demonic character of diseases."[140] Somehow Edelstein has chosen to ignore the grand and opaque jumble of opinions characteristic of his "great pharmacologists," including many authors represented in the Hippocratic corpus, Theophrastus, Dioscorides, Soranus, Rufus, and Galen. Edelstein seems to be aware of the artificial and imposed manner of his precise categories of "pharmacological thinking" in classical antiquity, since he slithers into a supporting argument that begins, "But all the pharmacologists, nay almost all the physicians, believed in sympathetic remedies."[141] In turn this "proves" the rarity of pure empiricism and pure magic in ancient pharmacy, proof that collapses when one reads the admixture of scorn and praise by Galen for amulets and medical astrology,[142] the quotation of local customs attached to various herbs by Dioscorides,[143] or the painstaking evidence of botanical lore sandwiched with magicoreligious observances as assembled by Theophrastus. One cannot—as Edelstein does—dismiss medical magic as "found only in the magical papyri which contain not the knowledge of physicians but prescriptions of folklore"[144] any more than one can assert truthfully that Pliny the Elder is a "superstitious layman"[145] without carefully qualifying what this might mean among Roman aristocrats in the early Roman Empire.

Edelstein's "rationalistic supernaturalism"[146] unwittingly signals why almost all pharmaceutical texts in classical antiquity—from Homer and Theophrastus to Galen and the *Papyri Graecae Magicae*—encompass aspects of magic, empiricism in its strictest sense, religion as understood in its context of historical observation, and the constant shifting and mingling with one another of these three broad approaches. Sometimes there is the clear imprint of philosophy as *it* wrestled with definitions and attempted clarity, and sometimes witchcraft is incorporated as a meaningful part of man's investigation into the world of nature and its mysteries. To call some drugs "divine remedies" or "sacred stuff,"[147] as does Galen in his *Compound Drugs according to Place in the Body,* says no more and no less than that this particular substance acts in a godlike manner. Quite openly, Galen (quoting Asclepiades) can write that certain drugs have the "property (δύναμις) of Asclepius,"[148] not too distant from common, modern perceptions of aspirin and sulfonamides as "wonder drugs." Not comprehending molecular chemistry or the physiology of drug action does not make one a "superstitious layman."

DEFINITIONS, HERBS, AND THEIR PROPERTIES: GRECO-ROMAN INTERPRETATIONS

Theophrastus urges his readers to honor only certain of the magicoreligious traditions attached to a few plants, but once he has discarded fully rank superstitions as explanations for how some plants "work" as drugs, he attempts to fuse the data of the ῥιζοτόμοι with a curiously muddled definition of an "herb." He is clearly uneasy regarding the specifics of an "herb" or an "herbal remedy" as provided by his oral and written sources, and the description of πόαι (herbs) includes substances that have "medicinal powers" (φαρμακώδεις δυνάμεις), comprising "juice" (χυλισμός), "fruits" (καρποί), "leaves" (φύλλα), and "roots" (ῥίζα), because the "rootcutters term an 'herb' certain of the medicinals,"[149] and an "herb" consists of one or all of these parts. Theophrastus' hesitant definition of *herb* may have been borrowed directly from the ῥιζοτόμοι, or perhaps from the medicobotanical works of Diocles of Carystos,[150] or Theophrastus may have invented it himself. The ῥιζοτόμοι certainly had given him a basis for this definition, because they did not call all roots "herbs" but only roots from a group of medicinal plants or healing parts of certain plants.[151] Theophrastus continues to display his uncertainty as he writes that such δυνάμεις (probably now a mix of "powers" and "properties") of medicinal roots are distinguished from the δυνάμεις of roots generally. Thereby the "roots" that are medicinal include all four parts of the plant, especially the leaves, as the ῥιζοτόμοι themselves say. Consequently, "herbs" are all four parts, not merely the roots.[152]

By attempting to impose the morphological approach as he struggles with the basic question of how one distinguishes an "herb" from other plants, Theophrastus has provided a particularly unsatisfactory substitute for the answer to the question of "how drugs work," especially in view of the firm answers given in traditional superstitions and some religious connotations. In his listing of the herbs that follows in *Inquiry* IX one must link his brief descriptions of drug action with previous morphologies of plants that form almost all of *Inquiry* I–VIII. It is significant that Theophrastus does not employ the theory of elements and qualities to explain "how drugs work," as does the Pseudo-Aristotelian *Problems:* "The [substance] that according to its own nature, is not overcome by the body's heat and that enters the veins, and because of its excess heat or cold [is not concocted]: this is the φύσις of a drug."[153] The unknown author has grappled rather well with the basic questions of delineating a theory to explain drugs and their actions, as contrasted to foods or plant derivatives without noted pharmacological properties. There seems, oddly enough, a better awareness in the philosophical speculation regarding plants by the author of *Problems* than in the murky groping toward a morphological definition of *herb* by Theophrastus. It is not surprising that many scholars have assumed that book IX of *Inquiry into Plants* is a later addition by unknown medical botanists, a hypothesis disproven by the manuscripts.[154]

Toxic substances were reasonably understood by the second century B.C., as suggested by the difficult poems of Nicander of Colophon (fl. c. 140 B.C.), the *Theriaca* and *Alexipharmaca*.[155] Nicander borrowed heavily from an earlier Greek toxicologist named Apollodorus,[156] who, in turn, may have inherited his concepts from Diocles of Carystos. Again agricultural lore shows vividly in Nicander's semiplagiarism of details on the black widow spider, various cobras, wasps, large centipedes, and millipedes,[157] as well as in his description of what one did to administer an antidote for an excessive consumption of the famous aphrodisiac made from blister beetles.[158] Greek pharmacy had a good botanical morphology from Theophrastus, a rough taxonomy of plants derived from folklore, some medical entomology and toxicology as recorded by Nicander, and a mass of details on herbs and herbal concoctions as revealed in the Hippocratic tracts on women's ailments.[159] An organizing principle, however, appeared in none of these authors regarding "how drugs worked," even though Theophrastus' basic botany was superb. In classical antiquity, pharmacy and toxicology remained aspects of medical practice that occasionally purloined venerated superstitions or religious customs or that with leaps of uncertainty adapted and adopted facets of philosophical physical theory (especially the concepts of elements, qualities, and humors) to account for observed actions of drugs. As Riddle has remarked, at the very least Greco-Roman pharmaceutical theory included the firm notion of a "drug poison,"[160] first discerned in Homer's poems. Most physicians and pharmacologists, however, sought to reject divine explanations for drugs and their effectiveness, and those of the so-called Hippocratic persuasion were sometimes emphatic in dismissing *both* the magicoreligious and philosophic interpretations,[161] leaving either a jumble of empirical observations or a strange and quixotic denial of the efficacy of any drugs. Folk medicine, however, continued to carry on with its venerated assumptions, and even when Dioscorides of Anazarbus (fl. c. 65 A.D.) was able to produce his magnificent summary of drugs in the famous *Materia Medica,*[162] he acknowledges quite frequently the staying power of numerous folk traditions linked with particular pharmaceuticals.

Dioscorides is a watershed in the development of Greco-Roman pharmacy, and his masterpiece of compression, *Materia Medica,* brilliantly demonstrates the efforts of a skilled physician and astute medical botanist to bring order out of the chaos that characterized drug lore up to his own day. As a young man, Dioscorides probably studied herbs and herbal pharmacology with resident experts in Tarsus,[163] and there is just sufficient evidence to indicate that he spent a portion of his mature medical practice in the context of a Roman legion.[164] Most importantly, however, Dioscorides went about the business of testing herbs and drugs with a precision that would be noteworthy in any century[165] and invented an entirely fresh method of classifying drugs by what they did or did not do when given to a patient—a system called by Riddle, "drug affinity."[166] The prevailing medical theories in the early Roman Empire were generally linked in some way with a debased form of "medical atomism,"[167] and in the remarkable preface to his *Materia Medica* Dioscorides flatly rejects this and all other philosophic explanations of drugs when he writes, "They have not measured the activities of drugs experimentally, and in their vain prating about causation they have explained the action of an individual drug by differences among particles, as well as confusing one drug for another."[168] At the same time, he says he has checked "what was universally accepted in the written records and [made] inquiries of natives in each botanical region."[169]

At the conclusion of his discussion of black hellebore (*Helleborus niger* L.),[170] Dioscorides furnishes a typical example of how he treats long-lived religious or quasi-superstitious customs connected to herbs. His farmer-informants have told him that if one plants vines close to the root of the hellebore, the wine from such grapes will make an excellent purgative, and Dioscorides adds, "they sprinkle it [the wine] around houses thinking it to be a purification (*καθάρσιος*) from defilement [in a religious sense]"; moreover, when his informants dig up the hellebore, they stand facing the sunrise and pray to Apollo and Asclepius, watching all the while for an eagle (*ἀετός*) in flight as an evil omen, since "the bird engenders death should it see the digging of the hellebore; at any rate, it has to be dug up very quickly, since drowsiness is caused from its exuded vapors." Dioscorides merely acknowledges the belief that black hellebore was sacred to a god[171] by describing a series of steps taken by his informants when they dig up the root, passing rapidly to a clear, natural reason why such religious customs and precautions became accretions upon this particular medicinal root. Since Dioscorides mentions hellebore's association with an apparently ancient rite of purification, it is probable that the plant had some local history of use in such rituals (no locale is specified), quite reminiscent of the widespread employment of squill.[172] Perhaps herbs like squill and the hellebores were early considered sacred plants due to their heavy pungency, especially when they were cut or bruised, much as Dioscorides notes the "exuded vapors" (*ἀποφο-ραί*) from the cut hellebore roots. Parker's comments on the corollaries of using the hellebores (there are two main types: a "white" and a "black") in the medical and mythological treatment of madness,[173] said to have been invented by Melampus for the daughters of Proetus, are pointed and relevant,[174] although Hippocratic physicians strove to perceive insanity quite apart from divine causes, as one reads in the eloquent presentation by the author of the Hippocratic *Sacred Disease*,[175] very often cited as *the* Greek medical work that provided the seedbed for medical approaches in the West. Dioscorides, however, carefully limits his account of black hellebore to verified uses in terms of his "drug affinity" system, uses that are pharmacological and classed by pharmacognostic means.[176]

As brilliant as was Dioscorides' new classification method for drugs, later writers on pharmacology chose to revert to other systems, alphabetizing plants and drugs in place of Dioscorides' precise groupings according to physiological action. Although Dioscorides' collected data were incorporated into almost all future tracts on pharmacy in the West until the Renaissance, those data were rearranged according to the predilections of later writers, including the polymathic and enormously influential Galen of Pergamon (129–after 210 A.D.).[177] Galen's drug lore is a gigantic potpourri of herbs, animal products, written sources quoted verbatim, and quasi-legendary and pseudofolkloristic facts all compacted into three separate systems of organization, as he attempted to bring some sort of harmony into the chaos of pharmaceutical data as he found them in the second century.[178] In spite of some oddly informal sources of pharmaceutical information, such as one Orion the Groom,[179] Galen's drug lore became a model for later Byzantine encyclopedists,[180] who took the explicit humoral theory (borrowed directly from the Hippocratic *Nature of Man*)[181] as the major explanation of "how drugs worked." Except for certain compound recipes, which had their own venerated pedigrees,[182] the "drugs-by-degrees" classification, as evolved from Galen through successive Byzantine medical writers to Paul of Aegina (fl. c. 640 A.D. in Egypt), was based on the ancient notions of humors and qualities, an outlook that dominated Western pharmacy until the mid-nineteenth century. Later Roman and Byzantine pharmacy and pharmacology refined Greco-Roman theory on drugs and occasionally added new substances, but there was always a powerful undercurrent of folk medicine displayed in cults to the saints in the Byzantine Empire,[183] as well as a tenacious survival of nonlearned conceptions about drugs.[184] Two aspects of how pharmacology was perceived for sacred plants are illustrated by pharmacological astrology and the widespread acceptance of magical properties exemplified by the formulas and doctrines of Thessalus of Tralles and the collection of texts in Greek, Coptic, and demotic known as the *Papyri Graecae Magicae*.

"HERMETIC" MEDICAL ASTROLOGY AND HERBAL PHARMACOLOGY

An important and often ignored facet of late hellenistic and Roman religion are the so-called Hermetic texts.[185] Having purported origins in particular revelations by a Hermes Trismegistus, Asclepius, and other pagan gods of Greco-Egyptian background, "Hermetic" medical revelations assume multiple aspects that attempt to reconcile several competing concepts of medicine and pharmacy, employing the device of an "archetypal" beginning. The "Hermetic" manner of medicine and pharmacology is sharply defined in an astrological-pharmaceutical text by a Thessalus, called *Powers of Herbs*.[186] The preface to Thessalus' *Powers of Herbs* was first published in the *Catalogus Codicum Astrologorum Graecorum*[187] and was simply called Thessalus' *Letter to Claudius* (or *Nero*). Publication of the full *Powers of Herbs* awaited proper collation of several Greek and Latin manuscripts, and the complete (and dual) texts were eventually published by Friedrich in 1968. In the preface (= *Letter*), Thessalus of Tralles informs the emperor about a revelation by Asclepius, earlier revealed to him by Hermes Trismegistus,[188] naming the herbs truly associated with the planets[189] and the signs of the zodiac.[190] The *Powers of Herbs* and its preface date to the reigns of Claudius and Nero (sometime between 41 and 68 A.D.),[191] and the preface precedes (in the *Catalogus* texts,) a tract on twelve plants linked to the twelve signs of the zodiac,[192] an exposition of seven planets associated with seven plants attributed to Thessalus of Tralles,[193] and a clipped treatise attributed to an "Alexander" that surveys the same seven plants. The *Catalogus* texts indicate the wealth of ancient works discussing plants and herbs and their relationships to the planets, the signs of the zodiac and the traditional three decans (10-degree units) into which each was divided, and plants linked with the "fifteen fixed stars."[194] Thessalus' preface (*Letter*) and the full text of his *Powers of Herbs* represent medical astrology and pharmacology, and there are many instances of Roman physicians and scientists who believed astrology was an important diagnostic tool: Galen's *Crisis Days* (III. 5–6)[195] indicates how astrology pinpoints both diagnosis and prognosis in diseases,[196] although his account is bereft of the mysticism permeating corollary "Hermetic" works.[197] If Galen thinks of medical astrology as a diagnostic technique, the so-called Hermetic writers believe that herbal and medical astrology are revelations, explaining why certain plants have healing powers and properties.

In place of Theophrastus' confusion, Dioscorides' brilliant but inapplicable "drug affinity" system, and Galen's basic uncertainty about classifications of drugs, authors in the Hermetic traditions could claim that their acceptance of divine power was an active manner of receiving it that thereby increased its strength, much as a person instructed by Poimandres can say he "has been invested with the power and instructed in the Nature of the Whole and in the Greatest Divinity."[198] Such herbal astrology is marked by a simplicity—a deceptive simplicity from the standpoint of modern pharmacology—illustrated by "A Plant of the Sun: Chicory," from Thessalus' *Powers of Herbs*:[199]

A Plant of the Sun: Chicory (*κιχώριον*)[200]

(1) First named among the plants of the sun is "heliotrope"; yet there are many kinds of "heliotropes," and of all these most efficacious[201] is the one called chicory. (2) Its juice mixed with oil of roses is an ointment.[202] (3) It is suitable for relieving heartburns,[203] and it releases tertians,[204] quartans, and intermittent fevers,[205] and mixed with an equal part of the oil of unripe olives, it stops headaches. (4) If someone looking toward the sunrise smears on the juice of the chicory, invoking the presence of the [god] Helios, and begs to give him praise, he will be most favored among all men on that day. (5) One prepares from the chicory's root little pills (*καταπότια*) for heartburns and disorders of the stomach, in

which the stomach is afflicted and will not accept foods, and for disorders of the stomach in which the stomach receives nourishment but does not promote digestion: downy woundwort,[206] 8 drachmas; saffron crocus,[207] 2 drachmas; Pontic honey, 14 drachmas; mastic flower,[208] 6 drachmas; ginger[root],[209] 4 drachmas; pepper,[210] 4 drachmas; Dead Sea bitumen/mineral pitch,[211] 2 drachmas; anise,[212] 4 drachmas; mastic gum/resin,[213] 4 drachmas; the root of the chicory, 24 drachmas; pound these ingredients in a mortar with very old mead (honey plus wine, *οἰνομέλι*)[214] and make lozenges of 1 drachma. Drink one with water for heartburns; drink one with the best wine for stomach ailments.

Parallels to Thessalus' recipe occur in Pliny, Columella, Dioscorides, Galen, the *Papyri Graecae Magicae,* and the Byzantine *Geoponica,*[215] but Thessalus' "chicory stimulant" seems unique among Roman prescriptions. Initially, the botany appears poor, with the "heliotrope" including the chicory; Thessalus, however, is not saying this is *the* heliotrope but rather that it belongs to a broad class of herbs "attracted to the sun," the literal meaning of the word. And excepting the self-anointing with chicory juice while invoking Helios, the recipe is a fairly straightforward listing of ingredients and preparation methods frequently encountered in the drug books of Galen and later Byzantine pharmacy. Moreover, folk medicine retains employment of mastic, anise, and chicory as stomach calmers, traditions backed to some extent by the physiological chemistry of the herbs. Thessalus' *Power of Herbs* encompasses a number of Greco-Roman pharmacological traditions, including the technical approaches seen in Dioscorides and Galen, as well as specifics of drugs and herbs found in the magical papyri. Hermetic astrological herb lore could claim it knew *why* such plants had pharmaceutical properties by their clear linkages to the divinities represented by constellations and planets (including the sun and moon) at various points in the zodiac. Festugière classes some of these texts of astrological herbalism according to which plants were associated with the sun, and he gives these plants separate status by means of determination of the contents of the list (categories labeled A, B, C, etc.).[216] For example Chicory is Festugière's type B, with a special listing of herbs for the moon, Saturn, Jupiter, Mars, Venus, and Mercury.[217]

The presence of pepper and ginger in Thessalus' *Powers of Herbs* shows that the text dates from the first half of the first century, on grounds independent of manuscript attribution. In full revival was a flourishing trade with Far Eastern markets,[218] and Dioscorides' *Materia Medica* shows fresh incorporation into the Roman pharmacopoeia of Indian, Malayan, and some Chinese spices.[219] Thessalus is well aware of good ingredients for his recipes, and there is a sophistication in *Powers of Herbs* reflecting formal drug lore of the day infused with the "Hermetic" assertions of special knowledge about herbal medicine. Such a pharmacology of sacred plants, organized along astrological lines, received a patina of acceptable "science" in the Roman Empire, although Roman law frowned upon the application of astrology to political ends.[220]

SACRED PLANTS IN THE MAGICAL PAPYRI

The Greek, Coptic, and demotic texts known as the *Papyri Graecae Magicae* mention over 450 plants, minerals, animal products, herbs, and other substances as presumably "pharmaceutically active" in the recorded spells, incantations, formulas, and imprecations. The texts in the collection are dated generally to Roman and Byzantine Egypt (c. 30 B.C.–c. 600 A.D.), but several instances of drug lore (e.g., the multiingredient incense called κῦφι) indicate a heritage going back many centuries, probably to dynastic Egypt. One cannot dismiss the documents contained in the *Papyri Graecae Magicae* as simply written forms of jumbled superstitions with a pseudopatina of Egyptian learning:[221] the *Papyri Graecae Magicae* have yielded a trove of insights on how Jew, Christian, and pagan perceived their world; and an important facet of our fresh understanding of these precious documents that emerge from beliefs of "common people" is the pharmacology of magical and sacred plants and drugs.

Ritual is essential for the Greco-Egyptian ῥιζοτόμος (here generally "herbalist"), much as suggested by Pfister in his basic essay on magical conceptions of plants,[222] a suggestion nicely illustrated by the following from the *Papyri Graecae Magicae:* "**Spell for picking a plant:** Use it before sunrise. *The spell to be spoken:* I am picking you, such and such a plant, with my five-fingered hand, I, NN, and I am bringing you home so that you may work for me for a certain purpose. I adjure you by the undefiled name of the god: if you pay no heed to me, the earth which produced you will no longer be watered as far as you are concerned—ever in life again, if I fail in this operation [then follow magical words]; fulfil for me the perfect charm."[223] One presumes the herbalist recited this spell as he went out to gather plants, and his work began before dawn, much as suggested in context by Thessalus of Tralles in *Powers of Herbs* and by Dioscorides when he mentions folk customs concerning certain powerful roots and their collection. Specifics followed by an herbalist come a little later in the magical papyri: "Among the Egyptians herbs are always obtained like this: the herbalist first purifies his own body, then sprinkles with natron and fumigates the herb with resin from a pine tree after carrying it around the place 3 times. Then after burning κῦφι and pouring the libation of milk as he prays, he pulls up the plant while invoking by name the daimon to whom the herb is being dedicated and calling upon him to be more effective for the use for which it is being acquired."[224] Natron and its use is very ancient, indeed,[225] and the full ritual includes burning pine resin and κῦφι and an offering of milk, each an aspect of hoary religious observances in many cultures, not only ancient Greek and Egyptian.[226] And, as the papyrus text continues, one soon comprehends that *all* herbs are sacred, since the invocation uttered at the moment of picking the plant is as follows: "You were sown by Kronos, you were conceived by Hera, you were maintained by Ammon, you were given birth by Isis, you were nourished by Zeus the god of rain, you were given growth by Helios and dew. You [are] the dew of all the gods, you [are] the heart of Hermes, you are the seed of all the primordial gods, you are the eye of Helios, you are the light of Selene, you are the zeal of Osiris, you are the beauty and glory of Ouranos, you are the soul of Osiris' daimon which revels in every place, you are the spirit of Ammon."[227] The invocation proceeds by naming a number of gods, goddesses, powers, and properties, thus assuring the herbalist of his sacred function as well as the primary acknowledgement of awe regarding the god-given properties of such plants. In many respects the act of collecting herbs is an act of worship, and the herbalist understands that the powers contained in these plants emerge from the divinity within each part collected (this is perhaps similar to the *ῥιζοτόμοι* and their definition of *herb* as recounted by Theophrastus) and that the plants conversion into "drugs" simply extends their powers for the kindly benefit of man.

With the natural awe in mind, as well as Betz's thought that "magic is the art that makes people who practice it feel better rather than worse,"[228] one may contemplate the pharmacology of a few representative texts from the magical papyri, for instance, this contraceptive recipe:

> **A contraceptive, the only one in the world:** take as many bittervetch seeds as you want for the number of years you wish to remain sterile. Steep them in the blood of a menstruating woman. Let them steep in her own genitals. And take a frog that is alive and throw the bittervetch seeds into its mouth so that the frog swallows them, and release the frog alive at the place where you captured him. And take a seed of henbane, steep it in mare's milk; and take the nasal mucus of a cow with grains of barley, put these into a [piece of] leather skin made from a fawn and on the outside bind it up with a mulehide skin, and attach it as an amulet during the waning of the moon [which is] in a female sign of the zodiac on a day of Kronos or Hermes. Mix in also with the barley grains cerumen from the ear of a mule.[229]

This contraceptive recipe certainly displays the typical ingredients expected in magical concoctions (nasal mucus, a mule's earwax, menstrual blood, etc.), but sandwiched within are two pharmacologically potent herbs, bittervetch (*ὄροβος* = *Vicia ervilia* [L.] Willd.) and henbane (the papyrus has *ὑοσκύεμος* as contrasted to the usual *ὑοσκύαμος*,[230] *Hyoscyamus niger*), both widely known in Greco-Roman pharmacy and in modern folk medicine and pharmacognosy. Dioscorides notes the use of the vetch in the treatment of skin diseases of several kinds[231] and gives warning regarding its ingestion since it engenders headaches, disturbs the bowels, and draws down blood in the urine. As Riddle notes,[232] vetch was used to treat cancer until the early nineteenth century, and active principles isolated from the herb include vicianin (a cyanogenetic glycoside), guanidine, and xanthine. Overdosage would resemble cyanide poisoning, much as Dioscorides suggests. Henbane is a rather poisonous plant from which hyoscyamine is obtained and was well known in classical antiquity as a drug that could cause madness or—as Dioscorides writes—"frenzies."[233] Preparations made from the henbane have atropinelike effects[234] and have some limited utility in modern therapeutics in the relief of spasms in the urinary tract. The scribe may indicate that the concoction is to be swallowed through the device of having the frog ingest the seeds (frogs and fertility were intimately linked in Egyptian and Greek lore),[235] but the basic function of the recipe is in its amuletic powers, a common approach as seen in the following text from the magical papyri. Yet muffled among the earwax, blood, milk, and nasal mucus is a record of the potency of two drugs, fully understood in the formal pharmacy of classical antiquity.

Even a short text from the magical papyri will yield numerous insights:

> Carried [with a magnetic] stone, or even spoken, [this verse] serves as a *contraceptive:* "Would that you be fated to be unborn and to die unmarried." Write this on a piece of new [papyrus] and tie it up with hairs of a mule.[236]

Apart from the quotation of Homer (*Il.* III.40), this small bit from late Roman or Byzantine Egypt does not appear to contain anything of particular interest until one checks the ancient references on the "lodestone" in the context of obstetrics and gynecology. The scribe has compacted an enormous lore on magnetic stones and "women's problems" into a single line recommending the lodestone as an amulet, although the use of quasi-Homeric lines as spells is also very common in the magical papyri. The lodestone charm appears in the writings of Soranus of Ephesus (fl. c. 98–117 A.D.), whose *Gynecology* was the finest tract of its kind before the European Enlightenment. Soranus' mention of the lodestone amulet suggests how common such charms must have been in the first and second centuries, and he writes that some individuals believe "some things" are useful according to their "antipathy," for example the lodestone, the "Stone of Assius," the stomach curd from a rabbit, and "some other amulets to which we ourselves pay no heed. But one must not forbid their use: even if the amulet has no real effect, it will possibly gain a cheerful attitude in the woman"[237] being treated for uterine hemorrhage. Soranus thus provides a listing of some charm ingredients while disapproving them, even though his patients obviously valued them all. Moreover, the lodestone was deemed very useful in ancient medicine as a "blood assimilator," as recorded by Dioscorides,[238] Galen,[239] Pliny the Elder,[240] and the Byzantine compilation of farm lore, the *Geoponica*.[241] Similar in properties was the "Stone of Assius" (probably some kind of pumice), which was an effective styptic esteemed by Dioscorides and other Roman medical writers.[242] Physicians and pharmacologists thought the lodestone was useful in treatment for uterine bleeding and other similar ailments, and it appears the magical papyrus records a logical assumption by the common folk: one could, indeed, wear the saying on a lodestone, since the magnetite already had styptic and presumably divine powers to prevent bleeding; such logic would also proceed to the next step, which meant that a contraceptive power was likewise provided in the stone.

Herbs, drugs, medicinal minerals, animal products used as medicines, and insects appear frequently in varying contexts in the magical papyri.[243] Particularly fascinating is a "substitution list" of names of herbs and other substances given as "code names," as explained by the scribes:

> **Interpretations** which the temple scribes employed, from the holy writings, in translation. Because of the curiosity of the masses they [i.e., the scribes] inscribed the names of the herbs and other things which they employed on the statues of the gods, so that they [i.e., the masses], since they do not take precaution, might not practice magic, [being prevented] by the consequence of their misunderstanding. But we have collected the explanations [of these names] from many copies [of the sacred writings], all of them secret. *Here they are:*
>
> A snake's head: a leech
> A snake's "ball of thread": this means soapstone
> Blood of a snake: hematite
> A bone of an ibis: this is buckthorn
> Blood of a hyrax: truly of a hyrax
> "Tears" of a Hamadryas baboon: dill juice
> Crocodile dung: Ethiopian soil
> Blood of a Hamadryas baboon: blood of a spotted gecko
> Lion semen: human semen
> Blood of Hephaistos: wormwood
> Hairs of a Hamadryas baboon: dill seed

Semen of Hermes: dill
Blood of Ares: purslane
Blood of an eye: tamarisk gall
Blood from a shoulder: bear's breach
From the loins: camomile
A man's bile: turnip sap
A pig's tail: leopard's bane
A physician's bone: sandstone
Blood of Hestia: camomile
An eagle: wild garlic(?)
Blood of a goose: a mulberry tree's "milk"
Kronos' spice: piglet's milk
A lion's hairs: "tongue" of a turnip
Kronos' blood: . . . of cedar
Semen of Helios: white hellebore
Semen of Herakles: this is mustard-rocket
[A Titan's] blood: wild lettuce
Blood from a head: lupine
A bull's semen: egg of a blister beetle
A hawk's heart: heart of wormwood
Semen of Hephaistos: this is fleabane
Semen of Ammon: houseleek
Semen of Ares: clover
Fat from a head: spurge
From the belly: earth-apple
From the foot: houseleek.[244]

This list of "interpretations" is striking in what it tells us about divine names given to ordinary herbs—divine names that the scribes insist are not understood by "the masses." The preface to the listing also shows a professional pride by the unknown priests who "know the codes" and suggests that the practice of magic had adopted as part of its skills and techniques the lore of herbalism.

If the "interpretations" of *PGM* XII. 401–44, are intended to explain code names taken for granted by priests, scribes, and probably the common people, there are many more instances in the magical papyri in which substances are merely named without further ado, and some of these terms disguise multiingredient drugs. In four (probably five) different passages of the *Papyri Graecae Magicae* one reads of the use of something called *κῦφι*,[245] which is a very ancient Egyptian incense, ointment, and edible drug,[246] containing up to thirty-six ingredients, all pharmacologically active.[247] If formal sources in ancient and Byzantine pharmacy—for instance, Dioscorides, Oribasius, and Paul of Aegina—did not specify ingredients in the various formulas for *κῦφι*, we would be reduced to learned speculation regarding just what this "sacred incense of Egypt" *was*. The *κῦφι* recipes show a sophistication of drug compounding among the common folk and a close study of the historical evolution of *κῦφι* indicates a slow "improvement of the product" from dynastic Egypt through the seventh-century texts of Paul of Aegina. In itself, the history of *κῦφι* destroys an accepted mythology of modern medical historians, who assume ancient medicine and pharmacy developed to a certain point, then remained utterly static for countless centuries. If *κῦφι* originated from the magicomedicine and pharmaceutics of dynastic Egypt, it became not only a substance that would be part of a magical tradition but also a drug adapted into the pharmaceutics of learned medicine, suggested by the ten-ingredient κῦφι given by Dioscorides.[248] The magical papyri certainly take κῦφι "for granted," offering no detailed explanation of either the ingredients or its significance other than to imply its importance in such procedures as *The Bear-charm which accomplishes everything,*[249] in which κῦφι is one item in a list offered to the "Bear" (constellation).

The cacophony of competing claims, apologies, and vitriolic criticism of magicians and magic (including purported drug lore) in the second century is dramatically illustrated by a masterful diatribe of Hippolytus of Rome (c. 170–c. 236 A.D.). As part of an ongoing and antagonistic free-for-all among Christians and non-Christians alike, Hippolytus' rhetorical fury is directed, in part, at the blatant charlatanism displayed by the practitioners of magic who "offer the Egyptian magicians' incense called κῦφι,"[250] accompanied by loud and noisy manipulation of gawking crowds. Hippolytus knows his drugs and simulates deep anger at the degradation of herbalism he has observed, remarking that all of the hocus-pocus of invisible writing produced by the magicians of Rome is easily understood with the use of malachite, powdered galls, milk, fish sauce (*garum*), spurge (τιθύμαλος), and fig juice.[251] Hippolytus' feigned tirade against practicing magicians delineates how popular such professionals were in the early Roman Empire.

PHARMACOLOGY AND MAGICORELIGIOUS ASSUMPTIONS: SOME CONCLUSIONS

Recent research on occult doctrines and texts in the Renaissance has convincingly demonstrated that underneath the bland labels of *Hermeticism, magic, astrology, witchcraft,* and similar terms, there is an enormous mélange of views and practices, a hodgepodge of concepts often in open conflict with one another. Vickers writes that "the influence of the hermetic texts was small in comparison with that of the main occult sciences, and their presence in the Renaissance . . . makes for just one more syncretic ingredient in an already syncretic mixture."[252] A comparable pasticcio characterized Greco-Egyptian "Hermeticism," as indicated by Fowden in *The Egyptian Hermes;* and once Fowden has led us through the curiously interwoven, yet conflicting, documents of this ancient Hermeticism, a clear awareness emerges of the patchwork quality in thinking, even with such presumably unified subjects like herbalism.[253] An analogous miscellany characterizes the data in the magical papyri, the erudite authors on matters pharmacognostic like Dioscorides, and almost all other extant Greek and Roman texts—from poetry to philosophy—that take up or mention the myriad of different facts gleaned from folk customs presuming divine or magicomedical properties in plants and herbs.

Even with the evidence of continually shifting and fluctuating hybrids of assumptions about φάρμακα, which seem to wax and then fade into composite forms through the centuries, a few limited conclusions can be drawn. At first glance, the presence of women as "experts" in drug lore, as seen first in Homer, might suggest that such expertise emerged solely from the arcane knowledge of sex and birthing, perhaps guarded, if not hoarded, by females. To be sure, many of the rituals that festoon rootcutting and the gathering of herbs do traditionally incorporate women as the "main characters," but the generally cited evidence of poetry ignores the plentiful and contrary evidence of male expertise in aphrodisiacs and matters obstetric as found in such authors as Theophrastus. In certain respects, Greek and Roman men were as fond of creating fantasies about women as are modern males, so that when one notes the listing of men who were learned in sexual and erotic magic, as provided by Galen, one is also reminded of the male practioners of erotic bewitchments as they appear in the *Papyri Graecae Magicae*. Modern psychology also has demonstrated the strong presence of a generalized masculine fear of women, particularly in the basic consideration of sex, so that the student of ancient erotic magic may be forewarned not to accept the notion of a special feminine expertise in drugs.

Magicoreligious concepts about drugs may probably be linked with very old traditions ultimately stemming from the civilizations of the ancient Near East. The curious presence of the *δημιουργός* in the pages of Greek literature, from Homer to Aristotle, may indicate a survival, if not an infusion, of Assyrian or Babylonian medical customs; and further study of Egyptian texts of magicopharmacy and rituals associated with the prescribing of drugs most likely will demonstrate further links with the later practices of Greek medicine. At the very least, one must conclude that the magicoreligious documents of the Greeks, which contain data on potent herbs, have very apparent links with cultures and religions of the ancient Near East.

Greek and Roman perceptions of the basic causes of pharmaceutical properties—in particular those of plants—continually fused religious and empirical data; and the pattern of thought in its multiple levels on the actions of drugs, first enunciated by Homer, remained fairly consistent throughout the centuries of Greek, Roman, and Byzantine pharmacology. This pattern combined the conviction of divine powers of drugs—whether beneficial or deleterious—with deeply rooted observations gathered by farmers over hundreds of generations; and properties (*δυνάμεις*) attributed to varying *φάρμακα* quite frequently were amalgams of venerated rituals fused with carefully deduced pharmaceutical effects, for instance, the association of squill with purification ceremonies and its treatment in the *Materia Medica* of Dioscorides. The *δυνάμεις* of herbs and drugs could be viewed "rationally" through magicoreligious means, by one who also assumed the basic divinity of the world at large (and therefore of the men and the plants that lived in that world) or accepted explanations of botanical astrology. All these and other approaches touched on above were quite acceptable to most thinkers in classical antiquity so that Edelstein is almost certainly incorrect to argue the utter rarity of pure magic or pure empiricism in ancient pharmacy and medicine. Even as Theophrastus muddles his definition of "herb" thanks to the folk customs of his oral sources, the *ῥιζοτόμοι*, he not only applies the precepts of Aristotelian morphology to his herbal botany but also acknowledges the prevalence of a belief in sympathetic pharmacology, a "doctrine of signatures" applied to orchid bulbs, a belief paralleled in spells of the *Papyri Graecae Magicae* that note the efficacy of "mule products" as contraceptives.

Theophrastus' painstaking considerations of religious and folkloristic practices concerning plants and herbal preparations provide priceless details about the so-called unlettered levels of Greek society as they conceived herbal medicine and its functions. The magical papyri, in spite of priestly attempts to restrict knowledge of herbal lore, show quite vividly the ordinary and sophisticated command of drug compounding by the common people, a command not surprising in view of their usual rural upbringing. They would know the plants from childhood, and they would also know the appropriate magicoreligious connotations and their proper interpretations. Drugs were indeed "the hands of the gods."

Notes

1. O. Regenbogen, "Theophrastos," in *RE* Suppl. VII (1940), 1354–1562 (esp. 1453–79: botany). John Scarborough, "Theophrastus on Herbals and Herbal Remedies," *Journal of the History of Biology* 11 (1978): 353–85. G. E. R. Lloyd, "Theophrastus, the Hippocratics, and the Root-Cutters," in *Science, Folklore and Ideology: Studies in the Life Sciences in Ancient Greece* (Cambridge, 1983), 119–35. Almost all plants are identified according to international systematics in botany, with the usual abbreviations of authors' names (e.g., L. for Carl von Linnaeus [1707–1778]). The Latin binomials employed sometimes also contain indications of priorities. For instance "L" sometimes appears as "[L]," indicating that Linnaeus was the original author of a name, revised by some more modern botanist, whose naming had been accepted by the international community of botanists, who, in turn, wished to retain Linnaeus' priority, even with the "new" name. One can gain insight into the formation of modern botanical names by consulting William T. Stearn, *Botanical Latin,* 2d ed. (Newton Abbot, England, 1973). A standard listing of priorities and author's names for plants is the *Draft Index of Author Abbreviations Compiled at the Herbarium, Royal Botanic Gardens Kew* (London, 1984), followed faithfully by D. J. Mabberley, *The Plant-Book* (Cambridge, 1987). Numerous botanical equivalents are conveniently provided in George Usher's *Dictionary of Plants Used by Man* (London, 1974) and Oleg Polunin's *Flowers of Europe* (London, 1969) and *Flowers of Greece and the Balkans* (Oxford, 1980). Various "flora," however, differ as to which binomials are listed for particular genera and species, so that nonbotanists often become confused regarding the "real" name. See the lists "Floras and Handbooks" and "Periodicals" in Mabberley, *Plant-Book,* pp. 637–49.

2. Theophrastus' *Inquiry into Plants* is frequently cited by its Latinized title, *Historia plantarum* or simply *HP;* the most commonly used Greek text of *Inquiry* is that edited and translated in two volumes by Arthur Hort in the Loeb Classical Library (London, 1916). Victorian prudery, however, caused Hort to omit certain passages from both the Greek text and English translation (viz., *Inquiry,* IX.18.3–11 on ὄρχις as an aphrodisiac and other plants as anaphrodisiacs; see also nn. 124–32 below), so that one must still employ F. Wimmer, ed., *Theophrasti Eresii Opera* (Paris, 1866; rpt. Frankfurt, 1964) for the complete Greek text. (*Inquiry,* IX.18.3–11 = Wimmer, pp. 159–61, with Latin translation). An English translation of these "missing lines" was published as Chalmers L. Gemmil, "The Missing Passage in Hort's Translation of Theophrastus," *Bulletin of the New York Academy of Medicine* 49 (1973): 127–29 (a spare translation with no commentary). One must also be extremely wary of accepting many of the proposed nomenclatures for plants as listed in the "Index of Plants" of the Hort edition. The identifications and binomial names were provided by William Thiselton-Dyer, and many are no more than wildly inappropriate guesses. Unfortunately, many of these false nomenclatures have entered *LSJ,* where they mislead novitiates unacquainted with Thiselton-Dyer's less-than-skilled botany. Far better are the nomenclatures in the *Oxford Latin Dictionary,* and best are listings in Jacques André, *Les noms de plantes dans la Rome antique* (Paris, 1985); useful among older lexicons are A. Carnoy, *Dictionnaire*

étymologique des noms grecs de plantes (Louvain, 1959) and R. Strömberg, *Griechische Pflanzennamen* (Göteborg, 1940). Had John Raven of Cambridge been able to complete and publish his projected *History of Botany in Ancient Greece* before his death in 1980, we might, indeed, have a reliable guide to the plant lore in English. Colleagues were treated to a foretaste of the study in Raven's 1976 Gray Lectures, with the subject on "Plants and Plant Lore in Ancient Greece." Nick Jardine, "Ancient Greek Botany," in *John Raven by his Friends* ed. John Lipscomb and R. W. David [Bath, 1981; privately printed], 88–90) writes the following about one aspect of the 1976 Gray Lectures: "Throughout the lectures John succeeded in communicating a specific enthusiasm, an enthusiasm which gave life to all his interests, classical, philosophical and botanical, the enthusiasm of the hunt. Sometimes the quarry was a person. In the opening lecture the hapless Sir William Thiselton-Dyer, F. R. S., successor to Hooker in the directorate of Kew, and author of the diagnoses of plant names that enliven the standard Greek dictionary, Liddell and Scott, is hunted down. With unholy glee the superficiality of his scholarship and the absurdity of his diagnoses are gently but remorselessly exposed" (p. 89).

3. Books I and II only of *Causes* have appeared in the Loeb series: *Theophrastus "De Causis Plantarum"* ed. and trans. Benedict Einarson and G. K. K. Link (Cambridge, Mass., 1976); for the remainder (*Causes*, III-VI) cf. Wimmer (see n. 2), 218–319.

4. On Theophrastus as a botanist, see Ernst H. F. Meyer, *Geschichte der Botanik,* (Königsberg, 1854–57; rpt. Amsterdam, 1965), I:146–88; A. G. Morton, *History of Botanical Science* (London, 1981), 29–43; and Edward Lee Greene, *Landmarks of Botanical History,* ed. Frank N. Egerton (Stanford, 1983) I:128–211.

5. Scarborough, "Theophrastus" (see n. 1), 353–55.

6. W. H. S. Jones, *The Medical Writings of Anonymus Londinensis* (Cambridge, 1947; rpt. Amsterdam, 1968), 5–8.

7. Exhaustively, P. J. Rhodes, *A Commentary on the Aristotelian Athenaion Politeia* (Oxford, 1981).

8. Scarborough, "Theophrastus" (see n. 1), 354–55, 359–60. Lloyd, *Ideology* (see n. 1), 120–22. Greene (see n. 4), I.120–27.

9. Very common in the modern literature is the use of some "school" of psychology or psychiatry, and studies are replete with "psychosexual" or "psychosomatic" explanations—as if the ancient Greeks thought in terms identical to those of modern Freudians, Jungians, Piaget, or even Eric Berne. Sometimes, however, ancient Greek drug lore becomes part of a debased structural anthropology in modern scholarship, especially prominent in works by followers of Claude Lévi-Strauss, e.g., Marcel Detienne, *The Gardens of Adonis: Spices in Greek Mythology,* trans. Janet Lloyd (London, 1977). See my review and comments on Detienne's book in *Classical Journal* 76 (1981): 175–78.

10. Two books stand out in this respect, whose authors attempt to interpret early Greek thinking in ancient terms, not modern ones: Richard Broxton Onians, *The Origins of European Thought* (Cambridge, 1951) and Robert Parker, *Miasma: Pollution and Purification in Early Greek Religion* (Oxford, 1983).

11. Quoted in Galen, *Compound Drugs According to Place of Ailment* VI.8 (in C. G. Kühn, ed., *Galeni Opera Omnia* [Leipzig, 1821–33], XII:966).

12. P. M. Fraser, *Ptolemaic Alexandria* (Oxford, 1972), I:348–57.

13. Cynthia Wright Shelmerdine, *The Perfume Industry of Mycenaean Pylos* (Göteborg, 1985), 123–30.

14. *Od.* X.394, *οὐλόμενον*; II.329, *θυμοφθόρον*; I.261, *ἀνδροφόνον* (viz. *φάρμακον ἀνδροφόνον διζήμενος, ὄφρα οἱ εἴη / ἰοὺς χρεχρίεσθαι*); *Il.* XXII.94 and *Od.*X.213, *κακόν*; *Od.* IV.230, *λυγρόν* [cf. X.236]; IV.229 and X.287, *ἐσθλόν*; *Il.* V.401 and 900, *ὀδυνήφατα φάρμακα*; and IV.218, *ἤπιον*.

15. X.391–94.

16. *Od.* X.392.
17. *Od.* X.394.
18. One may observe how these stories degenerated as they were passed along over the centuries, e.g. Pliny, (*HN* XXV.10–12) says that the "origin of botany" was closely aligned with pure magic, and he notes that Medea and Circe were early investigators of plants—and that Orpheus was the first writer on the subject of botany. His reflections on the topic, from the vantage point of the "enlightened" first century, are instructive of the differences between Homeric Greece and Pliny's Roman intellectual world eight hundred years later: *Durat tamen tradita persuasio in magna parte vulgi veneficiis et herbis id cogi eamque unam feminarum scientiam praevalere* (XXV.10). The old "folk traditions" remained strong. Further degeneration of the Circe legend appears in Ael. *On Animals* I.54, in which the sorceress kills merely by touch. By the time Augustine wrote his *City of God,* folklore was current in Italy that women (esp. women who ran inns) gave drugs along with cheese to travelers and thereby changed their guests into beasts of burden; August., *City of God,* XVIII.18; cf. Apul., *Golden Ass* (*Metamorphoses*) III.24, and *Apologia* (*De Magia*), 31 (*Apuleius,* ed. R. Helm, vol. II. [Leipzig, 1972], 36–37).
19. *Od.* X.305.
20. *Od.* X.304: ῥίζῃ μὲν μέλαν ἔσκε, γάλακτι δὲ εἴκελον ἄνθος.
21. *Od.* X.316–17.
22. *Od.* X.305.
23. *Od.* X.305–06.
24. In some later Greek texts, μῶλυ means "garlic," viz., *Allium sativum.*, e.g., Theophr., *Inquiry into Plants,* IX.15.7; the Hippocratic *Nature of Women* 85 (*Oeuvres complètes d'Hippocrate* ed. E. Littré [Paris, 1839–61], VII:406); and Diosc. *Mat. Med.* III.47 (*Pedanii Dioscuridis Anazarbei "De Materia Medica,"* ed. M. Wellmann, [Berlin, 1906–14], II:60). E. Steier, "Moly," *RE,* vol. XVI (Stuttgart, 1933), cols. 29–33. Jerry Stannard, "The Plant Called Moly," *Osiris* 14 (1962): 254–307, esp. 255–56. G. M. Germani, "Ulisee, Circe, e l'erba moly nella tradizione Omerica e nella realità poetica," *Rassegna di clinica, terapia, e scienze affini* 14 [1934]: 168–172) offers mandrake (*Mandragora officinarum*), which can be rejected; likewise the identification of Homer's μῶλυ as black hellebore (*Helleborus niger*) is merely fanciful, e.g., in Oswald Schmiedeberg, *Über die Pharmaka in der Ilias und Odyssee* (Strassburg, 1918), 25–29, esp. 27. Cf. Carnoy, *Dictionnaire* (see n. 2), 180. Robert Graves (*The Greek Myths* [Harmondsworth, 1955], II:367) offers "wild cyclamen" after rejecting John Tzetzes' guess of "wild rue."
25. As does Walter Artelt, *Studien zur Geschichte der Begriffe, "Heilmittel," und "Gift"* (Leipzig, 1937; rpt. Darmstadt, 1966), 38–46, with minor modifications.
26. G. E. R. Lloyd, *Magic Reason and Experience: Studies in the Origins and Development of Greek Science* (Cambridge, 1979), 31.
27. D. Holwerda, ΦΥΣΙΣ (Groningen, 1955), 63.
28. Lloyd, *Magic* (see n. 26), 31, n. 106.
29. *Od.* IV.220–30. This seems to be the earliest reference to compound φάρμακα in the Greek world.
30. *Od.* IV.231–32; cf. Hdt. II.84.
31. *Il.* VIII.306–7.
32. Diosc. *Mat. Med.* IV.64.1 (Wellmann [see n. 24], II:218–19) and IV.64.7 (ibid., p. 221).
33. George Edward Trease and William Charles Evans, *Pharmacognosy,* 11th ed. (London, 1978), 570; Albert F. Hill, *Economic Botany,* 2d ed. (New York, 1952), 260; Varro E. Tyler, Lynn R. Brady, and James E. Robbers, *Pharmacognosy,* 8th ed. (Philadelphia, 1981), 226–27.
34. viz., *Od.* IV.220–32.

35. Guido Majno, *The Healing Hand: Man and Wound in the Ancient World* (Cambridge, Mass., 1975), 108–11; and A. Lucas, *Ancient Egyptian Materials and Industries,* 3d ed. (London, 1948), 401 (on poppy seed oil).

36. *Od.* IV.219–34.

37. *Od.* X.391–94; cf. *Il.,* XI.638–41 (and slightly earlier in line 623): Hecamede mixes a potion (*κυκεών*) for Nestor and his companions; the potion consisted of Pramnian wine, grated goat's cheese, and white barley meal. Delatte argues that this name was attached to the ritual drinking and initiations at Eleusis; see Armand Delatte, *Le Cycéon: Breuvage rituel des mystères d'Éleusis* (Paris, 1955); see also A. Delatte, *Herbarius* (Liège, 1938), 1–6 for women in legendary roles as drug compounders.

38. Macrob. *Sat.* V.19.9–10 (quotations from Sophocles' lost play, *Rootcutters*); Sen. *Medea,* passim, but esp. 705–70; Petron. *Sat.* 137–38; and Lloyd, *Science* (see n. 1), 76–79 for women as sources of medical information. But almost all of the names of "herbalists" and experts on folk medicines (including matters of sexual nature), as given by Theophrastus, Pliny, Galen, and Athenaeus are those of men, not women. Friedrich Pfister, in "Pflanzenaberglaube," *RE,* XIX, pt. 2 (1938), 1446–56, esp. 1448–49.

39. E.g., C. G. Jung, "Commentary on the Visions of Zosimus," in *Alchemical Studies,* trans. R. F. C. Hull (Princeton, 1967), 66–108, esp. 99 on Circe; Jolande Jacobi, "Symbols in an Individual Analysis" in *Man and His Symbols,* ed. C. G. Jung (London, 1964), 272–303, esp. 283: ("The pig is closely associated with dirty sexuality [Circe . . . changed the men who desired her into swine]"); Wolfgang Lederer, *The Fear of Women* (New York, 1968), 35 ("Odysseus . . . refused to yield to Circe's advances on the grounds that his vigor would be impaired"), 57 ("Passive innocence of the dangerous woman is far less common than her active destructiveness" [with the example of Circe]), and 126 (Circe as a death goddess). Cf. Robert Graves, *The White Goddess* (New York, 1966), 448–49; idem, *Greek Myths* (see n. 24), I.177, with sources. Lederer and Graves provide a full panoply of references to the vast literature in comparative mythology, anthropology, and psychoanalytic theory.

40. Trease and Evans, *Pharmacognosy* (see n. 33), 569; Tyler et al., *Pharmacognosy* (see n. 33), 225–26.

41. W. Burkert, "Itinerant Diviners and Magicians: A Neglected Element in Cultural Contacts," in *The Greek Renaissance in the Eighth Century* B.C., ed. R. Hägg (Stockholm, 1983), 115–20 and idem; *Die orientalisierende Epoche in der griechischen Religion und Literatur,* SBHeidelberg (Heidelberg, 1984), no. 1, 43–48. One may argue that a "familial control" over sacred drug lore could explain why Polydamnia is called the daughter of Thon, but Theophrastus (*Inquiry into Plants,* IX.15.1) expresses his basic skepticism regarding these tales.

42. Pl. *Symp.* 186d. The Hippocratic *Ancient Medicine,* I.11 (*Hippocrates,* ed. W. H. S. Jones [London, 1923–31], I:12).

43. Arist. *Pol.* 1282a3: "'Physician' [*ἰατρός*] is both the ordinary practitioner [*δημιουργός*] and the master of the art of medicine [*ἀρχιτεκτονικός*]."

44. See nn. 107–13.

45. esp. Pl. *Resp.* 406a, 408a and *Ion* 538c.

46. E.g., esp., Celsus, *Med.* proem 3 and Galen, *Thrasyb.* 33 (ed. G. Helmreich, in *Claudii Galeni Pergameni Scripta Minora* [Leipzig, 1884–93], III:78) after quoting *Od.* IV.230–31.

47. *Sed vulneribus tantummodo ferro et medicamentis mederi.*

48. Celsus, *Med.* proem 9, with 6.

49. Wesley D. Smith, "Physiology in the Homeric Poems," *TAPA* 97 (1966): 547–56, esp. 547.

50. Onians, *Origins* (see n. 10), 13–65.

51. Andreas Plaitakis and Roger C. Duvoisin, "Homer's Moly Identified As *Galanthus*

nivalis L.: Physiologic Antidote to Stramonium Poisoning," *Clinical Neuropharmacology* 6 (1983): 1–5. Cf. suggestions in the references in n. 24 above.

52. See, e.g., Clifford Geertz, "Curing, Sorcery, and Magic in a Javanese Town" in *Culture, Disease, and Healing*, ed. David Landy (New York, 1977), 146–54.

53. E.g., Melford E. Spiro, "The Exorcist," in *Burmese Supernaturalism* (Englewood Cliffs, N.J., 1967), 230–45.

54. *Il.* IV.189–219; V.401–2, 889–90; XI.828–47; *Od.* XIX.457.

55. Wesley D. Smith, "So-Called Possession in Pre-Christian Greece," *TAPA* 96 (1965): 403–26.

56. Onians, *Origins* (see n. 10), 489–90.

57. Among many examples, see Klaus Stopp, "Medicinal Plants of the Mt. Hagen People (Mbowamb) in New Guinea," *Economic Botany* 17 (1963): 16–22 and, generally, John Mitchell Watt, "Magic and Witchcraft in Relation to Plants and Folk Medicine," in *Plants in the Development of Modern Medicine*, ed. Tony Swain (Cambridge, Mass., 1972), 67–102.

58. See n. 14.

59. *Od.* IV.568. Cf. *Il.* V.795, X.575, and XIII.84.

60. M. R. Wright, *Empedocles: The Extant Fragments* (New Haven, 1981), 261, frag. 101 (= frag. 111 DK), as adapted and slightly altered from Wright's translation. The text (apud Diog. Laert. VIII.59), as edited by Wright (p. 133), is basically identical to that provided by Diels-Kranz, *Fragmente* I:353–54. "Drugs" as a translation of *φάρμακα* is more fitting than Wright's "remedies" and certainly less restrictive than the "Gifte" of Diels-Kranz.

61. Hes. *Works and Days* 41. (*Hesiod: Works and Days*, ed. M. L. West [Oxford, 1978], 153, commentary).

62. West on *Hesiod, Works and Days* 733–734. Cf. Plut. *Mor., Dinner of the Seven Wise Men*, 157D–158B for Hesiod as learned in medicine. The debate over the "magic food," *ἄλιμον*, occupied much learned discussion in antiquity: Walter Burkert, *Lore and Science in Ancient Pythagoreanism*, trans. Edwin L. Minar, Jr. (Cambridge, Mass., 1972), 151 with n. 174; to Burkert's list of refs. on *ἄλιμον* one can add Porph., *Life of Pythagoras* 34 (*Porphyre: Vie de Pythagore. Lettre à Marcella*, ed. Edouard des Places [Paris, 1982], 52), which gives a twelve-ingredient formula for *ἄλιμον*, including poppy seeds, squill, asphodel, and honey.

63. Lloyd, *Ideology* (see n. 1), 119–20.

64. Hesiod, *Works and Days* 824.

65. *Scholia to Apollonius of Rhodes* IV.156 (= Musaeus frag. 2 DK). *Juniperus oxycedrus* yields a distillation of the heartwood, an oil called Oil of Cade, used in the Middle East as an antiseptic and as an external parasite killer; see George Usher, *Dictionary of Plants Used by Man* (London, 1974), 329.

66. Apollonius of Rhodes, *Voyage of the Argo*, IV.156–58.

67. Musaeus frag. 19 DK.

68. R. Merkelbach and M. L. West, eds., *Fragmenta Hesiodea* (Oxford, 1967), p. 173, frag. 349 in "Fragmenta dubia."

69. Dioscorides IV.132 (Wellmann [see n. 24], II:277) not surprisingly retains some of the folklore about *τριπόλιον*, writing that "it is written that the flower changes its color three times a day." Dioscorides recommends its white root, mixed in two drachmas of wine, as a diuretic and that the root is "cut into antidotes." In southern European folk medicine, the root of *Aster tripolium* has been used to cure eye diseases. Oleg Polunin, *Flowers of Europe* (Oxford, 1969), no. 1365 (p. 427 with plate 142).

70. Alan Cameron, "The Date and Identity of Macrobius," *JRS* 56 (1966): 25–38.

71. Verg. *Aen.* IV.513–14; trans. C. Day Lewis, *The Aeneid of Virgil* (Oxford, 1952), 96.

72. Macrob. *Sat.* V.19.9 (*Macrobius*, ed. J. Willis [Leipzig, 1970], I:326).

73. Macrob. *Sat.* V.19.10 (= Soph. frag. 534 Pearson).

74. Theophr., *Inquiry into Plants* IX.8.5; Scarborough, "Theophrastus" (see n. 1), 359. Dioscorides IV.153.2 (Wellmann [see n. 24], II:298–99).

75. John Gerarde, *The Herball or Generall Historie of Plantes,* vol. 2 (London, 1633), p. 1030.

76. Theophr., *Inquiry into Plants,* IX.8.3.

77. Theophr., *Inquiry into Plants* IX.20.3. Apparently, Athenian cows were immune to its poisonous properties.

78. Soranus, *Gynecology,* II.2.2 (*Sorani Gynaeciorum,* ed. J. Ilberg Corpus Medicorum Graecorum [Leipzig, 1927], p. 51). Soranus flourished in the reign of Trajan (98–117 A.D.).

79. Jeffrey Henderson, *The Maculate Muse* (New Haven, 1975), 135: "Pubic hair is almost always conceived in agricultural terms as a flowering growth . . . . Pennyroyal is jokingly used by Lysistrata . . . to refer to the Boeotian girl's neatly dipilitated campus muliebris, with a clever reference to the smooth, fertile plains of that region . . . with neatly trimmed pennyroyal plots."

80. Ibid., 186 with n. 137. The pennyroyal potion is prescribed for too much *ὀπώρα* (fruit or sex). Apparently, pennyroyal was hawked in the city along with reed mats, eels, and the like. Aristoph. *Ach.* 861, 869, and 874.

81. John Scarborough, *Facets of Hellenic Life* (Boston, 1976), 179–85; idem, "Nicander's Toxicology," pt. II, "Spiders, Scorpions, Insects, and Myriapods," *Pharmacy in History* 21 (1979): 3–34 and 73–92 (esp. 74–75 with nn. 240–246 and 249–254).

82. E.g., *Nature of Woman* 32 (*Oeuvres complètes d'Hippocrate,* ed. E. Littré [Paris, 1839–61], 8:364) among many refs; see X:751 for index entries for *pouliot.*

83. Dioscorides III,31.1 (Wellmann [see n. 24], II:40).

84. Gal. *Properties and Mixtures of Simples* VI.3.7 (Kühn [see n. 11] 11:857). Cf. Gal. *Treatment by Venesection* 18 (Kühn, 11:304); trans. Peter Brain, *Galen on Bloodletting* (Cambridge, 1986), p. 93.

85. E. Steier, "Minze," *RE, XV.*2 (1931), 2020–28, esp. 2027–28. Nic. *Alex.* 128–29, links pennyroyal with the infamous aphrodisiac called "Spanish fly," made from the elytra (wing covers) of blister beetles; see Scarborough, "Nicander's Toxicology," pt. II (see n. 81), 73–74.

86. A. Tschirch (*Handbuch der Pharmakognosie,* vol. II, pt. 2 [Leipzig, 1917], 1107–08) summarizes the literature to the early twentieth century.

87. R. G. Todd, ed., *Extra Pharmacopoeia Martindale,* 25th ed. (London, 1967), 1544.

88. Malcolm Stuart, ed., *The Encyclopedia of Herbs and Herbalism* (London, 1979), 223–24. Cf. Trease and Evans, *Pharmacognosy* (see n. 33), 412–14 and R. D. Mann, *Modern Drug Use* (Boston, 1984), 138.

89. Gal. *Properties and Mixtures of Simples* III.23 (Kühn [see n. 11], XI:609); John Scarborough, "Some Beetles in Pliny's *Natural History,*" *Coleopterists Bulletin* 31 (1977): 293–96.

90. E. Rohde, *Psyche,* 2d ed., trans. W. B. Hillis (London, 1925), 198, n. 95; J. Murr, *Die Pflanzenwelt in der griechischen Mythologie* (Innsbruck, 1890), 31–35 and 104–6.

91. Theophr., *Inquiry into Plants* VII.13.4.

92. Theophr., *Characters* XVI.14.

93. Parker, *Miasma* (see n. 10), 307.

94. Theophr., *Characters* XVI.2.

95. Theophr., *Characters* XVI.10.

96. Jerry Stannard, "Squill in Ancient and Medieval Materia Medica, with Special Reference to its Employment for Dropsy," *Bulletin of the New York Academy of Medicine* 50 (1974): 684–713.

97. Theogn. 537 (in *Elegy and Iambus,* ed. J. M. Edmonds [London, 1931], I:292).

98. Hipponax frag. 48 in *Herodes, Cercidas, and the Greek Choliambic Poets,* ed. A. D. Knox (London, 1961; pt. 2 of vol. entitled *Theophrastus: Characters,* ed. J. M. Edmonds) 34) as translated by Knox.

99. John Tzetzes, *Chiliades* V.726–61 (*Ioannis Tzetzes Historiarum variarum Chiliades,* ed. T. Kiessling, [Leipzig, 1826], 185–86).

100. John Tzetzes, *Chiliades* 5.743 (Kiessling, p. 185). Although modern scholars debate the reliability of later compilations regarding the *φαρμακός* rituals, one may certainly trust the main lines (squill and its importance) in summaries by such as John Tzetzes. See Jan Bremmer, "Scapegoat Rituals in Ancient Greece," *HSCP* 87 (1983): 299–320. Bremmer's arguments, however, that all such plants are symbolic (e.g., esp. 302–13: such herbs had no "fruit," and thereby justified the brutal treatment of "nonfruitful" or "marginal" persons) ignores the plentiful evidence of squills and their use in medicine and herbal remedies.

101. John Tzetzes, *Chiliades* 734–36 (Kiessling, p. 185).

102. *Ἄγνος*, as in Plut. *Mor., Table Talk* 693F, in which we learn that at Chaeronea a slave is beaten with branches of the *Vitex agnus castus* to drive away a "great hunger" termed *βούλιμος*. The "honey" of *Vitex agnus castus* was famed in antiquity for its use as a wound healer (Arist., *HA* 627a7–9), and the fruit of the "tree" (*Vitex agnus castus* is more of a bush) was valued as an emmenagogue, contraceptive, and sedative; see Diosc. *Mat. Med.* I.103.1–2 (Wellmann [see n. 24], 1:95). Traditionally, the leaves and fruit ("Monks' peppers") are sexual suppressants. Usher (see n. 65), 602. Stuart, *Herbs* (see n. 88), 282–83. Walter H. Lewis and Memory P. F. Elvin-Lewis, *Medical Botany* (New York, 1977), 322.

103. Diosc. *Mat. Med.* II.171.3 (Wellmann [see n. 24], I:239).

104. *Septuagint: Exod.* VII.11 and *Malach.* III:5.

105. Theophr., *Inquiry into Plants,* VIII.1.

106. Scarborough, "Theophrastus" (see n. 1), 356–57.

107. Stannard, "Squill" (see n. 96), 689.

108. Theophr. *Inquiry into Plants* VII.12.1; Hort (see n. 2), II:477 s.v.

109. Apul., *Apol.* 27 (Helm [see n. 18], 31).

110. Einarson, Introduction to *De Causis Plantarum* (see n. 3), xix-xx.

111. Diog. Laert. *Lives of the Eminent Philosophers: Epimenides* I.112.

112. The texts give contradictory evidence: Pl., *Laws* I.642d4 indicates a floruit of c. 500 B.C.; Arist., *Constitution of the Athenians* 1 points to a floruit of c. 600 B.C.. See commentary by Rhodes (see n. 7), 81–82.

113. Stannard, "Squill" (see n. 96), 687–89.

114. Lucian, *Menippus, or Descent into Hell* 7. Cf. E. Steier, "Skilla," *RE* Suppl. III.1 (1927), 522–26, esp. 524.

115. E.g., *Geoponica* XV.1.6–7 (*Geoponica,* ed. H. Beckh [Leipzig, 1895], 432).

116. Parker, *Miasma* (see n. 10), 231–32.

117. Ibid., 232.

118. Theophr., *Inquiry into Plants* I.6.7–9; I.10.7; II.5.5; VII.2.2; VII.4.12; VII.9.4; and VII.13.1–7.

119. Scarborough, "Theophrastus" (see n. 1), 355–60.

120. In folk medicine, both the common mallow, *Malva rotundifolia* L., and high mallow, *Malva sylvestris* L., retain their usefulness; the dried leaves of both species contain mucilage and tannin, making them quite suitable for preparations to treat inflamed tissues and coughs. A decoction of the flowers (*Flores malvae*) is employed for gargling and as a mouthwash. Usher (see n. 65), 375; Todd, *Martindale* (see n. 87), 1533.

121. Lycoph. *Alex.* 582. Nic. *Ther.* 496 and 587; Nic. *Alex.* 112.

122. The Hippocratic *Fractures,* X.3 (*Hippocrates,* ed. E. T. Withington [see n. 42], III.120–21); Nic. *Ther.* 493.

123. Theophr., *Inquiry into Plants* IX.18.2.

124. Ibid. 3. With οὐ μόνον τῶν σωματικῶν ἀλλὰ καὶ τῶν τῆς ψυχῆς, the Greek text in Hort's edition breaks off, presumably due to its offensive content—offensive to a squeakily prudish mind of the late Victorian era (the omitted passages contain only the mildest references to aphrodisiacs and a single, rather bland notice of an "Indian drug" that promotes multiple erections in the male). Sarton rightly comments that "Such prudishness in a scientific book is truly shocking." George Sarton, *A History of Science,* vol. I (Cambridge, Mass., 1959), 555, n. 96. See also above, n. 2.

125. Polunin, *Flowers of Europe* (see n. 69), 577 no. 1894. Todd, *Martindale* (see n. 87), 81.

126. Theophr., *Inquiry into Plants* IX.18.3 (Wimmer [see n. 2], 160).

127. Jerry Stannard, "Medicinal Plants and Folk Remedies in Pliny, *Historia Naturalis,*" *History and Philosophy of the Life Sciences* 4 (1982): 3–23, esp. 14–15, with refs. to sources and secondary works.

128. Anthony Preus, "Drugs and Psychic States in Theophrastus' *Historia Plantarum* 9.8–20," in *Theophrastean Studies,* ed. W. W. Fortenbaugh and R. W. Sharples, Rutgers University Studies in Classical Humanities 3 (New Brunswick, 1988), 76–99.

129. Theophr., *Inquiry* IX.18.4 (Wimmer [see n. 2], 160).

130. Ibid.

131. Theophr., *Inquiry* IX.19.2–4.

132. The pharmacological effects of squill (*Urginea maritima* Baker) are fairly well understood. It has a digitalislike action on the heart, and in small amounts is employed occasionally as an expectorant. The outer scales are removed from the bulb. In Europe, a powdered form of rat poison is manufactured from squill. It contains not less than 70 percent alcohol[60%]-soluble extractive, as well as goodly percentages of scillarin A and scillarin B (A is a pure crystalline glycoside; B a mixture of glycosides). Todd, *Martindale* (see n. 87), 692. Trease and Evans, *Pharmacognosy* (see n. 33), 504–6. G. Baumgarten and W. Förster, *Die Herzwiksamen Glykoside* (Leipzig, 1963), 70–75.

133. John Scarborough, *Pharmacy's Ancient Heritage: Theophrastus, Nicander and Dioscorides* (Lexington, Ky., 1985), 19–20.

134. Ibid., 6, and Scarborough, "Theophrastus" (see n. 1), 355–59.

135. See nn. 74–77.

136. Theophr., *Inquiry into Plants* IX.8.6. The particular hellebore (white or black) is not distinguished.

137. Ludwig Edelstein, "Greek Medicine in Its Relation to Religion and Magic," in Owsei Temkin and C. Lilian Temkin, eds., *Ancient Medicine: Selected Papers of Ludwig Edelstein* (Baltimore, 1967), 205–46 (quotation from p. 230); orig. *Bulletin of the History of Medicine* 5 (1937): 201–46.

138. Ibid.

139. Ibid., 231.

140. Ibid., 231–32.

141. Ibid., 232.

142. John Scarborough, *Roman Medicine* (London, 1969), 119–20.

143. John M. Riddle, *Dioscorides on Pharmacy and Medicine* (Austin, 1985): 14–24.

144. Edelstein, "Magic," in *Ancient Medicine* (see n. 137), 232, n. 88.

145. Ibid., 230.

146. Ibid., 231.

147. Gal., *Compound Drugs According to Place in the Body* VIII.2 (Kühn [see n. 11], 13:126), quoting from the works of Andromachus on remedies for upset stomach.

148. Gal., *Compound Drugs According to Kind* VII.6 (Kühn, XIII:986).

149. Theophr., *Inquiry into Plants* IX.8.1.

150. Scarborough, "Theophrastus" (see n. 1), 355–56 with nn. 10–17.

151. Theophr., *Inquiry into Plants* IX.8.2.

152. Scarborough, "Theophrastus" (see n. 1), 356–57 with texts.

153. Pseudo-Aristotle, *Problems,* 864b9–12; John Scarborough, "Theoretical Assumptions in Hippocratic Pharmacology," in *Formes de pensées dans la collection hippocratique: Actes du IVe Colloque international hippocratique Lausanne . . . 1981*, ed. F. Lasserre and P. Mudry (Geneva, 1983), 307–25, esp. the section 'Aristotelian Drug Theory' with nn. 7–28.

154. Benedict Einarson, "The Manuscripts of Theophrastus' *Historia Plantarum,*" *CP* 71 (1976): 67–76, esp. 68–69 with the readings from manuscript U*; Scarborough, "Theophrastus" (see n. 1), 353–54, with summary of contrary views.

155. John Scarborough, "Nicander's Toxicology", pt. I, "Snakes," *Pharmacy in History* 19 (1977): 3–23 and pt. 2, "Spiders, Scorpions, Insects, and Myriapods," 21 (1979): 3–34, 73–92.

156. Max Wellmann, "Das älteste Kräuterbuch der Griechen," in *Festgabe für Franz Susemihl* (Leipzig, 1898): 1–31 and idem, "Apollodoros" [no. 69], in *RE*, I.2 (1894), 2895.

157. John Scarborough, "Nicander, *Theriaca,* 811," *CP* 75 (1980): 138–40.

158. Scarborough, "Beetles" (see n. 89), 293–96; idem, "Nicander's Toxicology", pt. II (see n. 81), 73–78.

159. See, e.g., the Hippocratic *Diseases of Women* I.75–79 (Littré [see n. 24], VIII:162–99).

160. Riddle, *Dioscorides* (see n. 143), 137–38.

161. See esp. the Hippocratic *Ancient Medicine* as contrasted with the Hippocratic *Nature of Man*.

162. See n. 24.

163. John Scarborough and Vivian Nutton, "The *Preface* of Dioscorides' *Materia Medica:* Introduction, Translation, Commentary," *Transactions and Studies of the College of Physicians of Philadelphia* n.s. 4 (1982): 187–227 (esp. 192–194).

164. Ibid., 213–17; Riddle, *Dioscorides* (see n. 143), 2–4.

165. Scarborough, *Pharmacy's Ancient Heritage* (see n. 133), 63–65, 72–76 on silphium; Scarborough and Nutton, "*Preface*" (see n. 163), 199–202.

166. Riddle, *Dioscorides* (see n. 143), 94–131.

167. John Scarborough, "The Drug Lore of Asclepiades of Bithynia," *Pharmacy in History* 17 (1975): 43–57; Scarborough and Nutton, "*Preface*" (see n. 163): 206–8.

168. Dios. *Mat. Med.* 2, Preface (Wellmann [see n. 24], 1:2); translation from Scarborough and Nutton, "*Preface*" (see n. 163), 196.

169. Ibid., 5 (Wellmann [see n. 24], 1:3); translation from Scarborough and Nutton, "*Preface*" (see n. 163), 196.

170. Diosc. *Mat. Med.* IV.162.4 (Wellman [see n. 24], II:308–9).

171. Most likely to Zeus, through the eagle, but "the mythology of the eagle baffles analysis"; D'Arcy W. Thompson, *A Glossary of Greek Birds* (Oxford, 1936), 12.

172. See above pp. 348–52 with notes 91–117.

173. Parker, *Miasma* (see n. 10), 215–216.

174. But cf. Bennett Simon, *Mind and Madness in Ancient Greece* (Ithaca, N.Y., 1978), 218–19, 226, and 229.

175. See esp. *Sacred Disease,* II-IV (Jones, *Hippocrates* [see n. 42], III:140–51). One passage suggests the tone and approach of the writer: "My own view is that those who first attributed a sacred character to this malady were like the magicians, purifiers, charlatans and quacks of our own day, men who claim great piety and superior knowledge" (*Sacred Disease* II.1–5; trans. Jones, p. 141). Black hellebore is specifically prescribed by a Hippocratic physician in treatment of delirium accompanied by visual hallucinations in the *On Internal Diseases* 48 (Littré [see n. 24], VII:284–89, esp. 286–87). See also Owsei Temkin, *The*

Falling Sickness, 2d ed. (Baltimore, 1971), 68, 73, 76, 79, and 162 on later uses of hellebores in treatment of epilepsies.

176. Riddle, *Dioscorides* (see n. 143), 124, no. 162 in table 8, with refs. Cf. Scarborough, "Theophrastus" (see n. 1), 361–62.

177. The research of Vivian Nutton has proven the traditional dates for Galen (130–200 A.D.) wrong. Vivian Nutton, "The Chronology of Galen's Early Career," *CQ* 23 (1973): 158–71; idem, "Galen in the Eyes of His Contemporaries," *Bulletin of the History of Medicine* 58 (1984): 315–24.

178. John Scarborough, "Early Byzantine Pharmacology," in *Symposium on Byzantine Medicine,* ed. John Scarborough, Dumbarton Oaks Papers 38 (Washington, D.C., 1985), 213–32 (esp. the section "Galen's Pharmacy," 215–21).

179. Vivian Nutton, "The Drug Trade in Antiquity," *Journal of the Royal Society of Medicine* 78 (1985): 138–45 (see esp. 145). A similar characteristic is observed in the drug recipes of Criton, a major source for Galen's drug lore. John Scarborough, "Criton, Physician to Trajan: Historian and Pharmacist," in *The Craft of the Ancient Historian: Essays in Honor of Chester G. Starr,* ed. John W. Eadie and Josiah Ober (Lanham, Md., 1985), 387–405 (see esp. 394–97).

180. Scarborough, "Byzantine Pharmacology" (see n. 178), 215–16.

181. See, esp., *Nature of Man* IV–V (Jones, *Hippocrates* [see n. 42], IV:10–15).

182. E.g., the Egyptian compound-ingredient incense called κῦφι; Scarborough, "Byzantine Pharmacology" (see n. 178), 229–32.

183. See, e.g., Michael J. Harstad, "Saints, Drugs, and Surgery: Byzantine Therapeutics for Breast Diseases," *Pharmacy in History* 28 (1986): 175–80.

184. Gary Vikan, "Art, Medicine, and Magic in Early Byzantium," in *Byzantine Medicine,* ed. Scarborough (see n. 178), 65–86.

185. John Scarborough, "Hermetic and Related Texts in Classical Antiquity," in I. Merkal and A. G. Debus, eds., *Hermeticism and the Renaissance: Intellectual History and the Occult in Early Modern Europe* (Cranbury, N.J., 1988), 19–44.

186. Hans-Viet Friedrich, ed., *Thessalos von Tralles: griechisch und lateinisch* (Meisenheim am Glan, 1968). The most recent discussion of Thessalus in his "Egyptian" and "Hermetic" setting is Garth Fowden, *The Egyptian Hermes* (Cambridge, 1986), 161–65.

187. P. Boudreaux, ed., *Catalogus Codicum Astrologorum Graecorum,* vol. VIII, *Codicum Parisinorum,* pt. 3 (Brussels, 1912), 132–65 ("Excerpta ex codice 41" [Paris. gr. 2256]).

188. A. J. Festugière, *Hermétisme et mystique païenne* (Paris, 1967) for a French translation of parts of the *Letter.*

189. A. J. Festugière, *La Révélation d'Hermès Trismégiste,* vol. I, *L'Astrologie et les sciences occultes* (Paris, 1950), 150.

190. Friedrich, *Thessalos* (see n. 186), 43.

191. R. Reitzenstein, *Die hellenistischen Mysterienreligionen,* 3d ed. (Leipzig, 1927), 127–28.

192. Boudreaux, *Catalogus* (see n. 187), 139–51.

193. Ibid., 153–63; but cf. Pfister, "Pflanzenaberglaube" (see n. 38), 1452.

194. Festugière, *Révélation* (see n. 189), 137–86.

195. Galen (Kühn [see n. 11]), IX:903–13.

196. Scarborough, *Roman Medicine* (see n. 142), 120.

197. Cf. Ptol. *Tetrab.* III.12 (*Ptolemy: Tetrabiblos,* ed. and trans. F. E. Robbins [London, 1940], 316–33).

198. *Poimandres* I.27 (*Corpus Hermeticum,* ed. A. D. Nock and A.-J. Festugière, vol. I [Paris, 1946], 16).

199. *Powers of Herbs* II.1 (Friedrich [see n. 186], 199 and 203.

200. *Cichorium intybus* Cf. Theophr., *Inquiry into Plants* VII.7.1; Diosc. *Mat. Med.*

II.132 (Wellmann [see n. 24], 1:205–6); Pliny, *Natural History* XIX.129, XX.73, and XXI.88; *Geoponica* XII.28 (Beckh [see n. 115], 375–76): from Didymus; Columella, *Agriculture,* XI.3.27 and XII.9.2.

201. Apparently "as a drug." Diosc. *Mat. Med.* I.98 (Wellmann [see n. 24], 1:89).

202. Cf. Diosc. *Mat. Med.* I.99 (Wellmann [see n. 24], 1:90–91).

203. Cf. Galen (Kühn [see n. 11]), VI:604 VIII:343.

204. Cf. *PGM* XXXIII.

205. Ibid., and *PGM* VII.211–12.

206. *Stachys germanica L.;* Diosc. *Mat. Med.* III.106 (Wellmann [see n. 24], II:118); Pliny, *Natural History* XXIV.136.

207. *Crocus sativus L.;* Diosc. *Mat. Med.* I.26 (Wellmann [see n. 24], I:29–31).

208. *Pistacia lentiscus L.;* Theophrastus, *Inquiry into Plants* IX.1.2.

209. *Zingiber officinale* Rosc.; Diosc. *Mat. Med.* II.160 (Wellmann [see n. 24], I:226).

210. *Piper nigrum L.;* Diosc. *Mat. Med.* II.159 (Wellmann [see n. 24], I:224–26).

211. This is the famous *ἄσφαλτος*; Diosc.*Mat. Med.* I.73 (Wellmann [see n. 24], I:72–73).

212. Here *γλυκάνισον* = *ἄννησον*, which is *Pimpinella anisum L.; Scholia on Theocritus,* VII.63d (*Scholia in Theocritum Vetera* , ed. C. Wendel [Leipzig, 1914; rpt. Stuttgart, 1967], 95).

213. Made from *Pistacia lentiscus L.;* (see n. 208).

214. Cf. Diosc. *Mat. Med.* V.8 (Wellmann [see n. 24], III:12).

215. Parallel texts listed in n. 200.

216. Festugière, *Révélation* (see n. 189), 146–60.

217. Ibid., 150–52.

218. John Scarborough, "Roman Pharmacy and the Eastern Drug Trade," *Pharmacy in History* 24 (1982): 135–43.

219. Ibid., 137, with nn. 20–31.

220. Frederick H. Cramer, *Astrology in Roman Law and Politics* (Philadelphia, 1954), 232–83 (on drugs and magic, see pp. 276–78). See also the chapter "Astrologers, Diviners, and Prophets," in R. MacMullen, *Enemies of the Roman Order* (Cambridge, Mass., 1967), 128–162.

221. See *PGM.* The magical tracts in Greek and Coptic (with occasional and clipped commentary) of the Preisendanz texts, have recently been restudied and translated into English by a team of scholars under the general direction of Hans Dieter Betz; added to the Greek and Coptic texts of the Preisendanz collection are important translations (by Janet H. Johnson) of demotic magical papyri; the first volume of translations of the *PGM* has been published as Hans Dieter Betz, ed., *The Greek Magical Papyri in Translation,* vol. I, *Texts* (Chicago, 1986) [= *GMPT*]. Volume II will be indexes to the English translations as well as the Greek and Coptic texts.

222. Pfister, "Pflanzenaberglaube" (see n. 38).

223. *PGM* IV.286–95, trans. E. N. O'Neil in *GMPT* (see n. 221), 1:43.

224. *PGM* IV.2967–75, trans. E. N. O'Neil in *GMPT* (see n. 221), 1:95.

225. Packets of natron were placed in Egyptian mummies from about 2000 B.C. "Natron" is a compound of sodium carbonate and sodium bicarbonate and, mixed with fats, forms sodium soaps; Majno, *Healing Hand* (see n. 35), 131–36. Resins were also essential in Egyptian mummification; ibid., 137–38.

226. J. G. Frazer, *The Golden Bough,* vol. IV, *Adonis Attis Osiris,* 2nd ed. (London, 1907), 74–75 on milk, 221–22, 231–33, and 340–43 on pine cones, etc.

227. *PGM* IV.2979–89, trans. E. N. O'Neil (with some clipped commentary) in *GMPT* (see n. 221), 1:95.

228. Betz, Introduction to *GMPT* (see n. 221) 1:xlviii.

229. *PGM* XXXVI.320–32, trans. John Scarborough in *GMPT* (see n. 221) 1:277.
230. Diosc. *Mat. Med.* IV.68 (Wellmann [see n. 24], II:224).
231. Diosc. *Mat. Med.* II.108 (Wellmann [see n. 24], I:182).
232. Riddle, *Dioscorides* (see n. 143), 57.
233. Diosc. *Mat. Med.* IV.68.2 (Wellmann [see n. 24], II:225).
234. Todd, *Martindale* (see n. 87), 170–71.
235. *PGM* XXXVI.324, n. 40 in *GMPT* (see n.221) 1: 277.
236. *PGM* XXIIa.11–14, trans. John Scarborough in *GMPT* (see n. 221) 1:260.
237. Soranus, *Gynecology,* III.42.3 (Ilberg [see n.78], 121).
238. Diosc. *Mat. Med.* V.126 and 130 (Wellmann [see n. 24], III:94–95 and 96–97).
239. Gal., *Mixtures and Properties of Simples* IX.11 (Kühn [see n.11], XII:204).
240. Pliny, *Natural History* XXVI.127.
241. *Geoponica* XV.1.28 (Beckh [see n.115] 435).
242. Dioscorides V.124 (Wellmann [see n. 24], III:92); Gal., *Mixtures and Properties of Simples* IX.9 (Kühn [see n. 11], XII:202); Celsus, *Med.* V.7. Pliny, *Natural History,* XXXVI.131.
243. When published, *GMPT* vol. II (see n. 221) will contain separate indexes of plants, animals, etc.
244. *PGM* XII.401–44, trans. H. D. Betz and John Scarborough (with some clipped commentary) in *GMPT,* (see n. 221), 1:167–69.
245. *PGM* IV.1313 ("priestly Egyptian incense" as trans. W. C. Grese in *GMPT* [see n. 221], 1:63); as *PGM* IV.2971 (see n. 224); *PGM* V.221 (transliterated by E. N. O'Neil as *kyphi* [*GMPT* I:204]); *PGM* VII.538 ("sacred incense" as trans. R. F. Hock [*GMPT* I:133]); and *PGM* VII.873 ("lunar ointment" as trans. E. N. O'Neil [*GMPT* I:141]).
246. G. Ebers, "Ein Kyphirecept aus dem Papyros Ebers," *Zeitschrift für ägyptische Sprache und Altertumskunde* 12 (1874): 106–11. Scarborough, "Byzantine Pharmacology" (see n. 178), 231.
247. Full discussion and identification of ingredients in a twenty-eight-substance formula in Scarborough, "Byzantine Pharmacology" (see n. 178), 230–32.
248. Diosc. *Mat. Med.* I.25 (Wellmann [see n. 24], I:28–29); translated, with ingredients identified, in Scarborough, "Byzantine Pharmacology" (see n. 178), 230.
249. *PGM* IV.1275–1322, trans. W. C. Grese in *GMPT* [see n.221], 62–63).
250. Hippol. *Refutation of all Heresies* IV.28 (*Origenis Philosophumena sive Omnium Haeresium Refutatio,* ed. E. Miller, Patrologiae Graecae, vol. XVI, pt. 3 [Paris, 1863], col. 3090). The *Refutation,* formerly attributed to Origen, is now firmly among the works of Hippolytus; F. L. Cross, ed., *Oxford Dictionary of the Christian Church* (London, 1957), 641–42.
251. Hippol. *Refutation* IV.28 (Miller, cols. 3090–91).
252. Brian Vickers, Introduction to *Occult and Scientific Mentalities in the Renaissance,* ed. Brian Vickers (Cambridge, 1984), 1–55, esp. 3, following Charles Schmitt.
253. Fowden, *Egyptian Hermes* (see n. 186), 79 with n. 19, 162–65, 168.

6

Dreams and Divination in Magical Ritual

† *Samson Eitrem*

When he died at age ninety-three on July 8, 1966, Samson Eitrem, professor emeritus of classical philology at Oslo University, left an unfinished manuscript of over seven hundred pages entitled *Magie und Mantik der Griechen und Römer,* written for the renowned *Handbuch der Altertumswissenschaft.*[1] Its intention was to give an exhaustive treatment of both magic and divination, topics that Martin P. Nilsson, Eitrem's contemporary (1874–1967), had touched upon in much shorter form in his *Geschichte der griechischen Religion* in the same *Handbuch* (3d ed., vol. 1 [1967]; 2d ed., vol. 2 [1961]). Eitrem was ideally suited for this task. Being a general classical philologist with an interest in, and knowledge of, archaeology as well, his scholarly activities were concerned especially with two fields: papyrology and the history of religion. The first manifested itself already in his first publication, an article on Bacchylides in the Oslo newspaper *Morgenbladet* in 1898 (a year after Kenyon had published his fundamental edition of the fragments); the second flourished early in the still-valuable monograph *Opferriten und Voropfer der Griechen und Römer* of 1915. The two fields merged in the study of ancient magic. From a trip to Egypt in 1920, Eitrem had brought back several papyri, among them magical ones, purchased from his own funds and donated to the Oslo University Library. After a thorough study of the major extant magical papyri in Paris, Berlin, and London, which yielded new readings and interpretations (1923), Eitrem edited the four Oslo magical papyri with translation and commentary (1925 and again for Preisendanz' *Papyri Graecae Magicae,* to which Eitrem was recruited as a collaborator shortly after World War I).

Eitrem's interest in the magical papyri stemmed from the same sources as his general interest in Greek and Roman religion—the tradition of German *Religionswissenschaft* as founded by Herman Usener (1834–1905) and continued by his pupil and son-in-law Albrecht Dieterich (1866–1904)—although Eitrem had never studied in Bonn (where Usener had taught) or in Heidelberg (where Dieterich taught) but in Berlin, Halle, and Göttingen with, among others, Wilamowitz, Diels, and Carl Robert (he dedicated *Opferriten* to Diels). Wilamowitz, for one, abhorred the "horrible superstitions of the magical papyri" as a sign of the decay of an old religion ("wenn die alte Religion in Verwesung ist und der wüste Aberglaube der Zauberpapyri sich an ihre Stelle drängt" [*Der Glaube der Hellenen,* vol. 1 , p. 10]). Albrecht Dieterich had not only edited one of the Leyden Papyri (*PGM* XII), he had demonstrated the relevance the papyri could and did have for the history of ancient religion (esp. in *Abraxas* of 1891 and *Eine Mithrasliturgie* of 1903). Hermann Usener had provided the theoretical framework by highlighting the importance of popular religion for an accurate understanding of Greek and Roman beliefs and ritual. (Magic formed a vital part of popular religion, thus the magical papyri were important documents.) The other leading figure of the period was, of course, Sir James Frazer (1854–1941) whose evolutionary view of magic and religion dominated the age. (Eitrem's Frazerian evolutionism is apparent, for example, below on p. 179 with nn. 37–38 where he refers to magical ritual as the nearly self-evident basis and background to the Homeric conception of dream apparitions.)

Frazerian evolutionism has been long since dismissed and superseded by other approaches to religion. Eitrem's work nevertheless remains in most part valid and nearly everywhere interesting. His was, fortunately, a philological and descriptive approach, a way of presenting the material that was only rarely affected by outmoded theories. As Festugière had put it in his obituary: "Ce qui ne passe pas, c'est l'exactitude dans l'édition des textes . . . et c'est la sûreté dans l'intepretation." Eitrem's magnum opus, to be published in a revised and completed form in the hopefully not-too-distant future, deserves some editorial care and scholarly attention. The chapter that follows (previously unpublished) has been excerpted from this work. It valuably reflects the state of the art at the time when Eitrem wrote it and for this reason has been printed as is, with only minor editorial additions and updated notes.

For Eitrem's bibliography, as far as scholarly works are concerned, see Leiv Amundsen in *Symbolae Osloenses* 43 (1968):110–23; for the obituary see Festugière, *CRAI* (1966):413–17.

Fritz Graf

MAGICAL DREAM POWER

New and abundant material regarding magical dream visions has been provided by Egyptian papyri. Here we find an astonishing wealth of practices for either inducing a dream (*ὀνειραιτητά*) or causing someone else to have a dream (*ὀνειροπομποί*).[2] Here Greek and Egyptian practices merge, as might be expected in this syncretistic milieu.[3] We find Apollo and Hermes side by side with Ra, Thoth, Bes, Isis, and every imaginable daemon—laurel and olive branches mixed with native Egyptian plants, and the tripod with magical dolls and magical songs. Christian angels make their first appearance in these texts. All the intellectual and material tools of coercion (*Zwangsmittel*) familiar to us from this brand of magic find their place here: the great name, the powerful names, magical formulae, letters, designs, and so on. Lamp or lantern magic (*Lampenzauber*) plays a major role here as well as generally in Egyptian magic—for light, the nocturnal sun, was something to be exploited. The night with its horde of dead spirits and eerie ways—the night through which the sun god navigated in his vessel to reach the east through the dark kingdom of the underworld while the moon shone or the heavens were starry—offered the magician the best opportunity for exercising his art or arts. We have very simple instructions as to how the desired dream might be had, then again we find extremely complicated practices devised with all the finesse of magical wisdom and requiring longer time and greater expenditures. Ritual "cleanliness" or "purity" is everywhere the overall important prerequisite; but here one may even intensify the demand for it.

The following examples will make this clear. The inscribed spell of "Pitys the Thessalian" for the "interrogation of a corpse" is very simple. Two magic words are written in magic ink on a flax leaf, and the leaf is then stuffed into the mouth of a corpse. The context makes clear that the dead body enabled the magician either to dream himself or to transmit dreams to others.[4] Hermes, the Greek dream sender and guide of the dead will then appear as called. In another prescription a Hermes is painted with the blood of a quail on a strip of linen; but here Hermes has the face of an ibis and is therefore identified with the Egyptian Thoth. The name is added in myrrh ink; then the god ("he whom the god of gods set above the spirits") is invoked briefly, together with his parents, Osiris and Isis. Mystical names known only to the practitioner are pronounced. The invocation ends, "Tell me about the matter at hand,[5] about everything that I wish to know." The practitioner then lies down to sleep in the belief that the god would make his appearance.[6]

In its main features the outline of this ritual remains the same nearly everywhere. In another spell an inscription including the magic name and presumably the entire invocation is to be written on a papyrus leaf and placed under a lamp. Then the practitioner is to go to sleep "in a pure state."[7] In another ritual, a tin tablet previously inscribed with the invocation *lords, gods* (of whom the dreamer is the slave) then crowned with myrtle and carried round a burnt offering of frankincense is placed beneath one's pillow[8] (this version thus omits the lamp). A good illustration of the sophisticated etiquette of "union" with a divinity (*systasis*) is provided by a partially hexametric invocation of Helios-Apollo that solemnly apostrophizes the god's soothsaying laurel. A *systasis* with the Moon completes this consecration of the dream-bringing night. Here precise timing is indicated: the prayer is said toward either the east or west on the second or, better, the fourth day of the new-moon period. One might imagine oneself in a purely Greek milieu if the magical formulae were not mixed in with it (there is even an invocation of Sabaoth).[9]

A very detailed ritual—involving an invocation of Apollo, the smoke of incense on an altar and a lamp ("that has not been colored red") placed on a wolf's head—shows how Apollo retains his position as a great divinatory god even in this Greco-Egyptian oriental magic.[10] A laurel branch is held in the right hand, an ebony staff in the left (the staffs are shifted to the other hands when one wishes to rid oneself of the divinity who appears). The "heavenly gods and the daemons of the earth" are called upon, "the holy and divine names" are pronounced, in order that they may send to the dreamer "the divine spirit"—once again in good hexameters mixed with magical formulae. This is not a common type of dream demand. But the text includes the claim that the god, for whom a throne is prepared, is able to provide information in the form of a general prediction "about dream sending, dream requests, and dream explanation (*ὀνειροκριϲία*)."[11]

We again recognize Apollo's tripod in a "dream vision" in which there is no mention of Apollo's name.[12] Three reeds are plucked from the ground while a magic formula is recited,[13] the [particular] purpose of the oracle request is stated as the third one is plucked. Then they are written upon with a magic ink compounded of seven substances; at the same time, the practitioner recites what he requests. Then a lamp "that has not been colored red" is filled with pure oil. A wick is fashioned from pure material and magic names are inscribed on it, while they are pronounced seven times facing the lamp; then the lamp is turned toward the east. Lumps of frankincense are offered up in a censer; then, finally, the reeds are put together to form the tripod (date palm fibers should be used in its construction). Then one crowns one's head with olive branches. Although this can hardly be called an "autopsy", the heading might well be *Charm for seeing Apollo with one's own eyes,* as at *PGM* VII.727–39, where one is to sleep in a room on flat ground and *without light,* even though Helios is invoked with the relevant magical formulas.[14] Ancient Greek magical practice seems preserved in an invocation of the dangerous Hecate[15] (now named Hecate Ereschigal[16]) who is summoned at night at an intersection of three roads: "She will give you *in a dream* all the information you desire, even if you are in the face of death." Then one must leave the intersection quickly.

The following recipes or formulas are very complicated.[17] One is entitled *Pythagoras' request for a dream oracle and Democritos' mathematical* (i.e., *astrological*) *dream divination.*[18] Here one invokes a star angel (named Zizaubio) of the "all-ruling Pleiades." This angel is subordinated to Helios and appears in the form of a friend of the dreamer.[19] A laurel branch with twelve leaves is used in the invocation. On each leaf a sign of the zodiac is traced with a magic word and character (each leaf is numbered). The name of the god is written on a special laurel leaf. The practitioner wraps these in a new "sweat cloth," which is placed under his head[20] for three nights while he sleeps. On the last day, facing west, he offers frankincense, invoking the angel and the twelve other angels of the Pleiades and Helios. Finally the laurel branch is held over burning incense and then bent around one's head as a crown; this phylactery with its power-charged name should then remain near the head of the sleeper.

In a double version of this rite there is a dream request addressed to "the weak-sighted Bes."[21] This popular god of unusual form and horrible appearance gives protection against everything evil, against the evil eye, and in particular against everything that disturbs the sleeper.[22] The invocation, which is written in an ink composed of seven (in the other version, nine) ingredients[23] is pronounced while facing the lamplight and is addressed first to a particularly power-charged daemon, "the headless god whose countenance is at his feet."[24] After the invocation comes the conjuration wherein the "two (secret) names of Bes" (Anouth Anouth) are recited solemnly, followed by the command to "predict without deception, without treachery."[25] During the performance of the ritual the magician holds a black "Isis cloth" (i.e., from the garment of an Isis statue) in the left hand and also places such a cloth around his neck "so that the god may not strike (him)" (*PGM* VII.232). A figure of Bes is drawn on papyrus with the remaining ink by the left hand;[26] when Bes is ordered to leave, the drawing is erased with the Isis cloth.[27] The expectation is that the god will appear only toward morning.[28]

Two interesting dream requests are addressed to Hermes, invoked as early as *Homeric Hymn* 4.14 as ἡγήτορ' ὀνείρων (bringer of dreams). Both spells are very elaborate. In the first[29] a figure of Hermes is fashioned from a specially blended dough.[30] The time for preparing the figure is also indicated (when the moon is in

Aries, Leo, Virgo, or Sagittarius). This figure of Hermes is brought to life by a prayer written on hieratic papyrus (or, according to the rival version of the spell, on the windpipe of a goose!) that has been slipped inside the figurine. Then the Hermes figure (wearing a mantle and holding a herald's staff) is placed in a small shrine made of linden wood. If a dream prediction is desired, the prayer and the question about the future are written out once again (along with powerful magic formula accompanied by magic words) and placed at the feet of the statuette.

This is presumably a simple, miniature replica of the normal rite of temple incubation. A special incense offering[31] is also to be made before one goes to sleep. The accompanying hymn, worthy of a theurgy having the all-encompassing power and wisdom of Hermes, is of remarkable interest, since it refers to the god as the "eye of Helios," the divine Oneiros[32] (the oracle speaks by day and night), the elder son of Mneme, and invokes him with titles familiar to us from the teachings of Asclepius Soter.[33] For a comparison with such magical prescriptions we have the theurgic procedure reported by Porphyry as a Hecate oracle, which is relatively simple. A figure of Hecate is fashioned from rue (*Ruta graveolens*) and this is "purified" in a special way and placed in a small laurel-wood shrine. Consecration with an accompanying prayer takes place at night by the waxing moon. The prayer is repeated over and over and concludes, "Appear to me in sleep."[34]

DREAM TRANSMISSION AND THE PAPYRI

At least as significant as the dream request was dream transmission (*ὀνειροπομποί, ὀνειροπομπία*). The antiquity of this type of dream magic must have been very great indeed in Greece. When Zeus as Lord of Dreams in Homer (*Il.* 2.63) sends Oneiros to Agamemnon (2.6), or Athena sends an *eidōlon* in the form of a friend to Penelope (*Od.* 4.795f.), these Olympian divinities are only doing what an experienced practitioner of magic had been doing for ages. The fact that dreams could be altered at will by others and that dream images could assume the likeness of this or that person who would awaken the deepest trust in the dreamer (the dream image appeared to Agamemnon as Nestor), only shows that this dream technique known from later sources[35] went back very far in Greece.[36] It suits Olympian religion that dreams (the significance of which affected both high and low alike, both god and poet) were under the control of the Olympian deities; in other words the entire dream technique with its coercion of spirits and magical offerings underwent restriction and modification in Olympian religion and yet a certain recollection of magical practice remained.[37] The Olympians did not make an effort to bring dreams themselves (Hermes *ὀνειροπομπῶν*[38] is of a later date). The many dreams became, in Homer's graphic clarity, the personified "dream," Oneiros, who is a divinity adapted to Olympian society. The Homeric poet also uses Hypnos (*Il.* 14.231), whereas the practitioner of magic preferred Eros as *ὀνειροπομπῶν*. The matters with which Zeus and Athena concerned themselves are of an entirely different level than the egoistic trivialities with which the magician dealt to satisfy the wishes of his clients. The practitioners of magic even pressed the Olympians into service in order that they might direct dreams in the proper direction, namely, according to the wishes of the master magician. As might be expected, most of the gods and daemons who were used for the purpose of dream requests were also used in transmitting dreams even when this is not stated specifically in the prescriptions.[39]

The moon goddess in all her different aspects is found active in both kinds of procedures.[40] In the transmission of dreams she appears to the individual concerned in the form of that divinity or daemon to whom "NN" habitually prayed (*PGM* IV.2500). In order that the moon goddess be of service, a magic likeness of the "Egyptian Lady Selene" can be fashioned from a magic mixture (potter's clay, sulfur, and the blood of a spotted goat, which is the mount of Selene-Hecate[41]). The figure is anointed and crowned and is placed in a small shrine made of olive wood, erected late at night in the fifth hour, facing the moon. An offering of incense is made to Selene and a coercive prayer is addressed to her. The moon goddess is asked to send a different angel at each of the twelve hours of the night (it is easily understandable why it is the moon who rules these nocturnal hour-angels).[42] The entire procedure, entitled *The lunar spell of Claudianus,* has the objective of leading the loved one to the practitioner (i.e., it is an ἀγωγή), but the text includes the claim that the practice is also useful in magical binding and dream transmission.[43]

The power of an "attendant" (πάρεδρος, *PGM* I.1 and 37) or the use of an "assistant" (παραστάτης, *PGM* IV.1849f.)[44] is also extensive. The "assistant," as one of the many spirits or stellar angels or daemons of the dead,[45] may contribute anything, including dream transmissions and revelation by dreams. A recipe is recorded in the name of the magician Agathocles that works by means of a "violently slain" (drowned?) tomcat. A small piece of papyrus on which the magical formula and the oracular request are written (preferably in myrrh ink) is put into the mouth of the cat, and the ritual formula with the "great name" Aoth[46] is enchanted. The cat represents Helios-Osiris. Its body is used in another detailed description of a magical procedure that asks Helios (in particular) at sundown and sunrise for dream transmission, among other things.[47] This is followed in our papyrus by another spell ascribed to a certain Zminis of Tentyra (i.e., Dendera on the Nile).[48] A winged daemon with the horns of a bull and the tail of a bird,[49] with a diadem on its head and swords at its feet, is used forcibly for dream transmission. The daemon is drawn on a piece of linen and the powerful name is added. Other "sacred names" of the god are uttered into a lamp filled with cedar oil; in addition, the Agathos Daemon is apostrophized and Seth is invoked. The hour of birth and the 365 names of the "great god" are pronounced under a grim threat of severe punishment "so that I not be forced to say this twice": "Tell him (NN) such and such" is the order, when the god appears in a dream before the given individual.[50]

In another spell[51] a hippopotamus—the beast of Typhon-Seth—is fashioned from reddish wax; gold, silver, and iron are inserted in its belly and the figure is placed at a clean window. Here the dream that one wishes to transmit is written on hieratic papyrus,[52] which is then rolled into a wick placed in a new lamp. The foot of the hippopotamus is placed on the lamp, the name is pronounced, and the dream is transmitted. So also the Ouroboros, the serpent that bites its own tail and that is carved on a heliotrope stone and worn as a ring, makes its wearer capable of all types of magic; consequently the wearer also masters ὀνειροπομπία.[53]

There is also an isolated practice whereby the practitioner can himself appear to a woman in her sleep (*PGM* VII.407f.). Reference is made to magic words spoken to the "lamp used daily" and repeated several times over (CHEIAMOPSEI EREPEBOTH), and one says briefly, "Now let NN see me in sleep, now, quick, quick" and adds whatever one habitually wishes.[54]

Most often the transmission of the dream is entrusted to other, more powerful dream images. In one of the spells for a direct revelation mentioned above (*PGM* IV.3205) the daemon (a lamp daemon) even enters into the practitioner (crowned with olive branches) and reveals everything to him—an extraordinary compromise between the independently active dream soul and the magical daemon world, comparable with "possession."[55]

The instructions handed down in the papyri for the transmission of dreams may be supplemented by the description of the use of a dream-sending sympathy doll that prefaces the Alexander-romance (*Historia Alexandri Magni*).[56] Here Nectanebos, a former Pharaoh and traditionally a master in all kinds of magic[57] appears in the form of Ammon to the Macedonian queen Olympias, not only in her nocturnal dreams but also by day. He convinces the queen about the compatibility of their horoscopes and tells her in advance what will happen to her in her dreams; then he departs, gathers quickly the requisite magic plants, makes a magic doll out of wax, naming it Olympias, and a small bed. He lights a lamp filled with an oil into which he has blended the sap of plants and utters the necessary invocations into the lamp for *ὀνειροπομπία* (in one variation he summons daemons). Since the queen wishes that her nocturnal experience be repeated by day, the prophet grants her wish: Alexander the Great (the issue of this union) is consequently born divine as a son of Ammon. Just as Nectanebos had predicted, he now transforms himself into a hissing serpent, into Ammon, Heracles, Dionysus (all of whose powers are thereby transmitted to the yet-to-be-conceived Alexander). Naturally, there was no place here for the usual *ἀγωγή*.[58] It is perhaps possible that the sympathy doll might not have been present in the original magical operation.[59] But a medieval treatise substantiates that it did indeed have a place in the transmission of dreams:[60] a doll of that type was fashioned after the investigation of the planetary constellation, the doll being given the name of the dreamer and adorned with the symbol of Hermes, and the names of both Hermes and Selene. The doll is then told what the individual should dream about. It is further remarked that the content of the dream truly came to pass, both Olympias' and many others' dreamed in the *Asclepieia* and later Christian healing shrines.

Also in the Alexander-romance (*Historia Alexandri Magni*) there is a remarkable dream transmission whereby Nectanebos calms the suspicious King Philip (chap. 8 p. 8 Kroll). He takes a falcon (*ἱέρακα πελάγιον*, probably a sea hawk),[61] performs his magical arts on the bird (i.e., he kills it, as was the usual practice in similar instances; see Porph. *Abst.* 4.9), and sends it through the night to "bring the dream" to Philip. Here a dramatic scene is enacted in the dream for the king. As spectator of the action in the dream (compare Hom. *Od.* 19.535f.), Philip sees how the god Ammon embraces the queen and also receives the verbal explanation from the bird.[62] Here again Olympias' Egyptian prophet predicted the confirmation of Philip's dream.[63] The dual task of the deified bird must be due to the fabular nature of the story.

Christians were horrified by the transmission of dreams and the related magic of sorcerers. Accordingly, in their view evil daemons were invented to beguile the weak and subjugate them to their power. Justin Martyr (*Apol.* 18.3) and Irenaios (*C. Haer.* 16.3) protested such daemon-mania. The latter cited *ὀνειροπομποὺς δαίμονας* one after the other and *ὀνειροπομποὶ καὶ πάρεδροι δαίμονες*. A general, detailed discussion on the value of visions and dreams takes place in Pseudo-Clement (*Hom.* 17.13f.) between Peter and Simon Magus. Here the dream as a source of truth and spiritual enlightenment is emphatically denied by Peter: an evil daemon had in this [situation] the best opportunity to pass himself off as being sent by God. Previously Peter had expounded to his listeners that it was precisely in dreams that the daemons assumed the likeness of gods, in order to receive the adoration and offerings accruing to those same gods (9.15).[64] Simon counters these arguments by saying that a pious man might see truth in a dream while anyone who was not so pious would not see truth in anything (17.15). It was by such reasoning that religious belief was substituted for psychic gifts and somatic conditions, whereby dreams took on apocalyptic significance in matters of religion. We see how the disputes among the ancients on the subject of dream interpretation continued in a lively fashion in Christian circles. Peter maintained that fears and desires call forth dreams, which are then shaped either by a daemon or one's own psyche.[65] He knew that no god appeared in dreams to Jews, simply because they did not believe in such gods.[66] It should be pointed out that Simon and Carpocrates with their disciples were discredited precisely because of such magical arts, which involved daemons and dream transmission with erotic overtones. Hippolytos drew upon Irenaios,[67] and Eusebius relied upon the authority of Irenaios.[68] Tertullian, too, is outraged by mischief that involved the invocation of angels and daemons (*Apol.* 23.1). One understands the indignation of the apologists when one considers that there were Christians who believed in mantic dreams without reservation. On the other hand Christians were convinced that God could reveal his will and his counsel to men in dreams. This is taught already in the Old Testament: God bestowed his exceptional grace by this means on the God-fearing. The mother of Augustine received comfort and sound counsel in this way at times of extreme spiritual need (*Conf.* 3.11, 5.9, 6.1). But Monica thought she knew exactly which dreams were of divine origin and which found their cause in her human, sinful soul (*Conf.* 6.13).[69]

Notes

English translation by D. Obbink of "Magischer Traumzwang" and "Traumsendung und die Papyri," two chapters from an unpublished monograph by S. Eitrem, with a preface by F. Graf.

1. For notices of the projected volume see H. G. Gundel and W. Gundel, *Astrologumena,* Sudhoffs Archiv 6 (Wiesbaden, 1966), ixf.; and Zeph Stewart in *La società ellenistica: Economia, diritto, religione,* Storia e civiltà dei Greci 8, ed. R. Bianchi Bandinelli (Milano, 1977), 509, n. 8.

2. In *PGM* I.329 the "Divine Spirit" invoked is much vaunted, because he is extraordinarily helpful in *ὀνειροπομπία*, *ὀνειραιτησία* and *ὀνειροκρισία* (sic) and in general in all magical experience. *ὀνειραιτητά* are dream requests [i.e., for revelations in dreams]; *ὀνειροπομποί* are spells for transmitting dreams (see index in *PGM,* vol. 3).

3. The relevant Greek papyrus texts, in addition to three demotic ones were published, translated, and discussed by Th. Hopfner, *Griechisch-ägyptischer Offenbarungszauber* [*OZ*], 2 vols., Studien zur Palaeographie und Papyruskunde 21, 23 (vol. 1: Leipzig, 1921, repr. Amsterdam, 1974; vol. 2: Leipzig 1924, partial repr. Amsterdam 1983) vol. 2 , sec. 162–211.

4. *PGM* IV.2140ff., see also 1950, 2078. The "Demotic magical papyrus of Leiden and London" (*The Leyden Papyrus,* ed. F. L. Griffith and H. Thompson [New York, 1974], 113–117 at verso col. 17. 1f.) says that a sedge leaf inscribed with magic symbols is placed under the head and "calls forth dreams (in the magician) and transmits dreams." If the leaf is placed in the mouth of a mummy, the mummy will transmit the dream.

5. I.e., either the matter already mentioned earlier in the invocation or "such and such," meaning that the practitioner here supplies his request; cf. *PGM* XXIIb.35.

6. *PGM* XII.144–51. Another ibis-faced Hermes appears at VIII.10. On Hermes in these contexts see G. Fowden, *The Egyptian Hermes: A Historical Approach to the Late Pagan Mind* (Cambridge, 1986), 22–31, esp. 25–26. Another short dream request is addressed to the constellation of the Bear. With oil in one's left hand, her secret names are pronounced; then one goes to sleep, facing the sunrise (*PGM* XII.190–92). (*Jesus* heads the list of magic names.) Still shorter are two other practices (both from the fourth century A.D.): In *PGM* XXIIb.27–31 one repeats several times over in "whatever light is in daily use" (thus no "new lamp" or other is required) a short invocation until the light is extinguished. In 32–35 the last morsel of bread or meat is shown to the light, a brief *logos* is recited, the morsel is eaten, and a little wine is drunk; then one lies down to sleep "without speaking to anyone" (μηδενὶ λαλήσας). Otherwise one might read "without answering to anyone" (as, e.g., at VII.440, 1011). In this way one keeps the curious from disturbing one in order to preserve the sacred stillness or taboo of the dream (Hopfner, *OZ,* vol. 2, sec. 171) states incorrectly that the dreamer should not answer any question put by the visiting divinity). On ritual silence see, further, O. Casel, *De Philosophorum Graecorum Silentio Mystico,* RGVV vol. 16, pt. 2 (Giessen, 1919); G. Mensching, *Das Heilige Schweigen* (Giessen, 1926); F. Sokolowski, *Lois sacrées des cités grecques,* Supplement (Paris, 1962), 115 B 54 (Cyrene); W. Burkert, *Homo Necans,* trans. P. Bing. (Berkeley, 1983), 220, 223, 290.

7. *PGM* VII.703–26.

8. *PGM* VII.740–55 (cf. 1016, where the multiple invocation "Michael, Raphael, Gabriel" remains uncertain). At VII.843ff. a laurel branch as amulet is to be incensed and placed by the head; the practitioner is similarly instructed to "sleep pure" and admonished that the place of performance must be "absolutely pure."

9. *PGM* VI.1–47; for Sabaoth, see VI.33; see also S. Eitrem, "Die *systasis* und die Lichtzauber in der Magie," *SO* 8 (1929): 49–51. For further details see Hopfner, *OZ* vol. 2 sec. 171.

10. *PGM* I.263–347. On the cult of Apollo in Greek magic see S. Eitrem, "Apollo in der Magie," in *Orakel und Mysterien am Ausgang der Antike,* Albae Vigiliae 5 (Zürich, 1947), 47–52. The laurel branch should have seven leaves, and a magic symbol is to be inscribed on all of them. This branch, otherwise an attribute of the god and his ἱκέτης, is here described as "the body's greatest protective charm" (1.272). One should keep oneself free of all uncleanliness, abstaining from fish eating and cohabitation ("in order to excite the god into the greatest possible desire for you," I.290) and be robed in a prophet's apparel (I.278). In addition to offering incense (with a wolf's eye—a plant?—and various spices as a burnt offering) there is a libation of wine, honey, milk, and rain water and two sets of seven cakes. The linen cloth that serves as a wick for the lamp is to be inscribed with magic symbols.

11. *PGM* I.329ff. The hexametrical part is reconstructed as *PGM* Hymn 23; not all the hexameters are defective.

12. *PGM* IV.3172f., esp. 3197, "Make the three reeds into a kind of tripod." The papyrus calls this practice an ὀνειροθαυπτάνη (cf. IV.2624–25), i.e., ὀνειραυτοπτική, al-

though in the related invocation at IV.3206 it is said that the divinity addressed (ὑπαρέτης, instead of ὑπηρέτης?) will enter into the dream and reveal all. The numbers three and seven have particular power and must be observed throughout (the four cardinal points toward which one turns did nothing to change this).

13. The MASKELLI formula with the seven vowels and magic words is spoken while facing east, south, and west (and probably to the north by turning around). For the significance of the cardinal points in uprooting a plant see A. Delatte, *Herbarius,* 3d ed. (Bruxelles, 1961), 68.

14. Here the wearing of sandals made of wolf's leather is stipulated, together with a crown of marjoram (see Hopfner, *OZ* vol. 1, sec. 494). At his appearance Apollo comes already bearing a cup for a drink offering; he does not receive a gift but, if so asked, gives one to the dreamer/petitioner (VII.736: "if you ask, he will let you drink from his cup").

15. *Pap. Michigan* III.154 (third to fourth century A.D.) (= *PGM* LXX.4–25).

16. Ereshigal = Erisch-ki-gal, the Babylonian goddess of the dead: see M. Jastrow, *Die Religions Babyloniens und Assyriens* vol. II, pt. 2 (Giessen, 1912), 712, n. 3; Hopfner, *OZ* vol. 1, sec. 177.

17. All of these are contained in the long, important *Pap.* CXXI of the British Museum (third to fourth century A.D.) (= *PGM* VII).

18. *PGM* VII.795–845.

19. Asclepius appears in the dream to the tutor of Aristeides in the form of the Roman Consul Salvius (Aristeid. *Or.* LVIII.9 Keil). The saints Cosmas and Damian appear as physicians (p. 173, line 2 Deubner), as priests (p. 145, lines 42f. Deubner), as other individuals (pp. 187, line 8; 188, line 25 Deubner). They may also appear to physicians treating the patient (e.g., p. 178, line 7; 171, line 11 Deubner).

20. Cf. *PGM* VII.748.

21. *PGM* VII.222–49; VIII.64–110.

22. See A. Erman, *Die Religion der Ägypter* (Berlin, 1934), index s.v. *Bes* and pp.147, 395 (with plates). Also Hopfner *OZ*, vol. 2, sec.185 (with plate); A. Delatte, "Akephalos," *BCH* 38 (1914): 201f.; idem, *Musée Belge* 18 (1914) 53; (with plate 2).

23. *PGM* VII.226. The ordinary ink for writing is used.

24. See K. Preisendanz, *Akephalos. Der kopflose Gott,* Beihefte zum Alten Orient 6 (Leipzig, 1926), 44–50; A. Delatte, "Akephalos," *BCH* 38 (1914):221–32; K. Abel, "Akephalos," in *RE* Suppl. vol. 12 (1970), p. 13.

25. The prayer calling on the divinity or daemon concerned to reveal the truth is repeated, for example, in calling on Hermes (*PGM* V.431) or Helios (XIVa.6, where it is emphasized that the oracle should answer "without equivocation" [ἀναμφόγως]) or in a lamp divination (IV.1034). The fear of *misleading* spirits was common and widespread, particularly in connection with dreams.

26. *PGM* VIII.65f.

27. An extension and greater refinement of this Bes ritual is to be found at *PGM* VIII.64f. At the beginning there is an impressive hexametric appeal to Helios—that he send forth from the kingdom of the dead "the sacred daemon Anouth Anouth"—followed by an request for a direct revelation (by means of a lamp). One sleeps on a rush mat. A small tablet for writing convenience ("lest after going to sleep you forget") should be available.

28. *PGM* VII.229.

29. *PGM* V.370–446.

30. Twenty-eight leaves (i.e., 7 × 4) of a laurel bush ("though a man from Heracleopolis recommended the olive tree to me"), virgin soil (i.e., soil that has not yet been used), wormwood seeds, wheat meal, dogbane grass (suitable for Thoth-Hermes, on whom see n. 6), and the liquid of an ibis egg are blended (5.370ff.).

31. *PGM* V.394f.: to incense is added soil from a field where wheat has been grown, together with a handful of ammonium salt (three ingredients). On the significance of these elements, see the essay by F. Graf (chap. 7, p. 196 with nn. 64–65. On temple incubation in general, see the treatment of the Dream of Nectanebos by L. Koenen in *BASP* 22 (1985): 171ff.

32. Hermes is likewise referred to as Oneiros in the parallel versions of this hymn (*PGM* VII.675 and XVIIb.10).

33. *PGM* V.413: "You heal all pain of mortals," and 416: ἱλαρὸς φάνηθι. The recipe is repeated in VII.664–85 but in a much simpler form. The shrine for Hermes is omitted; one speaks (seven times) into the light of a lamp. Instead of papyrus the writing is done on a strip of linen (with myrrh ink); this is then wound around an olive branch and is placed on the head. The magical formulas vary and are shorter (there is no compulsory conclusion to the invocation). It is likely that an "uncorrupted youth" is to be employed as a medium: at VII.679f. the "uncorrupted youth" is said to be used by the divinity (see, e.g., II.56, VII.554, XIV.68 and 287; but compare V.416ff.). In the parallel recipe at V.375f. appropriate magic substances are to be used by a παῖς ἄφθορος for creating the magic figure. See Th. Hopfner, "Die Kindermedien in den griechisch-ägyptischen Zauberpapyri," in *Recueil d'études dédiées à la mémoire de N. P. Kondakov* (Prague, 1926), 650–74; R. Ganschinietz, *Hippolytos' Capitel gegen die Magier, Refut. haer. IV.28–42*, Texte und Untersuchungen vol. 39, pt. 3, (Leipzig, 1913), 30, 32–33.

34. Euseb. *Praep. Evang.* VII; G. Wolff, *Porphyrii De Philosophia ex Oraculis Haurienda* (Berlin, 1856), 130f.; on the substance rue, see p. 195f. For a discussion of the many parallels between the activities of the theurges and those described in *PGM*, see S. Eitrem, "Die *systasis*" (see n. 9) and idem, "La théurgie chez les Neoplatoniciens et dans les papyrus magiques," *SO* 22 (1942): 49–79.

35. *Il.* 2.6, 63.

36. Preisendanz judges otherwise: "Oneiropompeia," *RE* vol. 18, pt. 1 (1939), esp. pp. 440f.

37. In Homer as in the later magic texts the dream image or likeness appears "above the head" of the dreamer: see *PGM* IV.2335, with verbal repetition of the Homeric expression (Preisendanz [see n. 36], 441). It depended on the dreamer "seeing" the dream vision, as the Greeks always stated. Sleep closes the eyes of those who rest, as Homer says, e.g., *Il.* 14.236.

38. First perhaps in *Schol. Od.* 7.38; cf. *PGM* 1.98. Galen uses the word at XII.251 Kühn.

39. At *PGM* XVIIa.15 Anubis is induced by means of a magical spell of attraction. He is called upon to exercise his power over the object of desire; he also has power to show ἐνύπνια (nighttime visions) and ὄνειροι (dreams).

40. *PGM* VII.862–918.

41. W. H. Roscher, *Selene und Verwandtes*, 2d ed. (Leipzig, 1895), 43; Hopfner *OZ* vol. 1, sec. 423.

42. A magic name was given to each "hour-angel," who necessarily had power only for a limited period of time: Menebain, Lemnei, Nouphier, etc. (*PGM* VII.900ff.).

43. *PGM* VII.878.

44. Here the attendant-assistant is a winged Eros fashioned from the wood of a mulberry tree. In the hollow back a small gold leaf is inserted with the inscription "MARSABOUTARTHE— be my assistant and aid and sender of dreams." With the statuette of Eros one knocks on the door of the beloved (*PGM* IV.1854). For a parallel instance see the charm entitled *Eros as assistant: consecration and preparation* at *PGM* XII.14–95.

45. I.e., necromancy; see for example *PGM* IV.2076f., from the *Spell of attraction of King Pitys* (IV.1928–2005).

46. *PGM* XII.107–21. At the end of the text we are told that this formula was "also used by Apollobex"—another well-known sorcerer, Apollobex of Coptos (Pliny, *HN* 30.9, Apul. *Apol.* 90); cf. Hopfner, *OZ* vol. 2 sec. 210 with p. 250.

47. *PGM* III.1–164. At the end it reads, "good for charioteers in races, transmission of dreams, love magic, and magic to cause separation (διάκοπος) and to arouse hatred."

48. *PGM* XII.121–43: "Draw on pure linen—according to Ostanes—with myrrh ink," etc. Commonly Ostanes is thought the source for the entire prescription, which thus relies on two authorities: see Bidez-Cumont, *Les mages héllenisés,* vol. 2 (Paris, 1938), 307 and K. Preisendanz, "Zminis," *Roscher* 6 (1936) 762.

49. For comparable magical figures see Hopfner, *OZ* vol. 1, sec. 212f.

50. *PGM* XII.136.

51. *PGM* XII.308–18.

52. *PGM* III.314. Here the ink is a myrrh solution with the blood of a baboon (i.e., the sacred beast of the moon god Thoth-Hermes: Hopfner, *OZ* vol. 1, sec. 429). On the hippopotamus, see Hopfner, *Der Tierkult der alten Ägypter nach den griechisch-römischen Berichten und den wichtigeren Denkmälern,* Denkschr Wein, Philosophisch-historische Classe 57.2 (Vienna, 1913).

53. *PGM* XII.271f, 305. (Preisendanz, [see n. 36], 444 is of the opinion that this is rather more an ὀνειροτησία.) On the consecration of the heliotrope stone see S. Eitrem, "Die magischen Gemmen und ihre Weihe," *SO* 19 (1939): 66f.

54. Here the magical procedure enables the lover performing the rite (and who presumably then goes to sleep) to enact the role for which a daemon was usually invoked (as opposed to Preisendanz [see n. 36], 444). Nowadays one would speak of a telepathic dream.

55. In the *Charm of Solomon* (*PGM* IV.850–929) a boy (or adult) is similarly ordered into a medium-like state for the purpose of revelation from Hesies, i.e., the dead Osiris (IV.897). The magical prayer is repeated seven times into the ear of the medium who then collapses in an ecstatic seizure.

56. *Historia Alexandri Magni* (Ps.-Gallisthenes) chap. 1, p. 3f. Kroll. See also M. Pieper, "Nektanebos," *RE* vol.16, pt. 2 (1935), 2238f.; O. Weinreich, *Der Trug des Nektanebos* (Leipzig, 1911); A. Ausfeld, *Der griechische Alexanderroman* (Leipzig, 1907); S. J. Storost, *Studien zur Alexandersage in der älteren italienischen Literatur,* Romanistische Arbeiten 23 (Halle [Saale], 1935); Fr. Pfister, *Kleine Schriften zum Alexanderroman,* Beiträge zur klassischen Philologie 61 (Meisenheim am Glan, 1976), in addition to the works of Nöldeke and Paul Meyer. The Nectanebos episode may be traced back to an Alexandrian novelist using Egyptian elaboration: for this argument, see, e.g., R. Merkelbach, *Die Quellen des Alexanderromans,* 2d ed., Zetemata 9 (Munich, 1977), 77–81; B. Berg, "An Early Source of the *Alexander Romance,*" *GRBS* 14 (1973): 381–87.

57. So also in *PGM* IV.156 Nephotes hails King Psammetichos as "the best σοφιστής" (i.e., professional magician).

58. See S. Eitrem, *Papyri Osloensis,* fasc. 1, *Magical Papyri* (Oslo, 1925), 49–51 for further discussion of the ἀγωγή ritual that aims at "leading" the victim out of her house and into the embraces of the practitioner.

59. In the earliest textual witness (a text exceedingly inaccurate and lacunate) the possibility of transmitting the dream is attributed entirely to magical plants.

60. *Catalogus Codicum Astrologorum Graecorum,* ed. F. Cumont et al., vol. 3 (Brussels 1912), 41 (cf. Preisendanz, [see n. 36], 446).

61. On falcons consecrated to Egyptian deities see Hopfner *OZ* vol.1 sec. 457; the dead falcon is ὀνειροπομπός according to Ael. *NA* 11.39 (though see Preisendanz [see n. 36], 446); cf. S. Eitrem, "Sonnenkäfer und Falke in der syncretischen Magie," in *Pisciculi: Festschrift für F. Dölger* (Münster, 1939), 94–101; H. Bonnet, *Reallexikon der ägyptischen Religionsgeschichte* (Berlin, 1952), 178–180, s.v. *Falke.*

62. Philip is awakened by the sea hawk, which strikes him with its wings (chap. 8.2) in accordance with other magical practices. In the Typhon magic of Nephotes (*PGM* IV.211f.) the sea hawk hits the practitioner with its wings to indicate that he should get up.

63. What Ammon was to predict in a dream about the birth of Alexander was an oracular decree, which Philip had his "oneirocritic" explain to him (chap. 3).

64. Peter substantiates that God might reveal himself in dreams or visions to the ungodly in anger (Ps.-Clem. *Hom.* 17.17). According to Peter, for the godly man true understanding and perception are revelation enough.

65. Ps.-Clem. *Hom.* 9.15.

66. Ps.-Clem. *Hom.* 9.16. Here it is also stated that in every healing process the explanation is either through a dream or "as of daemonic origin."

67. Hippol. Haer. 6.26; Wendland op. cit. 148.

68. Euseb. *Hist. Eccl.* 4.7, 9.

69. On this rather abrupt ending see n. 1 and Preface—Ed.

7

Prayer in Magic and Religious Ritual

Fritz Graf

The relationship between magic and religion has long been a problem widely discussed among historians of religion. Opinions ranged from the one extreme position that magic is different from and in strict opposition to religion if not its most dangerous opponent to the other where the term *magic* is denounced as a semantic trap and altogether expelled from scientific vocabulary.[1] Rather than revive this debate here, I propose to survey some prayers contained in the collection of the *Papyri Graecae Magicae* and ask in what respect, if at all, they are different from prayers in contexts traditionally regarded as non-magical. I shall also ask whether such differences as can be detected confirm the most widespread theory about the difference between magic and religion (at least among classicists), the one made famous by Sir James Frazer, namely, that the magician constrains, coerces, and forces the divinity to do his will, whereas religious man meekly submits himself to God's overpowering will. (The slight denigration of religion is Sir James's.)[2] Among anthropologists, this Frazerian dichotomy is long dead and buried. In classical scholarship however, it loomed very large and still is among us, explicitly[3] or, more often nowadays, implicitly.[4] I choose prayer as the focus of interest because it was, and still is, regarded as the quintessence of religion.[5] Its occurrence in a magical context is liable to pose some problems to those who support the traditional dichotomy.

When the major magical papyri were published during the latter part of the nineteenth century, scholars began to recognize spells, prayers, and hymns whose religious tone was not to be ignored. Partly they could be isolated as ritual texts of mystery cults that the magicians had appropriated, as they had Homeric verses as well; in the case of many prayers and hymns where no internal features provided clear indications as to their provenance but where one thought to detect religious (as opposed to magic) feelings, this one-sided borrowing was unhesitatingly assumed.[6] At least in the generation following the pioneering work of Albrecht Dieterich and Richard Reitzenstein, the Frazerian dichotomy, implicitly devaluating magic (at least from a Christiano-centric point of view), helped to facilitate this interpretation and prevented the apparent permeability between magic and religion from becoming a problem.[7]

It should have been otherwise. To the Greeks, a magician not only uttered spells, he also prayed to the gods: Plato, for one, connects the ἐπῳδαί (spells) and the εὐχαί (prayers) of the magician, both of which helped him to persuade (πείθειν) the gods.[8] In the magical papyri themselves, the usual term for the spoken part of the magical action is λόγος (formula), but the word *εὐχή* (prayer) occurs several times, as do the verb *εὔχομαι* and kindred terms.[9] Although the question is not confined to the formulas labeled *εὐχαί*, as we shall see presently, it seems safer to start the present investigation with them.

I count five instances where *εὐχή* occurs as an actual title of a spell (excluding the strongly Jewish-influenced *εὐχὴ* 'Ιακώβ, *PGM* XII b):[10] two hexametrical hymns[11] and three formulae in prose.[12] Furthermore, there are seven instances where either the formula itself or its context uses this term.[13] In each instance, the overall structure of the text conforms to the general structure of a Greek prayer, like Sappho's famous prayer to Aphrodite.[14] These prayers are tripartite, have an invocation (*invocatio* in the terminology of C. Ausfeld), a narrative middle part (*pars epica* or, as J. M. Bremer terms it, *argumentum*),[15] and a final section that contains the actual wish addressed to the divinity (*preces*).

This formal arrangement is important, since the tripartite structure is functional. The invocation calls the attention of the divinity (most often with the catchphrase *κλῦθί μου*) and invites it to come and participate in the ritual (the usual catchword is *δεῦρο*; Sappho's formula is *τυίδ' ἔλθε*, line 5). The meticulous listing of cult-places, myths, and epithets that follows assures that the divinity is addressed in all its relevant aspects, so that it will feel a real obligation to come.[16] The narrative in the second part gives the credentials of the persons who pray, establishes their right to ask something from the divinity: they refer either to the sacrifices they have performed earlier or to the one they are presently performing. (Both oblige the divinity to come to their aid.) They may also refer to earlier occasions where the divinity had helped, as Sappho does. This establishes the solvency, so to speak, of the petitioner. Finally, after they have caught the attention of the divinity and established their credentials, they may state their specific wishes. This is the standard form; there exist inversions where the wish immediately follows the invocation, and the *pars epica* rounds out the prayer. In this case, the wish is so urgent that it is brought forward as soon as the attention of the divinity is caught. The necessity to establish the credentials of the petitioner, though, persists.

The *εὐχαί* in the papyri conform to this formal pattern, as I said. To give only one example, the hexametrical *εὐχὴ πρὸς Σελήνην* in *PGM* IV.2785[17] begins with the invocation

> *ἐλθέ μοι, ὦ δέσποινα φίλη, τριπρόσωπε Σελήνη*
>
> Come to me, O beloved mistress, three-faced Selene (2786f.)

and asks her to listen to these incantations:

> *εὐμενίηι δ' ἐπάκουσον ἐμῶν ἱερῶν ἐπαοιδῶν*
>
> kindly hear my sacred chants (2787)

—incidentally confirming that (at least here) the magician felt no difference between *εὐχή* (prayer) and *ἐπῳδή* (incantation). There follows a plethora of epithets and circumscriptions of her power, then her identification with Dike, Moira, Persephone, Megaira, Allecto, Hecate, and Artemis; she is addressed as the mistress of the whole cosmos. This part ends with an allusion to an otherwise unknown myth that Cronos had handed over his scepter to her;[18] there follows an address and again the request to listen:

χαῖρε, θεά, καὶ σαῖς ἐπωνυμίαις ἐπάκουσον

Hail, goddess, and attend your epithets (2850)

Then follows a reference to the actual sacrifice:

θύω σοι τόδ' ἄρωμα

I burn for you this spice (2851)

—only to be followed by another list of epicleses that now center around the dark and harmful aspects of the goddess. Finally, the actual wish is brought forward:

ἐλθέ ἐπ' ἐμαῖς θυσίαις καί μοι τόδε πρᾶγμα ποίησον

Come to my sacrifices and fulfill for me this matter (2868)

The nature of "this matter" is left open, since the papyrus, as all these collections of magical recipes,[19] only gives a general instruction that has to be adapted to the individual case at hand. Except the general outline, the *preces* have to be left blank, so to speak. Similarly, personal names referring to the person praying or the other persons concerned are often left open by means of the formula *ὁ δεῖνα* (NN).[20]

Up to this point, it is impossible to describe the difference between magical and religious prayer in Frazerian terms: the two seem interchangeable. Structurally, all the canonical parts are there. It is not peculiar to magical prayers that the invocation contains a long list of epicleses and conflates scores of divinities and that the *pars epica* is rather short: similar features appear in other late religious texts, for instance, the so-called Orphic hymns.[21] More peculiar are the *voces magicae*, not mentioned so far. The hexameter immediately preceding the final *χαῖρε, θεά* of the invocation runs thus

σύ δὲ χαοῦς μεδέεις, αραχαρα ηφθισικηρε

Chaos, too, you rule, ARACHARA EPHTHISIKĒRE (2849)

The second part of this verse consists of a palindrome and a word that could be written as *ἠ φθισίκηρε*, containing the vocative of the epiclesis *destroyer of evil demons*—a word confined to magical texts. But in another instance it belongs to a much longer palindrome without clear significance, *ερηκισιθφε αραχαρα εφθισικηρε*.[22]

As to the content, the sinister aspect of Selene-Hecate is stressed in the second list of epithets, where we find gruesome names like

αἱμαπότι, θανατηγέ, φθορήγονε, καρδιόδαιτε

(O you) who drink blood, who breed death and destruction, who feast on hearts (2864)

σαρκόφαγε, καπετόκτυπε, ἀωρόβορε

flesh eater, who strike the graves,[23] who devour those dead untimely (2867)

—which seem to correspond to more popular ideas of magic. But this tendency is balanced by the identification of Selene with divinities like Dike or Physis and by statements like

> *ἀρχὴ καὶ τέλος εἶ, πάντων δὲ σὺ μούνη ἀνάσσεις,*
> *ἐκ σέο γὰρ πάντ' ἐστὶ καὶ εἴς σε ἅπαντα τελευτᾶι.*
>
> Beginning and end are you, and you alone rule all,
> for all things are from you, and in you do all things come to their end. (2836–37)

The dark aspects are but one side of her all-embracing nature.[24]

As Greek prayer nearly always does, the magical prayer accompanies a sacrifice[25]—in the case of our hymn a burnt offering (*θυμίαμα*). Again, there is no essential difference between magic and religion. The ingredients are given in detail: if the spell should do good, different sorts of spice (storax, myrrh, sage, frankincense) and a fruit pit are prescribed. If, however, it is intended to do harm, the "magical material of a dog and a dappled goat, as well as of a virgin untimely dead" is called for.[26] At least the last-mentioned items again seem to conform to popular ideas of black magic and are unheard of in other rituals beside magic. Therefore they need some explanation.

So far, then, magical prayer in general structure, content, and context is not different from religious prayer, two peculiarities excepted: the *voces magicae* in the prayer, the *materia magica* in the harmful, black version of the ritual. What are the functions of these peculiarities?

Not all the magical words are understood, or even understandable.[27] Where we think to see through them, they derive from Near Eastern languages, especially Egyptian, and are names or epithets of divinities. The magician thought them all to be names, *ὀνόματα* or *ἐπωνυμίαι*. One of their functions is indicated by the way the prayer to Selene uses the palindrome: it forms part of the *invocatio* as another name of the divinity invoked. By using it the magician makes certain that the god would listen, since he had embraced the widest possible sphere of the god's activities and characteristics—a strategy well known from religious prayer.[28] A second function is well illustrated by a prose prayer belonging to a spell of Astrapsoukos, which aims at providing general success and well-being (*PGM* VIII.1–60). It begins with a triple invocation; I separate the cola in print to indicate the stylistic features:[29]

> *ἐλθέ μοι, κύριε Ἑρμῆ,*
> *ὡς τὰ βρέφη εἰς τὰς κοιλίας τῶν γυναικῶν,*[30]
> *ἐλθέ μοι, κύριε Ἑρμῆ,*
> *συνάγων τὰς τροφὰς τῶν θεῶν καὶ ἀνθρώπων,*
> *ἐλθέ μοι, τῶι δεῖνα, κύριε Ἑρμῆ.*

The *preces,* obviously urgent ones, follow immediately:

> *καὶ δὸς μοι χάριν, τροφὴν, νίκην, εὐημερίαν, ἐπαφροδισίαν,*
> *προσώπου εἶδος, ἀλκὴν ἁπάντων καὶ πασῶν.*

The next section, in this case, should contain the *pars epica,* the credentials:

> *ὀνόματά σου ἐν οὐρανῶι • Λαμφθεν Οὐωθι • Οὐασθεν Οὐωθι • Ὀαμενωθ •*
> *Ἐνθουμουχ • ταῦτά εἰσι τὰ ἐν ταῖς δ' γωνίαις τοῦ οὐρανοῦ ⟨ὀνόματα⟩.*
> *οἶδά σου καὶ τὰς μορφάς, αἵ εἰσι . . . .*

The list of his forms follows. The list of *voces magicae,* the celestial names of Hermes, is a demonstration of the superior knowledge the magician can display, as the continuation shows: the celestial names are not those ordinarily used and known. The list continues with the sacred forms, then with items of Hermes's sacred biography (again note the triple anaphora):

> οἶδά σου καὶ τὸ ξυλόν· τὸ ἐβεννίον.
> οἶδά σε Ἑρμῆ, τίς εἶ καὶ πόθεν εἶ καὶ τίς ἡ πόλις σου·
> Ἑρμούπολις.

This knowledge, then, serves as the credentials, and justifies a new invocation and repeated *preces:*

> ἐλθέ μοι, κύρι, Ἑρμῆ πολυώνυμε,
> εἰδὼς τὰ κρύφιμα τὰ ὑπὸ τὸν πόλον καὶ τὴν γῆν,
> ἐλθέ μοι, κύρι, Ἑρμῆ, τῶι δεῖνα,
> εὐεργέτησον, ἀγαθοποιὲ τῆς οἰκουμένης.[31]
> ἐπάκουσόν μου,
> καὶ χάρισόν με πρὸς πάντα τὰ κατὰ τὴν γῆν οἰκουμένην εἴδη.
> ἀμοίξας μοι τὰς χεῖρας πάντων συνδω⟨ροδο⟩κο⟨ύν⟩των
> ἐπανάγκασον αὐτοὺς δοῦναί μοι, ἃ ἔχουσιν ἐν ταῖς χερσίν.

With this generous wish, the prayer could end. Instead, it continues,

> οἶδά σου καὶ τὰ βαρβαρικὰ ὀνόματα • Φαρναθαρ • Βαραχηλ • Χθα • ταῦτά σοί ἐστιν τὰ βαρβαρικὰ ὀνόματα.

The rest of this prayer can be left aside. (There follows, as a further *argumentum,* the narration of a myth of how Hermes always helps Isis when she calls: so why not likewise help the magician?) The function of the *voces magicae,* at least, is clear. They are not used, as some have claimed,[32] to force the divinity: they take the place of, and serve as, the credentials, an ample display of knowledge. In several instances, the papyri state that these names were secret,[33] that the god enjoys being called by them and helps out of joy:[34] it was the gods themselves who had revealed them.[35] The magician behaves not very differently from an initiate of a mystery cult: both claim a special relationship with their respective gods, based on revealed knowledge—this can explain why parts of mystery rituals were taken over into the prayers of the magical papyri.[36]

Another text may be adduced as a further illustration and confirmation, a prose prayer (termed εὐχή in the text) to Selene in the usual structure (*PGM* VII.756).[37] After the invocation with an account of how Selene was created and of her powers, there follows a *pars epica* with two lists: the first gives the acoustic signs that are termed her "companions" and must form part of the magical ritual,[38] the second her animal symbols. This part closes with the statement,

> I have said your signs and symbols of your name, so that you might hear me, because I pray to you, mistress of the whole world.

This very ample display of secret knowledge again serves to recommend the magician to his divinity.

More can be learned from another set of documents. There exist three versions of a hymn to Hermes, one preserved on a papyrus in Strasbourg without any context, the other two forming part of a magical prescription to obtain a dream. The Strasbourg version, by far the longest and only fragmentarily preserved, was thought to be a prayer without any use in magic, "the private copy of a pious believer."[39] The absence of context, namely, the prescription for the accompanying ritual, is not wholly conclusive, as in some cases (e.g., *PGM* XIV) we have many spells or formulae in the collection of the magical papyri that contain no directions for the ritual. Preisendanz thought that a mutilated hexameter might possibly contain magical words.[40] Whatever the answer, the controversy as such is significant.

The three texts are similar in structure; they all show the canonical three parts of a Greek prayer. They begin with the invocation to Hermes, followed by a plethora of epithets and a list of functions, expressed by participles. The *argumentum* stresses Hermes's power as a healer and a giver of oracles. The hymn on the Strasbourg papyrus connects it with the *preces* proper:

> For you cure, too, man's every ailment, who send [him] oracles by day and night:
> send me, I pray, your form, for I am a pious man, a suppliant.

The other two versions add, after the *argumentum,* a very short invocation, then the wish, in a very similar form. The version in *PGM* VII.678 has

> *δεῦρο, μάκαρ, θεᾶς τελεσίφρονος υἱὲ μέγιστε,*
> *σῆι μορφῆι ἱλαρῶι τε νοῶι· ἀνθεῖς δὲ ἀφθάρτωι*
> *κούρωι μαντοσύνην ἔπεμψον ἀληθῆ.*

> Hither, O blessed one, O mighty son of the goddess who brings full mental powers, by your own form and gracious mind. And to an uncorrupted youth reveal a sign and send him your true skill of prophecy.

There follow two lines of *voces magicae*. This and the "uncorrupted youth" who is to receive the oracle are the main difference from the Strasbourg version. The medium of such a boy is well attested in magic, even outside antiquity;[41] but divination through a medium is, of course, not typically "magical," even though only virgins, not unspoilt boys, are attested in Greek cults.[42] The third version (*PGM* V.400) differs but slightly. It begins with similar and, at least at the beginning, better hexameters:

> *δεῦρο, μάκαρ, Μνήμης τελεσίφρονος υἱὲ μέγιστε,*
> *σῆι μορφῆι ἱλαρός τε φάνηθ' ἱλαρός τ' ἐπίτειλον*
> *ἀνθρώπωι ὁσίωι, μορφήν θ' ἱλαρὰν ἐπίτειλον*
> (in prose) *ἐμοί, τῶι δεῖνα,*
> *ὄφρα σε μαντοσύναις, ταῖς σαῖς ἀρεταῖσι, λάβοιμι.*

> Hither, O blessed one, O mighty son of Memory, who brings full mental powers, in your own form both graciously appear and graciously render the task for me, a pious man, and render your form gracious to me, NN, that I may comprehend you by your skills of prophecy, by your own wond'rous deeds.

This time, there follow no magical words but a final prose prayer:

> *δέομαι, κύριε· ἵλεώς μοι γενοῦ καὶ ἀψευδῶς μοι φανεὶς χρημάτισον*

> I ask you, lord: be gracious and without deceit appear and prophesy to me.

The fact that in this prayer the *voces magicae* are wholly absent confirms our previous finding: they are part of the whole complex of credentials, but they are not essential. The final prayer, however, would alone serve to demolish the Frazerian dichotomy: the submissive tone—"I ask you, lord, be gracious"—could not be better expressed and does not stand alone in the papyri.[43]

Two points now have become clear. The *εὐχαί* in the papyri are in no way different from many other texts, labeled either *λόγοι* (e.g., VII.668) or some type of *λόγοι*, or more specifically as, for instance, *φίλτροκατάδεσμος* (VIII.1). And they do not essentially differ from prayers outside the magical praxis: religion and magic, at least with regard to prayer, are coterminus.

This is not to say that coercion, the idea of magic power over superhuman beings, is absent from magic or at least the magical papyri. That would be ludicrous. But coercion is not omnipresent in the spells and prayers in a manner that would justify taking it as a—or the—*differentia specifica* of magic from religion. The prayers analyzed so far are proof enough.[44] There also exists a separate class of spells and rituals labeled *ἐπάναγκοι* (coercive procedures). They are not very frequent, and already their existence as a distinct class of spells could argue against making coercion the distinctive characteristic of magic. In most instances, the coercive spell or ritual is not the rule but a sort of last straw for the magician—when the invoked divinity does not arrive quickly enough,[45] when the praxis after several repetitions brings no result,[46] when the divinity appears threatening and dangerous.[47] But its use is not without severe risks:

> Use this for a spell of coercion, for it can accomplish everything, but do not use it frequently against Selene, unless the procedure which you are performing is worthy of its power.[48]

Coercion, then, should be used but reluctantly. The gods are not that easily managed.

There are only a few unambiguous instances where the coercive spell is integrated into a praxis and is to be used without "the necessity for the compulsive procedure arising" (*τῆς χρείας τοῦ ἐπανάγκου καλούσης*, *PGM* II.63). The clearest case is a ritual forming part of the magician's initiation that is described on the opening pages of the *Eighth Book of Moses* (*PGM* XIII). Towards the end of this ritual, the would-be magician has to eat three figures made from flour, "saying the spell for the gods of the hours (which is in the *Key*) and the compulsive formula for them and the names of the gods set over the weeks".[49] The divinities concerned are the seven planetary gods already mentioned earlier in the same ritual, who preside not only over the days of the week, but also over the hours.[50]

But even here, precision is necessary. The supernatural world according to the magical papyri has a clear hierarchy, expressed, for instance, in another coercive spell (*PGM* XII.117):[51]

> Hear me, because I am going to say the great name AŌTH, before whom every god prostrates himself and every daimon shudders, for whom every angel completes those things which are assigned.

At the highest point is the supreme god, the "great name"; below him are the lesser gods, then the (evil) daemons and the helpful angels. The same hierarchy was a common belief in the society contemporary with the papyri, where the manifold world of supernatural powers had gained, under Platonist influence, a clear hierarchical structure. Magic especially used daemons and angels, whom the magician could command with the help of the supreme god.[52] Another coercive spell has, "If somehow he [sc. the god called for] delays, say in addition this following incantation . . . : the great, the living god commands you, he who lives for eons of eons, who shakes together, who thunders, who created every soul and race IAŌ AŌI ŌIA AIŌ IŌA ŌAI" (IV.1037).[53] Similarly, a love charm, that is to say, a spell from a subcategory of coercive spells, invokes Selene in the traditional forms of the prayer: "I call upon you, mistress of this entire world. . . . Give a sacred angel or a holy assistant who will serve this very night, in this very hour . . . and order the angel to go off with her, NN, to draw her by the hair."[54] Apuleius may have in mind such a relationship between the magician and the upper gods when he defines the magician, according to popular belief, as "someone who, by conversing with the immortal gods, gains an incredible power of charms for everything he wishes."[55] Planetary gods, on the other side, belonged to the world of the daemons;[56] here, coercion was not unusual.

Incidentally, we just learned that with regard to daemonology and the structure of the divine, the magicians' views were not different from those of religious individuals in the same period (and arguably before)—just as prayer, in form and intention, could be coterminous in magic and in religion. And it is not only in Frazerian terms that the difference between the two seems to disappear; the *voces magicae* can be reduced to a separate, but not essentially irreligious, phenomenon.

This leaves us with ritual. We have already noted that magical ritual could contain fumigations (ἐπιθύματα) not different from those accompanying, for instance the "Orphic" hymns: again, magic and religion appear coterminous. There are also libations of wine, honey, and milk, which belong to religious ritual as well,[57] and there is animal sacrifice. Here, at least, a more fundamental difference appears. Animal sacrifice occurs, as far as I can see, in the form of a holocaust (as in religious ritual) or strangulation—but never in the most usual and widespread form of the Olympian sacrifice, the killing of an animal followed by a common meal of the sacrificing group.[58] In the cases where the animal or parts of it are eaten, the magician always appears to be alone, in marked contrast to the ordinary sacrificial meal.[59] The difference is important: the community, which finds its identity and its feeling of *communitas* in the Olympian sacrifice and the ensuing meal, is absent from the magical praxis. The magician is an isolated individual—either an itinerant specialist working for a customer or an individual layperson (so to speak), practicing the ritual in his or her own interest. The sacrifices in *Papyri Graecae Magicae* are obviously not public festivals; but it remains debatable to what extent they differ from equivalent private small-scale ceremonies conducted by ordinary householders for their household gods.

Holocaust itself was viewed by ancient and modern theorists as standing in opposition to the Olympian sacrifice, as the sacrifice apt for the powers of the underworld.[60] It did stand in opposition to the Olympian sacrifice, but it was not addressed to underworldly or chthonic powers only.[61] In a similar way, the libation of (unmixed) wine, honey, milk, and oil stands in opposition to libations with mixed wine, the ordinary libation ritual in Greek sacrifice.[62] In both cases, the magician who performs this ritual puts himself in opposition to the more frequent way of Greek ritual practice—but not, it should be underlined, to religion as such. Both holocaust and libations with fluids other than mixed wine occur outside magical ritual, in "religious" rituals. Perhaps even the burnt offering without any ensuing (or accompanying) animal sacrifice can be understood in a similar way. It could be seen as the refusal of animal sacrifice, with the consequent marginalization of the performer, as M. Detienne showed for Pythagoreans and "Orphics."[63]

This same tendency of reversal or opposition to ordinary (civic) religion can be seen in smaller details, as in the ritual to which one of the hymns to Hermes (*PGM* V.401) belongs, a praxis to procure oracular dreams with the help of a small statue of Hermes.[64] Before he goes to bed and after all the preparations have been made, the performer burns incense, salt, and "soil from a wheat-field" on an altar and recites the hymn three times, facing east; then he goes to bed without reacting to other persons' questions.[65]

Unlike the incense, the traditional and well-known ingredient of burnt offerings, salt and soil are strange: they are not luxury goods imported from the east and used only in ritual but basics of human life. Precultural man, like the animals, had no use for them, since he lived on raw flesh and acorns.[66] Burning these substances thus could become a ritual sign for the alienation from the ordinary human world: accordingly, the magician no longer responds to attempts of human communication. The ritual moves him into a sphere removed from his fellow men, where he will converse with the divinity.[67]

A somewhat different explanation accounts for the strange burnt offering that accompanied the *Prayer to Selene* (*PGM* IV.2785) if one wanted to work harm. To do good, one had to burn spices and a fruit pit,[68] to work evil, the magical material of a dog, of a dappled goat, and of a virgin untimely dead (*PGM* IV.2875–78; see p. 191 and Appendix, pp. 199–202). These same substances appear (among many more) in a so-called *Slander spell to Selene,* a spell that ascribes unholy actions to an opponent in order to arouse Selene's wrath against her: "For you the woman NN burns hostile incense, goddess: the fat of a dappled goat, and blood, defilement, embryo of a dog, the bloody discharge of a virgin dead untimely" (*PGM* IV.2642)[69]. Obviously, these substances are things that should not be sacrificed to the goddess,[70] and such a sacrifice would alienate the sacrificer from her. The burnt offering works evil by a sort of abbreviated slander spell.

Thus, the main distinction of magic lies in the ritual, not in the prayers and not so much in the forms of the ritual—they are shared between magic and religion—as in the function. The rituals of the magician put him in opposition to ordinary, "religious" ritual and isolate him from his fellow man. The distinction, then, lies rather in social than in psychological factors. This, of course, would fit perfectly the social differentiation that made the magician an outsider and the outsider a potential magician—the best-known example being Apuleius, who had to defend himself against a charge of magic in the small town of Oia where he resided as a foreigner, scholar, and young husband of an elderly, rich widow.

There is a final conclusion to be drawn. Dieterich, Reitzenstein, and their contemporaries and pupils were convinced that the magicians had taken over entire mystery rituals, hymns, and prayers into their magical praxeis; the *Mithrasliturgie* seemed to confirm this view. But the spoken parts of mystery rituals on one side and hymns and prayers on the other are distinct. Since the magician felt himself to be an initiate, it is understandable that he appropriated parts of the *legomena* of mystery rituals that suited his purpose. As in the case of the catabasis ritual that H. D. Betz isolated, it was precisely by means of these appropriations that the magical performer could prove his special relationship with, and intimate knowledge of, the Lady of the Dead—just as any initiate could do, as Heracles did as an initiate of Eleusis, or as the persons buried with the so-called Orphic golden leaves did.[71] Hymns and prayers had no such function in mystery cults, and after a close scrutiny of some hymns, E. Heitsch concluded that there existed no *Ur-Fassung* that could be reconstructed:[72] the magicians used verses and formulas that came from a common stock of tradition, a stock that both magicians and non-magicians could use. In the world of syncretistic religion and magic, no provenances and clear borrowings of traditional formulae can be shown.

Appendix: The texts

Sappho Frag. 1

ποικιλόθρον' ἀθανάτ' Ἀφρόδιτα,
παῖ Δίος δολόπλοκε, λίσσομαί σε,
μή μ' ἄσαισι μηδ' ὀνίαισι δάμνα,
πότνια, θῦμον,

5 ἀλλὰ τυίδ' ἔλθ', αἴ ποτα κἀτέρωτα
τὰς ἔμας αὔδας ἀίοισα πήλοι
ἔκλυες, πάτρος δὲ δόμον λίποισα
χρύσιον ἦλθες

ἄρμ' ὐπασδεύξαισα· κάλοι δέ σ' ἆγον
10 ὤκεες στροῦθοι περὶ γᾶς μελαίνας
πύκνα δίννεντες πτέρ' ἀπ' ὠράνω αἴθε-
ρος διὰ μέσσω,

αἶψα δ' ἐξίκοντο· σὺ δ', ὦ μάκαιρα,
μειδιαίσαισ' ἀθανάτωι προσώπωι
15 ἤρε' ὄττι δηὖτε πέπονθα κὤττι
δηὖτε κάλημμι,

κὤττι μοι μάλιστα θέλω γένεσθαι
μαινόλαι θύμωι· 'τίνα δηὖτε πείθω
ἄψ σ' ἄγην ἐς Ϝὰν φιλότατα; τίς σ', ὦ
20 Ψάπφ', ἀδίκησι;

καὶ γὰρ αἰ φεύγει, ταχέως διώξει·
αἰ δὲ δῶρα μὴ δέκετ', ἀλλὰ δώσει·
αἰ δὲ μὴ φίλει, ταχέως φιλήσει
κωὐκ ἐθέλοισα.'

25 ἔλθε μοι καὶ νῦν, χαλέπαν δὲ λῦσον
ἐκ μερίμναν, ὄσσα δέ μοι τέλεσσαι
θῦμος ἰμέρρει, τέλεσον· σὺ δ' αὔτα
σύμμαχος ἔσσο.

Ornate-throned immortal Aphrodite, wile-weaving daughter of Zeus, I entreat you: do not overpower my heart, mistress, with ache and anguish, but come here, if ever in the past you heard my voice from afar and acquiesced and came, leaving your father's golden house, with chariot yoked: beautiful swift sparrows whirring fast-beating wings brought you above the dark earth down from heaven through the mid-air, and soon they arrived; and you, blessed one, with a smile on your immortal face asked what was the matter with me this time and why I was calling this time and what in my maddened heart I most wished to happen for myself: "Whom am I to persuade this time to lead you back to her love? Who wrongs you, Sappho? If she runs away, soon she shall pursue; if she does not accept gifts, why, she shall give them instead; if she does not love, soon she shall love even against her will." Come to me now again and deliver me from oppressive anxieties; fulfil all that my heart longs to fulfil, and you yourself be my fellow-fighter. (Text and translation by D. A. Campbell in *The Cambridge History of Classical Literature,* vol. I, *Greek Literature* [Cambridge 1985]), 203f.

Inscription from Phrygia, 175 A.D. (A. Körte, *AM* 25 (1900):421, n. 33)

[- - - - - - - - - - - - - - - *βρέχε γαῖ]αν,*
καρπῷ [ὅπ]ως βρί[θῃ καὶ ἐν]ὶ σταχύεσσι τεθήλῃ.
τ[αῦτ]ά [σε] Μητρεόδωρος ἐγὼ λίτομαι, Κρονίδα Ζεῦ,
ἀμφὶ τεοῖς βωμοῖσιν ἐπήρρατα(!) θύματα ῥέζων.

[Zeus . . . wet the ea]rth, that she becomes heavy with fruit and flowers with ears of corn. This I, Metrodoros, beg you, Zeus son of Cronos, while I am performing delightful sacrifice on your altars.

Magical Papyri (All of the Greek texts are from *PGM,* 2d ed.; the translations, with minor corrections, are from the English translation edited by H. D. Betz [*GMPT*].)

***PGM* IV. 2785–2879** Hymn 18: line divisions correspond to the metrical pattern reconstructed and set forth by E. Heitsch in *PGM* vol. 2, pp. 253–55.)

2785 **Εὐχὴ πρὸς Σελήνην ἐπὶ πάσης πράξεως ·|**
'ἐλθέ μοι, ὦ δέσποινα φίλη, τριπρόσω|πε Σελήνη,
εὐμενίῃ δ' ἐπάκου|σον ἐμῶν ἱερῶν ἐπαοιδῶν·|
2790 *νυκτὸς ἄγαλμα, νέα, φαεσίμβροτε, || ἠριγένεια,*
ἡ χαροποῖς ταύροισιν | ἐφεζομένη, βασίλεια,
Ἠελίου | δρόμον ἶσον ἐν ἅρμασιν ἱππεύ|ουσα,
ἣ Χαρίτων τρισσῶν τρισσαῖς | μορφαῖσι χορεύεις
2795 *ἀστράσιν κω||μάζουσα, Δίκη καὶ | νήματα Μοιρῶν, |*
Κλωθὼ καὶ Λάχεσις ἠδ' Ἄτροπος εἶ, τρικάρανε, ||
Περσεφόνη τε Μέγαιρα καὶ Ἀλληκτώ, | πολύμορφε,
2800 *ἡ χέρας ὁπλίζουσα || κελαιναῖς λαμπάσι δειναῖς,*
ἡ φο|βερῶν ὀφίων χαίτην σείουσα μετώ|ποις,
ἡ ταύρων μύκημα κατὰ στο|μάτων ἀνιεῖσα,
2805 *ἡ νηδὺν φολί|σιν πεπυκασμένη ἑρπυστήρων, ||*
ἰοβόλοις ταρσοῖσιν κατωμαδίοισι | δρακόντων,
σφιγγομένη κατὰ | νῶτα παλαμναίοις ὑπὸ δεσμοῖς, |
νυκτιβόη, ταυρῶπι, φιλήρεμε, ταυ|ροκάρηνε,
2810 *ὄμμα δέ σοι || ταυρωπόν, ἔχεις σκυλακώδεα φω|νήν,*
μορφὰς δ' ἐν κνήμασιν |ὑποσκεπάουσα λεόντων.
μορφό|λυκον σφυρόν ἐστιν, κύνες φίλοι | ἀγριόθυμοι·
2815 *τοὔνεκά σε κλήζουσι || Ἑκάτην, πολώνυμε, Μήνην, |*
ἀέρα μὲν τέμνουσαν, ἅτ' Ἄρτεμιν | ἰοχέαιραν,
τετραπρόσωπε θεά, | τετραώνυμε, τετραοδῖτι,
2820 *Ἄρτε|μι, Περσεφόνη, ἐλαφηβόλε, νυκτο||φάνεια,*
τρίκτυπε, τρίφθογγε, | τρικάρανε, τριώνυμε Σελήνη, |

θρινακία, τριπρόσωπε, τριαύχε|νε καὶ τριοδῖτι,
2825 *ἣ τρισσοῖς ταλά|ροισιν ἔχεις φλογὸς ⟨ἀκ⟩άματον πῦρ* ||
καὶ τριόδων μεδέεις τρισσῶν δ|εκάδων τε ἀνάσσεις·
ἵλαθί μοι κα|λέοντι καὶ εὐμενέως εἰσάκουσον, |
ἣ πολυχώρητον κόσμον νυκτὸς | *ἀμφιέπουσα,*
2830 *δαίμονες ἣν φρίσ||σουσιν καὶ ἀθάνατοι τρομέουσιν,* |
κυδιάνειρα θεά, πολυώνυμε, καλλι|γένεια,
ταυρῶπι, κερόεσσα, θεῶν | *γενέτειρα καὶ ἀνδρῶν*
2835 *καὶ* Φύσι | *παμμήτωρ • σὺ γὰρ φοιτᾷς ἐν* ᾿Ο||*λύμπῳ,*
εὐρεῖαν δέ τ᾿ ἄβυσσον | *ἀπείριτον ἀμφιπολεύεις.*
ἀρχὴ | *καὶ τέλος εἶ, πάντων δὲ σὺ μούνη* | *ἀνάσσεις·*
ἐκ σέο γὰρ πάντ᾿ ἐστὶ | *καὶ εἴς* ⟨σ᾿⟩, *αἰών⟨ι⟩ε, πάντα τελευτᾷ.*
2840 *ἀένα||ον διάδημα ἑοῖς φορέεις κροτά|φοισιν,*
δεσμοὺς ἀρρήκτους, ἀλύ|τους μεγάλοιο Κρόνοιο
καὶ χρύ|σεον σκῆπτρον ἑαῖς κατέχεις πα|λάμαισιν.
2845 *γράμματα σῷ σκήπτρῳ* || α[ὐ]τὸ[ς] Κρόνος *ἀμφεχάραξεν,*
δῶ|κε δέ σοι φορέειν, ὄφρ᾿ ἔμπεδα πάν|τα μένοιεν·
'Δαμνώ, Δαμνομέ|νεια • Δαμασάνδρα • Δαμνοδαμία.' |
σὺ δὲ χάους μεδέεις αραραχαρα|ραηφθισικηρε.
2850 *χαῖρε, θεά, καὶ* || *σαῖσιν ἐπωνυμίαις ἐπάκουσον.* |
θύω σοι τόδ᾿ ἄρωμα, Διὸς *τέκος,* | *ἰοχέαιρα,*
οὐρανία, λιμ⟨ε⟩νῖτι, | *ὀρίπλανε εἰνοδία τε,*
2855 *νερτε||ρία νυχία τε, ἀϊδωναία σκοτία τε,*|
ἥσυχε καὶ δασπλῆτι, τάφοις | *ἔνι δαῖτα ἔχουσα,*
Νύξ, Ἔρεβος, | Χάος *εὐρύ· σὺ γὰρ δυσάλυκτος·* | Ἀνάγκη,
2860 Μοῖρα *δ᾿ ἔφυς, σύ τ᾿* || Ἐρινύς, *βάσανος, ὀλέτις σύ,* Δίκη *σύ.* |
Κέρβερον *ἐν δεσμοῖσιν ἔχεις,* | *φολίσιν σὺ δρακόντων* |
κυανέα, ὀφεοπλόκαμε καὶ | *ζωνοδράκοντι·*
2865 *αἱμοπότι,* || *θανατηγέ, φθορηγενές, καρ|διόδαιτε,*
σαρκοφάγε καὶ | *ἀωροβόρε, καπετόκτυπε,* | *οἰστροπλάνεια·*
2870 *ἐλθὲ ἐπ᾿ ἐμαῖς* | *θυσίαις καί μοι τόδε πρᾶγμα* || *ποίησον.*' |

ἐπίθυμα τῆς πράξεως· *ἐπὶ μὲν τῶν* | *ἀγαθοποιῶν ἐπίθυε στύρακα,* |
2875 *ζμύρναν, σφάγνον, λίβανον,* | *πυρῆνα, ἐπὶ δὲ τῶν κακο||ποιῶν οὐσίαν*
κυνὸς καὶ αἰγὸς | *ποικίλης, ὁμοίως καὶ παρθένου* | *ἀώρου.*

2785 **Prayer to Selene for any spell:**
"Come to me, O beloved mistress, Three-faced
Selene; kindly hear my sacred chants;
Night's ornament, young, bringing light to mortals, /
2790 O child of morn who ride upon fierce bulls,
O queen who drive your car on equal course
With Helios, who with the triple forms
Of triple Graces dance in revel with /
2795 The stars. You're Justice and the Moira's threads:
Klotho and Lachesis and Atropos.
Three-headed, you're Persephone, Megaira,
Allekto, many-formed, who arm your hands /
2800 With dreaded, murky lamps, who shake your locks
Of fearful serpents on your brow, who sound

The roar of bulls out from your mouths, whose womb
Is decked out with the scales of creeping things, /
2805 With pois'nous rows of serpents down the back,
Bound down your backs with horrifying chains
Night-Crier, bull-faced, loving solitude,
2810 Bull-headed, you have eyes of bulls, / the voice
Of dogs; you hide your forms in shanks of lions.
Your ankle is wolf-shaped, fierce dogs are dear
2815 To you, wherefore they call you / Hekate,
Many-named, Mene, cleaving air just like
Dart-shooter Artemis, Persephone,
2820 Shooter of deer, night / shining, triple-sounding,
Triple-headed, triple-voiced Selene
Triple-pointed, triple-faced, triple-necked,
And goddess of the triple ways, who hold
Untiring flaming fire in triple baskets, /
2825 And you who oft frequent the triple way
And rule the triple decades, unto me
Who'm calling you be gracious and with kindness
Give heed, you who protect the spacious world
At night, before whom daimons quake in fear /
2830 And gods immortal tremble, goddess who
Exalt men, you of many names, who bear
Fair offspring, bull-eyed, horned, mother of gods
And men, and Nature, Mother of all things,
2835 For you frequent Olympos, / and the broad
And boundless chasm you traverse. Beginning
And end are you, and you alone rule all.
For all things are from you, and in you do
All things, Eternal one, come to their end.
2840 As everlasting / band around your temples
You wear great Kronos' chains, unbreakable
And unremovable, and you hold in
Your hands a golden scepter. Letters round
2845 Your scepter / Kronos wrote himself and gave
To you to wear that all things stay steadfast:
Subduer and subdued, mankind's subduer,
And force-subduer; Chaos, too, you rule.
2850 ARARACHARA / RA ĒPHTHISIKĒRE.
Hail, goddess, and attend your epithets,
I burn for you this spice, O child of Zeus,
Dart-shooter, heav'nly one, goddess of harbors,
Who roam the mountains, goddess of crossroads, /
2855 O nether and nocturnal, and infernal,
Goddess of dark, quiet and frightful one,
O you who have your meal amid the graves,
Night, Darkness, broad Chaos: Necessity
Hard to escape are you; you're Moira and /
2860 Erinys, torment, Justice and Destroyer,
And you keep Kerberos in chains, with scales

Of serpents are you dark, O you with hair
Of serpents, serpent-girded, who drink blood, /
2865 Who bring death and destruction, and who feast
On hearts, flesh eater, who devour those dead
Untimely, and you who strike the graves
And spread madness, come to my sacrifices,
2870 And now for me do you fulfill / this matter."

2875 **Offering for the rite:** *For doing good, offer storax, myrrh, sage, frankincense, a fruit pit. But for doing harm, offer magical / material of a dog and a dappled goat as well as of a virgin untimely dead.*

PGM VIII.1–26

Φιλτροκατάδεσμος Ἀστραψούκου. λόγος• | ʽἐλ[θ]έ μοι, κύριε Ἑρμῆ, ὡς τὰ βρέφη εἰς τὰ⟨ς⟩ κοιλίας τῶν γυναι[κ]ῶν. ἐλθέ μοι, κύριε Ἑρμῆ, συνάγων τὰς τροφὰς τῶν θεῶν | καὶ ἀνθρώπων, ⟨ἐλθ⟩έ μοι, τῷ δεῖνα, κύριε Ἑρμῆ, καὶ δός
5 μοι χάριν, τρο||φήν, νίκην, εὐημερίαν, ἐπαφροδισίαν, προσώ⟨π⟩ου εἶδος, | ἀλκὴν ἁπάντων καὶ πασῶν. ὀνόματά σοι ἐν οὐρανῷ· | Λαμφθεν Οὐωθι: Ο[ὐ]ασθεν Οὐωθι: Ὀαμενώθ: Ἐνθομουχ:ʾ | ταῦτά εἰσιν τὰ ἐν ταῖ⟨ς⟩ δ´ γωνίαις τοῦ οὐρανοῦ ⟨ὀνόματα⟩. οἶδά σου | καὶ τὰς μορφάς, αἵ εἰσι· ἐν τῷ
10 ἀπηλιώτῃ μορφὴν ἔχεις || ἴβεως, ἐν τῷ λιβὶ μορφὴν ἔχεις κυνοκεφάλου, ἐν τῷ βορέᾳ | μορφὴν ἔχεις ὄφεως, ἐν δέ τῷ νότῳ μορφὴν ἔχεις λύκου. | ἡ βοτάνη σου ηλολλα: ετεβεν θωητ: οἶδά σου καὶ τὸ ξύ|λον · τὸ ἐβεννίνου. οἶδά σε, Ἑρμῆ, τίς εἶ καὶ πόθεν εἶ, καὶ τίς ἡ | πόλις σου · Ἑρμούπολις. ἐλθέ μοι, κύριʾ
15 Ἑρμῆ, πολυώνυμε εἰδὼς || τὰ κρύφιμα τὰ ὑπὸ τὸν πόλον καὶ τήν γῆν. ἐλθέ ⟨μοι⟩, κύριʾἙρμῆ, || τῷ δεῖνα, εὐεργέτησον, ἀγαθοποιὲ τῆς οἰκουμένης. ἐπάκουσόν | [μ]ου καὶ χάρισόν με πρὸς πάντα τὰ κατὰ τὴν γῆν οἰκου|μένην εἴδη. ἀνοίξας μοι τὰς χεῖρας πάντων συνδω⟨ροδο⟩κο⟨ύν⟩|των, ἐπανάγκασον
20 αὐτοὺς δοῦναί μοι, ἃ ἔχουσιν ἐν ταῖς || χερσίν. οἶδά σου καὶ τὰ βαρβαρικὰ ὀνόματα· ʽΦαρναθαρ: | βαραχήλ: χθα:ʾ ταῦτά σοί ἐστιν τὰ βαρβαρικὰ ὀνόματα. | ἐὰν ἐπεκαλέσατό σε Ἶσις, μεγίστη τῶν θεῶν ἁπάντων, | ἐν πάσῃ κρίσει, ἐν πα⟨ν⟩τὶ τόπῳ πρὸς θεοὺς καὶ ἀνθρώπους | καὶ δαίμονας καὶ
25 ἔν⟨υ⟩δρα ζῷα καὶ ἐπί⟨γ⟩εια καὶ ἔσχεν τὴν χά||ριν, τὸ νῖκος πρὸς θεοὺς καὶ ἀνθρώπους κ[α]ὶ ⟨παρὰ⟩ πᾶσι τοῖς ὑπὸ τὸν | κόσμον ζῴοις, οὕτως κἀγώ, ὁ δεῖνα, ἐπικα[λ]οῦμαί σε.ʾ

Binding love spell of Astrapsoukos:

Spell: "Come to me, lord Hermes as fetuses do to the wombs of women. Come to me, lord Hermes, who collect the sustenance of gods and men; [come] to me, NN, lord
5 Hermes, and give me favor, sustenance, / victory, prosperity, elegance, beauty of face, strength of all men and women. Your names in heaven: LAMPHTHEN OUŌTHI OUASTHEN OUŌTHI OAMENŌTH ENTHOMOUCH. These are the [names] in the 4 quarters of heaven. I also
10 know what your forms are: in the east you have the form / of an ibis, in the west you have the form of a dog-faced baboon, in the north you have the form of a serpent, and in the south you have the form of a wolf. Your plant is the grape which is the olive. I also know your wood: ebony. I know you, Hermes, who you are and where you come from and
15 what your city is: Hermopolis. Come to me, lord Hermes, many-named one, who know / the things hidden beneath heaven and earth. Come [to me], NN, lord Hermes; serve well, benefactor of the world. Hear me and make me agreeable to all the forms

throughout the inhabited world. Open up for me the hands of everyone who [dispenses
20 gifts] and compel them to give me what they have in their / hands. I also know your foreign names: PHARNATHAR BARACHĒL CHTHA. These are your foreign names. Whereas Isis, the greatest of all the gods, invoked you in every crisis, in every district, against
25 gods and men and daimons, creatures of water and earth and held your favor, / victory against gods and men and [among] all the creatures beneath the world, so also I, NN invoke you.

PGM VII.756–94

εὐχή. | ʽ ἐπικαλοῦμαί σε, πάνμορφον καὶ πολυώνυ|μαν δικέρατον θεὰν Μήνην, ἧς τὴν μορφὴν | οὐδὲ εἷς ἐπίσταται πλὴν ὁ ποιήσας τὸν
760 σύμπαντα || κόσμον, Ἰάω, ὁ σχηματίσας ⟨σε⟩ εἰς τα εἴκοσι καὶ | ὀκτὼ σχήματα τοῦ κόσμου, ἵνα πᾶσαν ἰδέαν | ἀποτελέσῃς καὶ πνεῦμα ἑκάστῳ ζώῳ καὶ | φυτῷ νέμῃς, ἵν᾽ εὐερ⟨νὲς⟩ ᾖ, ἐξ ἀφανοῦς ἡ εἰς φῶς | αὐξανομένη καὶ
765 ἀπὸ φωτὸς εἰς σκότος || ἀποληγουσα (εἰς μείωσιν ἄρχουσα ἀπολήγειν).
καὶ ἔστιν σου | ὁ α´ σύντροφος τ[οῦ] ὀνόματος σιγή, | ὁ β´ ποππυσμός, | ὁ γ´
770 στεναγμός, | ὁ δ᾽ συριγμός, || ὁ ε´ ὀλολυγμός, | ὁ ς´ μυγμός, | ο᾽ ζ´
775 ὑλαγμός, | ὁ ἡ μυκηθμός, | ὁ θ´ χρεμετισμός, || ὁ ι´ φθόγγος ἐναρμόνιος, | ὁ ια´ πνεῦμα φωνᾶεν, | ὁ ιβ´ ἦχος [ἀ]νεμοποιός, | ὁ ιγ´ φθόγγος
780 ἀναγκαστικός, | ὁ ιδ´ τελειότητος ἀναγκαστικὴ ἀπόρροια. ||
βοῦς, γύψ, ταῦρος, κάνθαρος, ἱέραξ, καρκίνος, | κύων, λύκος, δράκων. ἵππος, χίμαιρα, || θέρμουθις, αἴξ, τράγος, κυνοκέφαλος, | αἴλουρος, λέων,
785 πάρδαλις, μυγαλός, | ἔλαφος, πολύμορφος, παρθένος, λαμπάς, || ἀστραπή, στέλμα, κηρύκειον, παῖς, κλείς. |
εἴρηκά σου τὰ σημεῖα καὶ τὰ σύμβολα | τοῦ ὀνόματος, ἵνα μοι ἐπακούσῃς, ὅτι σοι | ἐπεύχομαι, τῇ δεσποίνῃ τοῦ παντὸς κόσμου. | ἐπάκουσόν μου, ἡ
790 μόνιμος, ἡ κραταιά, || αφειβοηω μιντηρ οχαω πιζεφυδωρ | χανθαρ χαδηροζο: μοχθιον εοτνευ | φηρζον αινδης λαχαβοω πιττω | ριφθαμερ ζμομοχωλειε τιηδραντεια | οισοζοχοβηδωφρα.᾽ κοινόν.

Prayer: "I call upon you who have all forms and many names, double-horned
760 goddess, Mene, whose form no one knows except him who made the entire /world, IAŌ, the one who shaped [you] into the twenty-eight shapes of the world so that they might complete every figure and distribute breath to every animal and plant, that it might flourish, you who grow from obscurity into light and leave light for darkness" /
765 (*beginning to leave by waning*).
"And the first companion of your name is silence, the second a popping sound, the third
770 groaning, the fourth hissing, / the fifth a cry of joy, the sixth moaning, the seventh
775 barking, the eighth bellowing, the ninth neighing, / the tenth a musical sound, the eleventh a sounding wind, the twelfth a wind-creating sound, the thirteenth a coercive sound, the fourteenth a coercive emanation from perfection. /
780 "Ox, vulture, bull, beetle, falcon, crab, dog, wolf, serpent, horse, she-goat, asp, goat, he-goat, baboon, cat, lion, leopard, fieldmouse, deer, multiform, virgin, torch,/
785 lightning, garland, a herald's wand, child, key.
"I have said your signs and symbols of your name so that you might hear me, because I pray to you, mistress of the whole world. Hear me, you, the stable one, the mighty one,/
790 APHEIBOEŌ MINTĒR OCHAŌ PIZEPHYDŌR CHANTHAR CHADĒROZO MOCHTHION EOTNEU PHĒRZON AINDĒS LACHABOŌ PITTŌ RIPHTHAMER ZMOMOCHŌLEIE TIĒDRANTEIA OISOZOCHABĒDŌPHRA" (*add the usual*).

***PGM* XVII^b^** (Hymn 15/16 Heitsch, *PGM* vol. 2, p. 249)

[Ἑρμῆ κοσμοκρ]άτωρ, ἐνκάρδιε, κ[ύκλε σελήνης, |
στρογγύλε καὶ τ]ετράγωνε, λόγων [ἀρχηγέτα γλώσσης, |
πειθοδικαιόσυνε], χλαμυδηφόρε, [πτηνοπέδιλε, |
παμφώνου γ]λώσσης μεδέω[ν, θνητοῖσι προφῆτα, ||
5]..εἰσπνοιὴ γὰρ[|
]..παρων προει[|
]ἐν τυτθλῷ χρόνῳ[|
ὅτα]ν πάλι μόρσιμο[ν ἦμαρ ἐπέλθῃ, |
χρησμ]όν τιν' ἀληθέα [πέμπων· ||
10 Μοιρῶν τε κλωσ]τὴρ σὺ λέγῃ καὶ [θεῖος Ὄνειρος, |
πανδαμάτωρ, ἀδάμ]αστος, ἅπερ φε[|
....].α.[.]ρα ι[..]εμ[.]ν ἐπικρίνοιο[|
ἐσθλὰ μὲν ἐσ[θλο]ῖσιν παρέχεις, [δειλοῖσι δὲ λυγρά. |
σ[ο]ὶ δ' ἠὼς ἀνέ[τ]ειλε, θοὴ δ' ἐπελά[σσατό σοι νύξ. ||
15 στοιχείων αὐ κ[ρ]ατεῖς, πυρός, ἀέρο[ς, ὕδατος, αἴης, |
ἡνί⟨κ⟩α πηδαλιοῦχος ἔφυς κόσμοιο [ἅπαντος, |
ὧν δ' [ἐ]θέλεις ψυ[χ]ὰς προάγεις, | τοὺς δ' αὖτ' ἀνεγείρεις· |
κόσμος γὰρ κόσμου γεγαὼς[|
σὺ γαρ καὶ νούσους μερόπων [πάσας θεραπεύεις. ||
20 ἡμερινοὺς κα[ὶ] νυκτερινοὺ[ς] χρησμοὺς ἐπιπέμπων, |
καί μοι εὐχομέ[ν]ῳ τὴν σὴν [μορφὴν ἐπίτειλον, |
ἀνθρώπῳ, ὁσίῳ ἱκέτῃ καὶ σ[ῷ στρατιώτῃ, |
καὶ σὴν μαντοσύνην νη[μερτέα πέμψον ἐν ὕπνῳ.

"[Hermes, lord of the world], who're in the heart,
[O orbit of Selene, spherical]
[And] square, the founder of the words [of speech]
[Pleader of justice's cause,] garbed in a mantle,
[With winged sandals,] who rule [expressive] speech
5 [Prophet to mortals] / . . .
For he inspires . . .
. . . within a short time. . . .
[Whene'er] the fateful [day arrives] again
10 . . . [who send] some [oracle] that's true, / you're said
To be [the Moirai's thread] and [Dream divine],
[The all-subduer, Unsub]dued, just as
. . . may you judge . . .
You offer good things to the good, [but grief]
[To those who're worthless.] Dawn comes up for you,
15 For you swift [night draws] near. / You lord it o'er
The elements: fire, air, [water, and earth]
When you became the helmsman of [all the] world;
And you escort the souls of those you wish,
But some you rouse again. For you've become
The order of the world, for you [cure], too,
20 Man's [every] ailment, / [who send oracles]
By day and night; [send] me, I pray your [form],
For I'm a man, a pious suppliant,

And your [soldier]; and so, [while I'm asleep],
[Send to me your unerring] mantic skill."

***PGM* VII.668–85** (Cf. Hymn 15/16 Heitsch, *PGM* vol. 2, p. 249)

. . . λέγων τὸν λόγον ζʹ | πρὸς τὸν λύχνον·
'Ἑρμῆ, παντοκράτωρ, ἐνκάρ|διε, κύκλε Σελήνης,
670 στρογγύλε, τετράγωνε, λό||γων ἀρχηγέτα γλώσσης,
πειθοδικαιόσυνε, χλα|μυδηφόρε, χρυσοπέδιλε,
⟨αἰ⟩θέρι[ον] δρόμον εἱλίσ|σων ὑπὸ τάρταρα γαίης,
πνεύματος, ἠελίου | ἡνίοχε, ἀθαν⟨άτ⟩ων τε λαμπάσι τέρπων
675 τοὺς | ὑπὸ τάρταρα γαίης βροτοὺς β[ίο]ν ἐκτελέσαντας, ||
Μοιρῶν τε κλωστὴρ σὺ λέγῃ ⟨καὶ⟩ θεῖος Ὄνειρος, |
ἡμερινοὺς καὶ νυκτερινοὺ[ς χ]ρησμοὺς ἐπιπέμπων. |
ἰᾶσαι πάντων βροτῶν ἀλγήματα ⟨σαῖς⟩ θεραπείαις· |
δεῦρο, μάκαρ, θεῆς τελεσίφρονος υἱὲ μέγιστε, |
680 σῇ μορφῇ ἱλαρῷ τε νοῷ· δεῖγμ' ἀνθεὶς δὲ ἀφθάρτῳ ||
κούρῳ μαντοσύνην ⟨τὴν σὴν⟩ ἔκπεμψον ἀληθῆ.
Οιοσ|ενμιγαδων: Ὀρθώ: Βαυβώ: νιοηρε: κοδηρεθ | δοσηρε: συρε: συροε: σανκιστη: δωδεκακι[σ]τη:|ἀκρουροβόρε: κοδηρε: ρινωτον: κουμετανα: |
685 ρουβιθα: νουμιλα· περφερο[υ]: Ἀρουωρηρ: ||Ἀρουηρ:' λέγε ἑπτάκις καὶ κοινά, ὅσα θέλεις.

. . . *saying the spell 7 times to the lamp:*

"Hermes, lord of the world, who're in the heart,
O circle of Selene, spherical
670 And square, / the founder of the words of speech,
Pleader of Justice's cause, garbed in a mantle,
With golden sandals, turning airy course
Beneath earth's depths, who hold the spirit's reins,
The sun's and who with lamps of gods immortal
Give joy to those beneath earth's depths, to mortals
675 Who've finished life. / The Moirai's fatal thread
And dream divine you're said to be, who send
Forth oracles by day and by night; you cure
Pains of all mortals with your healing cares.
Hither, O blessed one, O mighty son
Of the goddess who brings full mental powers,
By your own form and gracious mind. And to
680 An uncorrupted youth / reveal a sign
And send him your true skill of prophecy,
OIOSENMIGADŌN ORTHŌ BAUBŌ NIOĒRE KODĒRETH DOSĒRE SURE SUROE SANKISTĒ DŌDEKAKISTĒ AKROUROBORE KODĒRE RINŌTON KOUMETANA ROUBITHA NOUMILA PERPHEROU AROUŌRĒR/
685 AROUER" (*say it seven times and add the usual, whatever you wish*).

***PGM* V.392–423** (Cf. Hymn 15/16 Heitsch, *PGM* vol. 2, p. 249)

. . . ὅταν δὲ βούλῃ χρᾶσθαι, | πρὸς κεφαλῆς σου τίθει τὸν ναὸν || σὺν τῷ
395 θεῷ καὶ δίωκε ἐπιθύω[ν] | λίβανον ἐπὶ βωμοῦ καὶ γῆν ἀπ[ὸ] || σιτοφόρου χωρίου καὶ βῶλον ἁλὸς | ἀμμωνιακοῦ αʹ. κείσθω πρὸς κε-| φαλήν σου, καὶ κοιμῶ μετὰ τὸ

400 εἰ|πεῖν μηδενὶ δοὺς ἀπόκρισιν. ||
‘Ἑρμῆ κοσμοκράτωρ, ἐνκάρδιε, κύ|κλε σελήνης,
στρονγύλε καὶ τε|τράγωνε, λόγων ἀρχηγέτα γλώσσης,|
πειθοδικαιόσυνε, χλαμυδηφόρε | πτηνοπέδιλε,
405 αἰθέριον δρόμον || εἰλίσσων ὑπὸ τάρταρα γαίης |
πνεύματος ἡνίοχε, Ἡλίου ὀφθαλμέ, μέγιστε |
παμφώνου γλώττης ἀρχηγέτα | λαμπάσι τέρπων
410 τοὺς ὑπὸ τάρ|ταρα γαίης τε βροτοὺς βίον ἐκτελέ||σαντας·
μοιρῶν προγνώστης σὺ | λέγῃ καὶ θεῖος Ὄνειρος,
ἡμερινοὺς [καὶ] | νυκτερινοὺς χρησμοὺς ἐπιπέμ|πων.
ἰᾶσαι πάντα βροτῶν ἀλγήμα|τα σαῖς θεραπείαις.
415 δεῦρο, μάκαρ, Μνή||μης τελεσίθρονος υἱὲ μέγιστε.
σῇ | μορφῇ ἱλαρός τε φάνηθι ἱλαρός τ' ἐπί|τειλον
ἀνθρώπῳ ὁσίῳ μορφὴν θ' ἱλα|ρὰν ἐπίτειλον ἐμοί, τῷ δεῖνα,
420 ὄφρα σε |μαντοσύναις, ταῖς σαῖς ἀρεταῖσι λάβοι||μι
δέομαι κύριε· ἵλεώς μοι γενοῦ |
καὶ ἀψευδῶς μοι φανεὶς χρημάτισον.’ |
διώκε καὶ πρὸς ἀνατολὰς ἡλίου καὶ σελήνης.

395 *But when you want to use it, place the shrine beside your head / along with the god and recite as on the altar you burn incense, earth from a grain-bearing field and one lump of rock salt. Let it rest beside your head, and go to sleep after saying the spell without giving an answer to anyone. /*

400 "Hermes, lord of the world, who're in the heart,
O circle of Selene, spherical
And square, the founder of the words of speech,
Pleader of justice's cause, garbed in a mantle,
With winged sandals, turning airy course /
405 Beneath earth's depths, who hold the spirit's reins,
O eye of Helios, O mighty one,
Founder of full-voiced speech, who with your lamps
Give joy to those beneath earth's depths, to mortals
410 Who've finished life. / The prophet of events
And Dream divine you're said to be, who send
Forth oracles by day and night; you cure
All pains of mortals with your healing cares.
Hither, O blessed one, O mighty son
415 Of Memory, / who brings full mental powers,
In your own form both graciously appear
And graciously render the task for me,
A pious man, and render your form gracious
To me, NN,
That I may comprehend you by your skills
Of prophecy, by your own wond'rous deeds. /
420 I ask you, lord, be gracious to me and
Without deceit appear and prophecy to me."

Recite this both at sunrise and moonrise.

Notes

1. Among the vast literature, I note O. Petterson, "Magic-Religion: Some Marginal Notes to an Old Problem," *Ethnos* 22 (1957): 109–19; J. Goody, "Religion and Ritual: The Definition Problem," *BritJournSoc* 12 (1961): 142–64; M. Wax and R. Wax, "The Notion of Magic," *CurrAnthr* 4 (1963): 495–518; D. Hammond, "Magic—A Problem in Semantics," *AmAnthr* 72 (1970): 1349–56; R. Grambo, "Models of Magic," *Norveg* 18 (1975): 77–109; D. E. Aune, "Magic in Early Christianity," in *ANRW* vol. II, pt. 23, sec. 2 (Berlin, 1980), 1507–23; M. Winkelmann, "Magic: A Theoretical Reassessment," *CurrAnthr* 23 (1982): 37–66.

2. O. Petterson (see n. 1) p. 111 points out that submission was characteristic for religion as early as Paul and was subsequently highlighted by Luther, Calvin, and, later, Schleiermacher—hence the Frazerian dichotomy still seems to have a strong grip on Christian, especially Protestant, minds.

3. To give a few instances through time (see also n. 7): L. Deubner, *Magie und Religion* (Freiburg, 1922), 16 (= *Kleine Schriften zur klassischen Altertumskunde* [Königstein, 1982], 290); S. Eitrem, *Orakel und Mysterien am Ausgang der Antike* (Zürich, 1947), 37–40; J. de Vries, "Magic and Religion," *HR* 1 (1962): 214–21; A. A. Barb, *The Survival of Magic,* in *The Conflict between Paganism and Christianity in the Fourth Century,* ed. A. Momigliano (Oxford, 1963), 101; G. Luck, *Hexen und Zauberei in der römischen Dichtung* (Zürich, 1962), 4 (in his recent book of texts *Arcana Mundi* [Baltimore, 1985], Luck concedes that there may exist "a religious mood in magical texts," still upholding the old dichotomy; his further attempt at differentiation—that magical prayer never contains "consciousness of sin and the prayer for forgiveness"—is even more Christiano-centric). R. MacMullen (*Paganism in the Roman Empire* (New Haven, 1981), 70) defines magic as "the art that brings about the intervention of supernatural powers in the material world": this sounds like another subtle variation of Frazer.

4. See, e.g., the attempts to discern magical elements in Roman religion, from E. E. Burriss, *Taboo, Magic, Spirits: A Study of Primitive Elements in Roman Religion,* 3d ed. (New York, 1972); cf. idem, "The Magical Element in Roman Prayer," *CPh* 25 (1930): 48–49 to H. H. Scullard, *Festivals and Ceremonies of the Roman Republic* (London, 1981), 15; or in Greek hymnic poetry, A. Burnet, *The Art of Bacchylides* (Cambridge, Mass., 1985), 6 with nn. 2f. (I owe the reference to Dirk Obbink.)

5. Bibliography: C. Ausfeld, *De Graecorum Precationibus Quaestiones,* Jahrbücher für classische Philologie Suppl. 28 (1903): 502–47; K. Ziegler, *De Precationum apud Graecos Formis Quaestiones* (Breslau, 1905); F. Schwenn, *Gebet und Opfer: Studien zum griechischen Kultus* (Heidelberg, 1927) on evolutionary derivation of prayer from spell; K. von Fritz, "Greek Prayer," *RevRel* 10 (1945/46): 5–39; E. des Places, "La prière cultuelle dans la Grèce ancienne," *RevSocRel* 33 (1959): 343–59; H. S. Versnel, "Religious Mentality in Ancient Prayer," in *Faith, Hope, and Worship: Aspects of Religious Mentality in the Ancient World,* ed. H. S. Versnel (Leiden, 1981), 1–64; a general introduction in W. Burkert, *Greek Religion* (Cambridge, Mass., 1985), 73–75.

6. Mystery rituals: A. Dieterich, *Eine Mithrasliturgie,* 3d ed. (Leipzig, 1923); H. D. Betz, "Fragments from a Catabasis Ritual in a Greek Magical Papyrus," *HR* 19 (1980): 287–95; also see n. 36. Prayers: the principle is stated by R. Reitzenstein, *Poimandres* (Leipzig, 1904), 14: "Die Gebete . . . die z.T. ohne Rücksicht auf den Zweck der magischen Handlung aus älteren Quellen übernommen und für sie nur durch die Aufnahme unverständlicher Formeln überarbeitet sind"; his own analysis on p. 20f. of *PGM*.VIII.1-12 only shows

that the prayer combines earlier material, not that this material was "religious." Th. Schermann (*Griechische Zauberpapyri und das Gemeinde- und Dankgebet im 1. Clemensbrief* [Leipzig, 1909]), shows the close parallels between prayers in *PGM* and early Christian prayer, without drawing conclusions as to borrowing processes.

7. For Deubner and Eitrem, see n. 3; a typical statement is that of M. P. Nilsson, "Die Religion in den griechischen Zauberpapyri," *Bull.Soc.R.Lettres Lund* no. 2 (1947–48): 60: "Es ist sehr richtig . . . dass die Magie in ihrer Haltung den Göttern gegenüber der Religion grundsätzlich entgegengesetzt ist [the old Frazerian dichotomy]. Die Zauberer haben aber immer und überall sich Brocken und gar grössere Stücke der Religion angeeignet"—without drawing any conclusions for the theoretical position. Nilsson refers to A. J. Festugière, *L'idéal religieux des Grecs et l'évangile* (Paris, 1932); the second edition (Paris, 1981) still has, on pp. 284–85, "Une action magique . . . est d'essence *magique*, c' est-à-dire qu' elle commande une attitude particulière à l'égard de la divinité. On ne la prie point pour lui exprimer de la révérance ou de l' amour. . . . Ce n' est pas une prière, une demande, mais une *sommation*. On force la divinité à agir" (italics Festugière's).

8. Pl. *Leg*. 10.909b (certain people seduce men) *τοὺς δὲ τεθνεῶτας φάσκοντες ψυχαγωγεῖν καὶ θεοὺς ὑπισχνούμενοι πείθειν, ὡς θυσίαις τε καὶ εὐχαῖς καὶ ἐπῳδαῖς γοητεύοντες* ("claiming that they could draw up the souls of the dead, and promising to persuade the gods, because they bewitch with sacrifices, prayers, and spells"). Cf. *Resp*. 2.364b, about the activity of beggar priests and seers (*ἀγύρται καὶ μάντεις*): *ἐάν τέ τινα ἐχθρὸν πημῆναι ἐθέλῃ, μετὰ σμικρῶν δαπανῶν ὁμοίως δίκαιον ἀδίκῳ βλάψει ἐπαγωγαῖς τισιν καὶ καταδεσμοῖς, τοὺς θεούς, ὥς φασιν, πείθοντές σφισιν ὑπηρετεῖν* ("If someone wishes to harm an enemy, he will with small expense harm him, be he just or unjust, with spells and enchantments, since they persuade, as they claim, the gods to help them").

9. We still are badly in need of an *index verborum* to the second edition of *PGM*; see, on the problem, H. D. Betz, ed., *The Greek Magical Papyri in Translation* (Chicago, 1986), [= *GMPT*] xliv.

10. I exclude *PGM* XXIIb from further discussion because in a Jewish spell submission under the One God is nearly inevitable (a fact, though, that already undermines the Frazerian categorization); see also *PGM* IV.3008–78, a slightly hellenized orthodox Jewish exorcism (W. L. Knox, "Jewish Liturgical Exorcism," *HTR* 31 [1938]:191–203, where the magician says he has power over the daemon *ὅτι ἐπεύχομαι ἅγιον θεόν* [*PGM* IV.3029] "because I pray to the holy god")—although such submission is not confined to Jewish spells, as will be shown presently.

11. *PGM* IV.2785–2870 (= hymn 18 to Hecate-Selene); VI.5–46 (the hexametrical parts 6–21 and 40–47 = hymns 13 and 14, both to Daphne). In structure and function, the hymn is a versified prayer or spell. For a good discussion and earlier literature, see J. M. Bremer, "Greek Hymns," in *Faith, Hope, and Worship*, ed. H. S. Versnel (Leiden, 1981), 193–214; cf. Th. Wolbergs, *Griechische religiöse Gedichte der ersten christlichen Jahrhunderte, vol. I, Psalmen und Hymnen der Gnosis und des frühen Christentums* (Meisenheim, 1971); K.-D. Dorsch, *Götterhymnen in den Chorliedern der griechischen Tragödie* (Münster, 1983). For the magical hymns, *PGM*, vol. 2, p. 264 gives a bibliography; add H. Riesenfeld, "Remarques sur les hymnes magiques," *Eranos* 46 (1946): 153–60. E. Szepes ("Magic Elements in the Prayers of the Magical Papyri," *AAHung* 24 [1976]: 205–25) begins with the definition of religion as submissive versus magic as coercive and making claims on the gods (a position that overstates the Frazerian dichotomy and is belied by most Greek prayers, except the one of Socrates) and tries to find elements of magic, thus defined, in the hymns—which he does, given the generous definition. Morton Smith, "The Hymn to the Moon, *PGM* IV.2242–

2355," in *Proc.XVIth Intern.Congr.Pap.* (Chico, Calif., 1981), 643–54.

12. *PGM* VII.756–94; XII.103–6; XXXVI.211–30.

13. *PGM* II.9 (formula) and 13 (context of the same formula); III.498 and 590 (in the same formula). IV.2545 (in a hexametrical hymn to Selene [= hymn 20.20], *εὐχαῖσίν ⟨τ'⟩ ἐπάκουσον ἐμαῖς, πολυώδυνε* Μήνη, cf. ibid. IV.2566 [= hymn 20.38], *εὐχομένωι τ' ἐπάκουσον ἐμοί, λίτομαί σε, ἄνασσα*), 2973 (context, no formula), and 2998 (formula). In two other instances, *PGM* III.174 and 176, the text is too fragmentary to be of any help.

14. Frag. 1 L.-P. (the text is given in the Appendix along with a metrical prayer where the *preces* follow immediately after the [lost] invocation).

15. J. M. Bremer (see n. 11), 196–97.

16. For more details on the very standardized forms of these parts see E. Norden, *Agnostos Theos: Untersuchungen zur Formengeschichte religiöser Rede,* 4th ed. (Leipzig, 1923) pt. II, esp. pp. 143–76.

17. For the text see the Appendix, pp. 199–202; for the date of *PGM* IV, see E. N. Lane, "On the Date of *PGM* IV," *SCent* 4 (1984): 25–27.

18. *PGM* IV. 2844, *γράμματα σῶι σκηπτρῶι αὐτὸς* Κρόνος *ἀμφεχάραξεν, δῶκε δέ σοι φόρεειν, ὄφρ' ἔμπεδα πάντα μένοιεν*: Hecate, not Zeus, is here the successor as lord of the universe. For Cronos see S. Eitrem, "Kronos in der Magie," in *Mélanges J. Bidez,* vol. 1 (Bruxelles, 1934), 351–60.

19. This is explicitly stated in, e.g., the introduction to the *Spell of Pnouthis* (the sacred scribe, *PGM* I.42), who speaks of "all the prescriptions bequeathed to us in countless books" (*τὰ πάντα καταλει[πόμενα ἡμῖν ἐν] βίβλοις μυρίαις συντάγματα*). After Paul's successful exorcism in Ephesus, a huge number of such books (obviously containing prescriptions for exorcism) were burnt by "those interested in officiousness" (*οἱ τὰ περίεργα πράξαντες, Acts* 19.19); see A. Deissmann, *Licht vom Osten,* 4th ed. (Tübingen, 1923), 216. See, further, Lucian *Philops*. 31, who mentions "very many Egyptian books" on this topic.

20. E.g., *PGM* I.254 and 261; III.341 and 567 (the magician); IV.3013 (the person exorcised).

21. "Orphic" hymns: see W. Quandt, *Orphei hymni,* 3rd ed. (Berlin, 1962); parallels between them and hymns in *PGM* IV were collected by B. Kuster, *De Tribus Carminibus Papyri Parisinae Magicae* (Königsberg, 1911). Other instances: *Anth.Pal.* 9.524f.

22. The longer form, e.g., IV.1794, the shorter in II.100; see, esp., R. Merkelbach, "*φθισίκηρε*," *ZPE* 47 (1982): 172; L. Koenen, *ZPE* 8 (1971): 206.

23. This translation, given by Preisendanz ("die die Gräber schlägt"), seems preferable to E. N. O'Neil's translation (in *GMPT*) "you who make grief resound": *κάπετος* is "grave" already in Homer (*LSJ* s.v.); the goddess is the one who disturbs the graves.

24. In *PGM* IV.2834 the papyrus has *οὐ γὰρ φοιτᾶις ἐν* 'Ολύμπωι, *εὐρείαν δὲ τ' ἄβυσσον ἀπείριτον ἀμφιπολεύεις*. Preisendanz follows E. Miller, *Mélanges de littérature grecque* (Paris, 1868), 453, in writing *σὺ γὰρ*, and the translator E. N. O'Neal follows; whereas A. Henrichs in hymn 18.33 (after E. Heitsch, *Die griechischen Dichterfragmente der Kaiserzeit,* 2d ed. vol. 1 [Göttingen, 1963], lix) keeps the transmitted text. The correction is easy (the addition of *σύ ⟨τε⟩ γὰρ*, proposed in A. Meineke, *Hermes* 4[1870]:66, gives the required metrical form) and necessary; the transmitted form excludes Selene from Olympus, whereas all other predications make her an all-embracing, omnipotent divinity). For a similar correction, see XII.254 and XIII.779.

25. See Iambl. *Myst.* 5.26, *ἔργον τε οὐδὲν ἱερατικὸν ἄνευ τῶν ἐν εὐχαῖς ἱκετειῶν γίγνεται*. See the remarks of W. Burkert, (see. n. 5), 73.

26. *PGM* IV.2874, *ἐπὶ δὲ τῶν κακοποιῶν· οὐσίαν κυνὸς καὶ αἰγὸς ποικίλης, ὁμοίως καὶ παρθένου ἀώρου*. Preisendanz translates the end "desgleichen von einer vor-

zeitig verstorbenen Jungfrau"; this seems preferable to O'Neil's version in *GMPT* "or in a similar way, of a virgin untimely dead."

27. As already Iamblichus said (*Myst.* 7.4), τὰ ἄσημα ὀνόματα . . . ἡμῖν μὲν ἄγνωστα ἔστω ἢ καὶ γνωστὰ ἔνια, περὶ ὧν παρεδεξάμεθα τὰς ἀναλύσεις παρὰ θεῶν. cf. A. Dieterich (see n. 6), 32–40; Th. Hopfner, *Griechisch-ägyptischer Offenbarungszauber,* vol. 1 (Leipzig, 1923; repr. Amsterdam, 1974 [= *OZ*]); detailed discussions in his analysis in *Arch.Orient.* 3 (1931): 119–55 and 327–58; 7 (1935): 90–120; 10 (1938): 128–48; *GöttNachr.* (1931):441–58 and (1937): 147–56; a list in *PGM,* vol. 3, pp. 243–78 (*voces magicae*) and 279–80 (palindromes); F. Maltomini, "Nuovi papiri magici," pt. II: "I papiri greci," *SCOr* 29 (1979): 55–124; on the alleged connection with Christian glossolalia see D. E. Aune, "Magic in Early Christianity," *ANRW* vol. II, pt. 23, sec. 2 (Berlin, 1980), 1549–51.

28. Cf. J. Alvar, "Matériaux pour l'etude de la formule *sive deus sive dea,*" *Numen* 32 (1985) 236–73.

29. See text in the Appendix, pp. 202–3. The title given in the papyrus, φιλτροκατάδεσμος, is too narrow and indicates only one (perhaps the most sought for) application. R. Reitzenstein (see n. 6) pp. 20–21 analyzes the different layers of this complex text. This does not affect our analysis. For triple expressions in religious language, so prominent in our text, cf. E. Norden (see n. 16), 348–54; for anaphora in prayer, ibid., 149–60 ("Du-Stil").

30. Explained by A. Dieterich (see n. 6) p. 97; provided with Egyptian background by R. Reitzenstein (see n. 6) p. 19, see 365 with more parallels in A. Abt, *Die Apologie des Apuleius von Madaura und die antike Zauberei* (Giessen, 1908), 36; and misunderstood in *GMPT,* p. 145, n. 2.

31. In itself a very short *argumentum:* Hermes helps the whole world, so why not me?

32. From A. Abt (see n. 30), 44–49 into the reference books—C. Zintzen, "Zauberei, Zauberer," *Der Kleine Pauly,* 5 (Stuttgart, 1975), 1467.

33. *PGM* IV.1266, Ἀφροδίτης τὸ ὄνομα τὸ μηδενὶ ταχέως γινωσκόμενον· Νεφεριηρι (an Egyptian epithet); XII.240 (=XIII.763) τὸ κρυπτὸν ὄνομα καὶ ἄρρητον.

34. The material in Hopfner *OZ* 1, 176.

35. lambl. *Myst.* 7.4f. (see also above, n. 27); cf. Porph. *De Philos. ex Or. Haur.* p. 157 Wolff. One of the consequences is that these names cannot be translated; Iambl. *Myst.* 7.5; Origen, *C. Cels.* 5.45.

36. The magician as a *mystagogos,* A. Abt, (see n. 30), 36f.; magic and mystery religions, R. Reitzenstein, *Die hellenistischen Mysterienreligionen,* 3d ed. (Leipzig, 1927), 128–31 et passim; fragments of mystery rituals, the famous *Mithrasliturgie* and a ritual of the Dactyls (H. D. Betz [see n. 6]); the claim to be "enthroned" (*PGM* VII.746, [see n. 43]) refers back to Meter mysteries; see *GMPT,* p. 139, n. 134.

37. See Appendix, p.203.

38. For the acoustic signs, cf. A. Dieterich (see n. 6), 40–43, 228f.; R. Reitzenstein (see n. 6), 264–65; R. Lasch, "Das Pfeifen und Schnalzen und seine Beziehung zu Dämonenglauben und Zauberei," *ARW* 18 (1915): 589–93. Already Dieterich (see n. 6) p. 69, noted that such signs are not confined to magical ritual, see Aristoph. *Vesp.* 626; Pliny *HN* 28.25; Sophr. *Vita Cyri et Ioann.* (Migne, *PG* 87. 3521B). For comparable lists of signs and symbols see, e.g., *PGM* III.499; IV.1646 (Helios); LXX.9 (Hecate).

39. The texts appear in the Appendix, pp.204–6 (*PGM* XVIIb, VII.668–85, and V.392–423). See R. Reitzenstein, Review of Berliner Klassiker Texte VI, *GGA* 173 (1911):564 (copied by "offenbar ein gläubiger Heide zum Privatgebrauch"); E. Heitsch, "Zu den Zauberhymnen," *Philologus* 103 (1959): 223–36.

40. In the *apparatus criticus* to *PGM* XVII b.11.

41. *PGM* II.56, V.87, VII.554; Apul. *Apol.* 43; Hippol. *Haer.* 4.28; cf. R. Ganschinietz, *Hippolytus' Capitel gegen die Magier* (Leipzig, 1913), 30–33; Th. Hopfner, "Die Kindermedien in den griechisch-ägyptischen Zauberpapyri," in *Recueil N. P. Kondakov* (Prague, 1926), 65–74; E. R. Dodds, *"The Ancient Conception of Progress" and Other Essays* (Oxford, 1973), 190 and 201.

42. The Pythia at Delphi is the obvious example. E. Fehrle, *Die kultische Keuschheit im Altertum* (Giessen, 1910), 75–76 gives the sources. For Argos, Apollo: Paus. 2.24.1; for Patara, Apollo: Hdt. 1.182.2. Cf. in Rome Cic. *De div.* 2.86, Tibull. 1.3.11 (a boy for Praeneste).

43. See, e.g., *PGM* VII.745, *πάντως δέομαι, ἱκετεύω, δοῦλος ὑμέτερος καὶ τεθρονισμένος ὑμῖν* ("Emphatically I beg, I supplicate, I your servant and enthroned" [i.e., initiated; see n. 36]); XIII. 72, *ἐπικαλοῦμαί σε, κύριε, ἵνα μοι φανῇς ἀγαθῇ μορφῇ, ὅτι δουλεύω ὑπὸ τὸν σὸν κόσμον τῶι σῶι ἀγγέλωι* ("I call upon you, lord, to appear to me in a good form, for under your order I serve your angel" [there follows a series of *voces magicae,* giving the name of the angel]).

44. See also the remark of A.-M. Tupet (*La magie dans la poésie latine,* vol. 1 [Paris, 1976], 16) that invocations in the form of a command instead of a prayer are more common in literary descriptions of magic than in the actual magical texts.

45. *PGM* IV.1035, *ἐπάναγκος • ἐάν πως βραδύνῃι, συνεπίλεγε τὸν λόγον τοῦτον ὕστερον τῆς θεολογίας* ("*Charm of compulsion:* If somehow he delays, say in addition this following incantation").

46. *PGM* II.45 (after two days of failure); IV.1295 (*ἐπάναγκος τῆς γ' ἡμέρας*); IV.1434 (after three days of failure); IV. 2676 (the third day).

47. *PGM* IV.3112.

48. *PGM* IV.2567, *τούτωι δὲ ἐπὶ τῶν ἐπ⟨αν⟩αγκαστικῶν χρῶ· δύναται γὰρ πάντα ἐπιτελεῖν· μηδὲ πυκνῶς δὲ ποίει πρὸς Σελήνην, εἰ μὴ ⟨ὃ⟩ ἐπάξιον ᾖν τῆς ἐνεργείας πράσσεις.* It does not appear from the context in which situation the spell was used; but since the charm is very close to the following one, IV.2622–2707, where the coercive spell clearly is the last try (see n. 45), the same should be assumed here.

49. *PGM* XIII.35, *λέγων τὸν ὡρογενῶν, τῶν ἐν τῆι* Κλειδί, *καὶ τὸν ἐπάναγκον αὐτῶν καὶ τοὺς ἑβδοματικοὺς τεταγμένους.* The second version, *PGM* XIII 381 is not very different: *εὑρήσεις καὶ τοὺς ὡρογενεῖς καὶ τοὺς ἡμερησίους καὶ τὸν ἐπάναγκον αὐτῶν ἐν τῆι* Κλειδὶ *τῆι* Μοϋσέως (an otherwise unknown book). For *PGM* VIII see Morton Smith, "The Eighth Book of Moses and How It Grew," *Atti XVII Congr. Intern.Pap.* (Naples, 1984), 683–93.

50. See F. Boll, C. Bezold, and W. Gundel, *Sternglaube und Sterndeutung,* 6th ed. (Darmstadt, 1974), 178–80.

51. *PGM* XII.120 (another coercive spell), *ἐπάκουσόν μου, ὅτι μέλλω τὸ μέγα ὄνομα λέγειν* ΑΩΘ *ὃν πᾶς θεὸς προσκυνεῖ καὶ πᾶς δαίμων φρίσσει, ὧι πᾶς ἄγγελος τὰ ἐπιτασσόμενα ἀποτελεῖ.*

52. Surveys on daemonology in later antiquity: M. P. Nilsson, *Geschichte der griechischen Religion,* vol. 2 (München, 1961), esp. 210–12 and 540–42; C. Colpe et al., "Geister (Dämonen)," *RAC* 9 (1975): 546–797, esp. 640–68 (C. Zintzen), with vast bibliography. See also O. Böcher, *Dämonenfurcht und Dämonenabwehr* (Stuttgart, 1970)—the subtitle, *Ein Beitrag zur Vorgeschichte der christlichen Taufe,* is not very relevant; J. den Boeft, *Calcidius on Demons* (Leiden, 1977); J. Z. Smith, "Towards Interpreting Demoniac Powers in Hellenistic and Roman Antiquity," in *ANRW* vol. II, pt. 16, sec. 1 (Berlin, 1978), 425–39 (important from the point of view of methodology).

53. *PGM* IV.1037 (the beginning in n. 45), *ἐπιτάσσει σοι ὁ μέγας θεός, ὁ εἰς τοὺς αἰῶνας τῶν αἰώνων, ὁ συνσείων, ὁ βροντάζων, ὁ πᾶσαν ψυχὴν καὶ γένεσιν κτίσας*.

54. *PGM* VII.881, *ἐπικαλοῦμαί σε, δέσποινα τοῦ συμπάντος κόσμου, . . . δὸς ἱερὸν ἄγγελον ἢ πάρεδρον ὅσιον διακονή[σον]τα τῇι σήμερον νυκτί, ἐν τῇι ἄρτι ὥραι, . . . καὶ κέλευσον ἀγγέλωι ἀπελθεῖν πρὸς τὴν δεῖνα ἄξαι αὐτὴν τῶν τρίχων*.

55. Apul. *Apol.* 26 *Magus est qui communione loquendi cum dis immortalibus ad omnia, quae velit, incredibili quadam vi cantaminum polleat.* There is no reason to see in the *communio loquendi* the uttering of coercive spells, despite A. Abt (see n. 30), 118–24; cf. the way Plato expresses the fact (see n. 8).

56. M. P. Nilsson (see n. 52), vol. 2, 540.

57. *PGM* IV.2191 (milk, honey, wine, and oil accompanying the sacrifice of a cock and a prayer to Aion for the consecration of a magical tablet); IV.3149 (milk of a black cow and holocaust); XII.214 (wine, honey, milk, and saffron accompanying a holocaust and a prayer for the consecration of a magical gem); XIII.1015 (milk, wine not mixed with seawater, water for the consecration of a magical tablet).

58. Holocaust: IV.40 (white cock); IV.2396, 3148 (falcon or onager); XII.35 (white chick); XII.213 (a pure white goose, three cocks, three pigeons). Strangling: XII.30 (seven birds). "Sacrifice as is customary" (I.25), whatever that was.

59. Explicit at *PGM* XII.36, *ποιῶν τὴν τελετὴν κατάφαγε τὸν νεοσσὸν μόνος, ἄλλος δὲ μηδεὶς συν[έστω* ("while conducting this portion of the ritual, consume the chick by yourself, allowing no one else to be present". Two other instances only prescribe eating, and the magician appears to be alone: IV.2398 (the *σπλάγχνα*, inward parts of an onager[?], otherwise burnt), and IV.3150 (*συνευωχοῦ*, "feast together" with a statue).

60. See W. Burkert (see n. 5), 60–64 for examples and bibliography.

61. E.g., the rituals that M. P. Nilsson labeled "Jahresfeuer," *Griechische Feste* (Leipzig, 1906), 223–25; more in F. Graf, *Nordionische Kulte* (Rome, 1985), 410–17.

62. See F. Graf, "Milch, Honig, und Wein," in *Perennitas: Studi in onore di Angelo Brelich* (Rome, 1980), 209–21.

63. M. Detienne, *Dionysos mis à mort* (Paris, 1977), 163–217.

64. The text in the Appendix, pp. 205–6 (*PGM* V.392–423).

65. There are ample parallels to the last prescription, e.g., *PGM* II.23, V.458, VII.1011, VIII.67, XXII b.33.

66. See. C. Segal, "The Raw and the Cooked in Greek Literature," *CJ* 69 (1974): 289–308; A. Henrichs, "The Sophists and Hellenistic Religion," *HSCP* 88 (1984): 139–58, esp. 142–43.

67. See the *communio loquendi cum dis* in Apul. *Apol.* 26 (see n. 55).

68. I feel unable to explain this detail.

69. IV.2642, *ἡ δεῖνά σοι ἐπιθύει, θεά, ἐχθρόν τι θυμίασμα· αἰγὸς στέαρ τῆς ποικίλης καὶ αἷμα καὶ μύσαγμα, κύνειον ἔμβρυον καὶ ἰχῶρα παρθένου ἀώρου.* Another slander spell with accompanying sacrifice: IV.2455.

70. In itself, the sacrifice of a dog was offered to divinities of marginality and reversal: Hecate, Enyalios, birth goddesses (F. Graf, *Nordionische Kulte* [*Rome,* 1985] 422); the goat was forbidden in some cults (ibid., 256); the usual animal had to be monochrome and pure (P. Stengel, *Die griechischen Kultusaltertümer,* 3d ed. (München, 1920), 151–52); pregnant animals, i.e., mother and embryo, were offered only in exceptional rituals (P. Stengel, ibid., 125 and 255).

71. H. D. Betz, (see n. 6); cf. also the simple reference to the enthronement (see n. 36). Heracles: Eur. *Heracl.* 613; H. Lloyd-Jones, "Heracles at Eleusis," *Maia* 19 (1967): 206–29.

The gold leaves prescribe the actual words one had to speak to the guards in the underworld; G. Zuntz, *Persephone* (Oxford, 1971), 358–70; W. Burkert, "Orphism and Bacchic Mysteries: New Evidence and Old Problems of Interpretation," in *Protocol of the Colloquy of the Center for Hermeneutical Studies,* Colloquy 28 (Berkeley, 1977), 1–8.

72. E. Heitsch, "Zu den Zauberhymnen," *Philologus* 103 (1959): 215–36, esp. 220; idem "Drei Helioshymnen," *Hermes* 88 (1960): 150–58. For the connections with the "Orphic" hymns, see n. 21.

8

The Constraints of Eros

John J. Winkler

Once on the island of Samos while a wedding procession moved through the streets to the groom's house, a bystander announced to his friends that the bride would be kidnapped before she reached her new home and that on that very night another man would make her his wife. And so it happened: a group of armed men descended on the crowd, killed some who resisted, and scattered the rest as they fought their way to the bride in the center and made off with her. The shrewd observer was no magician but Polemo, a second-century detective of the heart's secrets through the science of physiognomy, who continues his account, "I later learned from people discussing the events that it had happened with her consent. And now I will explain to you the signs on which I based my judgment."[1] Polemo had noticed near the bride a young man whose face and bearing revealed the forces that had mastered him: audible breathing, sweaty spots on his clothes, a palpitation in the nose, color shifting back and forth from pallor to blush and an overall trembling as if from fear of disgrace. The bride's eyes were unusually liquid, though her gaze was sharp, and a certain sorrow hovered over her features.

This scene contains most of the social and psychological traits that characterize relations of amicable association (φιλία, *philia*) and sexual desire (ἔρως, eros) in the Mediterranean family of cultures. Let us begin by looking at its telltale signs and, like Polemo, analyzing the character that informs its cultural surface. In this way we will be able to place the symbolic and tangible techniques for manipulating eros—known from a wide variety of practical handbooks as well as literary scenes—into a living context of intelligible purposes. To the original purveyors of and clients for "love spells" eros was no mystery at all—a problem, perhaps, but not a mystery. At the same time, though much of what we loosely call "erotic magic" can be seen to conform to the common sense of ancient Greek culture (and to pockets of our culture), the confrontation between actual erotic spells and the masculine literary fantasies about erotic witchcraft will illuminate some dark corners of personal anguish and interpersonal spite.

Indeed, the presence of such venomous and malicious feeling in many of the erotic magical rites offers twentieth-century readers quite a jolt. "Love" is certainly not *le mot juste* for the scenes of bondage and humiliation that are acted out in the central group of procedures aimed at bringing a desired person to one's bed; we can speak of this as "passion," "lust," or "desire" but hardly as "love." The vanilla connotations of "love" for us include mutual delight and consent, harmonious and balanced tenderness, perhaps a certain loss of self in the great mystery of the beloved other; they do not include wishing discomfort, annoyance, profound inner turmoil, and pain on the body and soul of one's beloved, as do the bulk of erotic incantations, both generic and prescription, found in *Papyri Graecae Magicae* and *Defixionum Tabellae*. When we further note that the norm for such procedures is male agency and female victimage, we clearly have much to be concerned about. Do we have here one more concerted assault on women as a group, comparable to foot binding, witch burning and similar examples of institutional misogyny?[2] The answer is a complex one, involving experiences of projection and desire symbolized in a Mediterranean setting where the whole question of "access"—to the divine or to anything else—had a rather different weight than it does in twentieth-century capital economies.

Each of the three sections of the present essay contains some ancient material, an argument about it, and a subtext on the relation of that ancient material to our conditions of understanding. I take it that an awareness of our "conditions of understanding" should be present in any hermeneutically sophisticated account, but it is particularly necessary in the case of a "distanced" subject like magic. *Magic* is a relative term: we call something "magic" only if we do not (or no longer) accept the premises of its meaning or operation.[3] The term thus reveals—or may be used to reveal—as much about the speaker as it does about the object. The perspective adopted here has something in common with Polemo's as a participant-observer in the wedding on Samos, that is, it neither exclusively identifies with the observed field nor pretends to stand wholly outside it but tries to understand what the actors themselves experience and intend and also, where possible, to see more than they do. Let me here make explicit the tendency of my subtexts.

First I present some of the ordinary technologies for managing eros and explain them as relatively unproblematic actions, given the social setting of Polemo's captured bride, with its cultural patterns of agonistic dramatization. The only reason to call these "magic" is that *modern* reliance on impersonal sciences and other centralized disciplines of the state and university has deprived *us* of the crafty resourcefulness in regard to available materials and symbols that can flourish in a *more* face-to-face and self-reliant society.

Note how that last sentence fictionalizes modernity and creates a role for the reader as a predictable and familiar contemporary, not a magically gifted or experienced individual. I enjoy the strangeness of erotic magic. I attempt to explain it only by invoking another, unarguable "strangeness"—that of cultural difference as studied by social anthropology. This tactic at its best may succeed in making the reader feel that such "magic" is not so strange after all, that it is not so much a queer custom of Others as it is a muted aspect of ourselves.

Risks of serious misrepresentation are built into the organized study of other cultures, which tends either to glamorize them or patronize them.[4] It is particularly great when there is not the safety mechanism of informants who can tell us when our reconstruction slips into fantasy. The temptation to discern an essentially fictional unity is increased by the fact that the material, though it is all drawn from ancient Greek texts, comes from many different locales, times, levels of society. The picture can at best be an impressionistic sketch, a series of experimental snaps, not an ethnographic report.

I then explore the twilit world of *agōgai* (ἀγωγαί), rituals designed to lead a desired person to one's house and bed. These provide an unusually intimate picture of private and heartfelt anxieties, as staged in one's psyche by persons more experienced in self-dramatization and in entertaining themselves than our relatively passive consumer culture. The bed at night, or the rooftop nearby, is the imagined location of most *agōgai,* their place of performance and the goal of the rite, and it is in the fantasy world of half-sleep that the desperate, sometimes suicidal, passions grow strong. Some of the violence of language and gesture in the *agōgai* is due to the projected intensity of the performer's own sense of victimization by a power he is helpless to control. My subtext takes the risk here of romanticizing Mediterranean passion (as Stendahl and Browning did for Italy) as something exquisite and vital, missing in our drier and paler culture. This too, of course, is a trick of sorts. It is not that modern culture is passionless, only that we are differently embarrassed about certain displays and expressions of it. If I were writing a novel rather than a scholarly essay, these scenes might be made to fit very well into a modern context.

Finally, I dwell on the implications of a disturbing terra-cotta statuette in the Louvre, using the anthropological common sense of the first section and the passionate psychology of the second to graph some of the deep tremors of hatred of women that seethe in the symbolic actions of the *agōgai.* It should not be possible in the late twentieth century to continue to ignore the institutions of terror that have circumscribed the experience of women over the centuries. But the more seriously we take this question, the harder it is to give a single answer. I will suggest, among other things, that the victimage models enacted in *agōgai* paradoxically incorporate rather than suppress women's desire but that they do so only within the models of family competition and male fantasy in which any desire is a dangerous irruption into one's autonomy and in which women's desire in particular must be thought of—that is, by men—as submitting to the pretensions of masculine control.[5]

HOME REMEDIES FOR EROS

A Public Drama

The Samian wedding witnessed by Polemo would not have been perceived as a tale of true love triumphing over opposition. It was rather an unmitigated social disaster for all concerned. The two families, putting their alliance on display in the open streets by a procession from house to house, have been shamed before an audience that watches, evaluates, spreads the tale and keeps it in long remembrance. In the zero-sum competition between families a fall in the stock of one group means a rise in the value of others.[6] The social force of prescribed enmity, manifested in competition, gossip, and envy is so strong that its deleterious effects can even be executed unconsciously and unwittingly, as in the case of the evil eye.[7] The two principal terms that articulate this perpetual jockeying for position are *honor* and *shame,* represented respectively by the men and the women of a family.[8] The behavior of men in controlling the reputation of a household and in prudently administering its resources has been analyzed from fourth-century sources by Michel Foucault.[9] The experience of Mediterranean women is considerably more difficult of access (appropriately enough) and falls outside the scope of Foucault's project, which is to trace the archaeology of certain practices of self-discipline, but modern analogies provide a consistent framework within which the ancient evidence fits snugly and comfortably.[10]

Young women are the actors (often treated as passive tokens in the social game) whose cooperation is essential in the highly unstable process of transmitting property through the recreation and redefinition of family units. As the bearers of a tremendous symbolic weight determining the good or bad reputation of two families, virgins, brides, and young wives are often perceived as the point of maximum vulnerability in a household's integrity. "A woman's status defines the status of all the men who are related to her in determinate ways. These men share the consequences of what happens to her, and share therefore the commitment to protect her virtue."[11]

An emphasis on protecting the vulnerable, of course, installs the vulnerability of women as a permanent and necessary part of the system—a bar to the crossover of women, should any aspire to do so, into activities of the public and male realms. This is the first key point at which we must invoke a sophisticated awareness of the interplay between our values and our evaluations. A naive reliance on the public pronouncements of ancient men would lead us to think that their wives and daughters were objectively weak in mind and body, both light-headed and prone to passion. Such statements should be taken seriously as a social move in the competitive game,[12] but should not be accepted at face value. An equally naive assumption that ancient women were much the same in character and aspirations as modern women would lead us to dismiss most ancient texts as patriarchal propaganda and even to resent their automatic classification of women as counters in an exclusively male game.

To counteract these temptations we must not only weave together various types of evidence—in this essay mainly literary productions and medical self-help procedures—but we must be aware of the historical, cultural, and material premises of their utterance.[13] Chief among these, and one very difficult for us to grasp, is the prevalence of lying.[14] Duplicity is not just a cultivated skill useful in special circumstances but a permanent state of defensiveness against intrusive enemies who will use any knowledge about the private affairs of a family to bring it down. Hence with every affirmation of family honor we must also posit the unspoken comments of neighbors doubting its truth; by the same token every aspersion on a family's integrity must be understood as an "interested" comment. In short, every statement implies a plethora of competing and opposing evaluations in a network wherein no allegation is disinterested and every assessment is a strategic move in the collective maneuvering of public opinion concerning the relative prestige of family units.

These comments on duplicity serve, first, to underscore the high stakes of a public performance like a wedding, which stages the outcome of delicate negotiations between families entering into an alliance in the presence of numerous well-wishers who undoubtedly entertain concealed thoughts of at least low-grade hostility. In 1968 Manoli, an eighteen-year-old young man from a socially prominent family on the island of "Nisi" fell in love with a girl from a poor family: "As Manoli had eaten at the girl's house, his family felt that someone there must have put a love potion into his food." Resisting all the advice and pressure from his family, Manoli eloped with the girl and after a few days returned and went through a formal wedding ceremony:

> The groom was unable to consummate the marriage. It was believed that when the best man passed the wedding crowns over the couple's heads three times, someone present had uttered magic words and had tied three knots in a string, thereby acquiring power over the couple. . . . The groom took ill and was bedridden for four months. He began to waste away, and the priest was summoned daily to bless him and the house. A relative suggested that a witch in Athens be consulted and he, his parents, and his wife went to Athens. The witch performed a curing ritual and instructed them to return to Nisi and be remarried at an outlying chapel. This was done. At this service, known as reversing the crowns, the magic was broken.

The intense disapproval of the bride's parents at her elopement and of the groom's family at Manoli's folly gives ample grounds to reconstruct the network of "social forces" at work—gossip, defamation, dirty looks, and other covert symbolizing activities such as knot tying and muttered words of power—influencing and correcting the young couple's improper behavior. A few years later the couple was divorced after Manoli became convinced that his original love "must have been witchcraft. Otherwise, he would have listened to his father and waited. He added, 'Why else would I have married during the best years of my life?' "[15] The transition along public thoroughfares from the protected inner sanctum of one house to that of another exposes the symbolically vulnerable family members, bride, and groom to malignancy that is both inevitable and invisible.[16] (Playing out this line of thought, we might maintain that many forms of what we call ancient "magic" are simply our way of representing "social forces," equally inevitable and invisible, "forces" that no longer prevail with us. In time to come what we assume to be powerful "social forces" may in turn be called our "magic" by a future "enlightened" society.)

Second, the assumption that hostility will be present but masked helps us to place the techniques of erotic influence where they belong—that is, in the extensive penumbra surrounding public events; they are usually unrecorded by conventional public discourse but undoubtedly present in unguessably large amounts. Thus it is only the more conversational or unpretentious or deliberately outrageous works of classical literature that allude with any regularity to the covert acts of symbolic influence that most citizens are constantly practicing—and always denying that they practice—on their fellows. How did Pericles come to have so many cooperating allies (φίλοι)? "One hears" and "they say" that it was due to spells (ἐπῳδαί, *epōidai*) and love charms (φίλτρα, *philtra*).[17] Playfully, Socrates can explain the faithful devotion of his friends Apollodoros and Antisthenes to *philtra* and *epōidai* and the drawing of Simias and Cebes all the way from Thebes to the even stronger attractive power of ἴυγγες (*iunges*) (Xen. *Mem.* 3.11.16–18).

Recipes for Success

The later handbooks of self-help procedures record many such devices for (in the words of a modern magus) "winning friends and influencing people." Those that are focused narrowly and exclusively on sex will be considered later. Here we must first note the texture of the social fabric into which such private practices are woven, for eros cannot be neatly detached from the total world in which these actors try to upstage each other. The ideal of personal success for men in an agonistic, duplicitous, self-dramatizing culture includes the items requested in a prayer to Helios (*PGM* III.494–611): "Come to me with your face gleaming, at the resting-place (κοίτῃ) you choose yourself, granting me, NN, life, health, safety, wealth, fine children, knowledge, ready hearing, good disposition, good counsel, good reputation, memory, charm (χάρις), looks (μορφή), beauty (κάλλος) before all persons who see me, and make my words persuasive, you who hear everything without exception, great god" (lines 575–81).

The petitioner would like to shine in his community not only with external marks of physical success but most particularly in personal qualities, because the truly significant interactions in which his value is continuously judged and rated by others are those where he shows his individual excellence. To get the edge on competitors often means charming or outfoxing them rather than fighting or insulting them, so generalized prayers for success frequently include what may seem to us a rather peacocklike pride in looking good and being seen as sexually appealing. The social implications of radiating χάρις (*charis*) in the eyes of the community are brought out in spells specifically directed to acquiring or reinforcing one's charisma, such as XII.397–400, which promises the bearer of an inscribed wormwood root, "You will be charming and befriended and admired by all who see you."[18] Part of such a person's total influence as a force in his society is his outright sexiness, as in the recommendation to carry the right eye and first tail-joint of a wolf in a gold container to make the bearer "well thought of and successful and honored and victorious and sweet and desirable in form and loved and desired by women."[19]

Competitive success is regularly joined not only with personal charisma but with power to soften and restrain the anger of one's enemies.[20] A prayer to Helios under the rubric *Spell to restrain anger, win victory, gain charisma* (XXXVI.211–30) asks not only for "sexiness (ἐπαφροδισία) and charm before all men and all women" and for "victory over all men and all women" but also for protection against failure, plots, harmful drugs, exile, and poverty. These latter eventualities are seen as the outcome of other people's θυμός (*thumos,* "anger, resentment"), which must be restrained, hence the name θυμοκάτοχον (*thumokatochon,* "anger-restraint"). Exuding charm and warding off anger are equally necessary and coimplicated strategies in the project of maximizing personal success.[21] Anger-restraining techniques range from reciting lines of Homer (IV.831–32)[22] to saying prayers (XXXVI.161–77) to carrying inscribed pieces of metal or papyrus (VII.940–68, IX, and XII.179–81).[23] The last three are designed so that the name of a particularly dangerous person may be inserted according to the needs of the wearer. We have one such prescription spell on an ostracon asking Cronos to check the anger of Hori, son of Maria, and not let him speak to Hatros, son of Taeses.[24] This is mild. Others beg the god to subject, silence, subordinate, enslave, and trample on NN (VII.940–68, IX); that intention is acted out by placing a tablet in one's right sandal inscribed with mystic vowels and angels' names: "Just as these sacred names are being trampled, so also let NN who hinders me be trampled" (X.36–50).

The systematic interlacing of violence and charm, which we may find puzzling and even repugnant, is simply the necessary shape given to aspirations for success in that agonistic, masked, and duplicitous society. The φιλτροκατάδεσμος (*philtrokatadesmos*) of Astrapsakos (VIII.1–63) is obviously "a spell for some shopkeeper to ensure good business," as its translator notes.[25] But the blessings prayed for NN and his workplace (ἐργαστήριον, line 63) include not only victory and wealth but charisma (4, 27, 36), sexiness (ἐπαφροδισία, 5, 62), and a handsome face and body (προσώπου εἶδος, 5; μορφή, 27, 30; κάλλος, 27). Our analytic surgery should not sever the nexus of triumph and seductiveness in this shopkeeper's social personality, for they are constantly juxtaposed in its litanies.[26] Both his charm and his strength serve the same goal in the wary game of life: "Humble all before me and give me power and beauty, etc." (30–31) Hence the title φιλτροκατάδεσμος, which equally invokes violent restraint of one's competitors (κατάδεσμος) and friendly alliance (φιλτρο–), is not misleading, as the translator claims, but expresses the diffusion of a visible, low-grade eros throughout the competitive structure.

The serviceableness of such eros is most vividly portrayed in the rite for constructing and empowering a wax image of Eros to be an all-purpose assistant (*parhedros*) in XII.14–95. When it has absorbed the life breath of seven strangled birds, the Eros statue will serve its master in bearing powerful messages of compulsion to all men and women whom the owner wishes to influence. "I call upon you, in your beautiful resting-place (κοίτη), in your house of desire, serve me and always convey whatever message I tell you" (40–41). Among other things, this assistant can "make all men and all women turn to desire me" (ἐπ' ἔρωτα μου, 61–62) and can "grant me charisma, sweet speech, sexiness towards all men and all women in creation, that they be subject to me in all things that I wish, for I am the servant of the most high god who controls the cosmos, the ruler of all" (69–72). The power of this Eros assistant to enter people's houses (82); to appear in their dreams or as a divine visitation in the midst of sleep (15–16, 41–42, 83–84); or to "afflict them with fear, trembling, anxiety, mental disturbance" (54–55, 84) assimilates this serviceable or "social success" eros to the more focussed *agōgai* of the next section. Before turning to those, however, we must round off this treatment of simple self-help therapies by taking a brief look at aphrodisiacs and antaphrodisiacs.

Erotic Pharmacology

The terms φιλτροκατάδεσμος (*philtrokatadesmos*) and χαριτήσιον (*charitēsion*) covered not only prayers and amulets but more directly material technologies for stimulating and managing sexual feelings, such as penis ointments[27] (VIII.191–92) and love potions. The penis of a lizard caught copulating produces indissoluble affection in the woman who unwittingly drinks it; and if you can throw your handkerchief over lizards copulating it will be a χαριτήσιον μέγα (great spell to produce charm); the tail worn as an amulet promotes erection (Cyranides 2.14.10–13). This is useful knowledge and it circulated in massive quantities along informal channels on the ground, leaving but a few traces in the stratosphere of dignified writing.[28] Before the compilation of handbooks it resided with families and individuals who might share it as they chose.[29] Plato's *Charmides* shows how a cure for headache (consisting of a leaf and recited spell) could be passed on;

Achilles Tatius' novel *Leucippe and Cleitophon* does the same for a bee sting remedy (2.6). But note that both cures are incorporated into the narrative as tricks to seduce a desirable person (*Charm.* 155B) rather than as straightforward home remedies.

It seems that one of the central institutions for promoting and sharing such knowledge was the symposium. At least, the antiquarian and medical literature that picks up information about aphrodisiacs tends to set it in a *symposiastic* context.[30] The Playful Tricks of Democritos includes lamp illusions, oinoprophylactics, and sexual stimulants.[31] Athenaios' experts on cultivated dining know a good deal about what foods rouse desire.[32] They cite treatises in prose and poetry from classical and hellenistic times (Philoxenos' *Dinner,* Heracleides of Tarenton's *Symposion,* Terpsicles' *Concerning Aphrodisiacs*) that included equally useful advice on which foods cause gas (Ath. 53C) and which promote sperm (bulbs, snails, and eggs [64A]). But this material is obviously much older and more widely available; most of Athenaeus' aphrodisiacs are cited from fourth-century Attic comedy (Ath. 63E–64B, 356E–F). He might have added Aristophanes' *Ecclesiazusai* 1092, where the crones demanding sex from a young man advise him to gulp down a potful of bulbs.[33] Theophrastos records, with varying degrees of skepticism, the claims of fourth-century herb merchants:[34] ointment of snapdragon produces good reputation (*Hist. Pl.* 9.19.3), mandrake root in vinegar is a love-potion (9.9.1).[35] Hipponax evidently referred to a love potion to be drunk when one saw the first swallow in the springtime (172 West).[36] The simplest ways are sometimes best: the Aristotelian *Problems* (3.33) recommend a big breakfast and light dinner to promote sexual desire.

Many of these ingredients are explicitly conceived in a system of balances, since the prudent householder needs to control eros both in its arousal and in its dispersal, both for male and for female. The plant "rocket" (*Eruca sativa, εὔζωμον*), eaten green, prevents erection and wet dreams,[37] but its seeds mixed with pepper and honey produces an erection two fingers long and is especially recommended for "older men whose part is relaxed."[38] The weasel's right testicle is conceptive, the left is contraceptive (Cyran. 2.7). The right molar of a small crocodile worn as a amulet guarantees erection in men, the left produces "equally powerful pleasure in women" (Cyran. 2.29).[39] The testiclelike double bulbs of the orchis and *satyrion* have opposite effects (*adversantur altera alteri* [Pliny *HN* 27.65]), stimulating or depressing desire,[40] producing male or female children.[41]

Some ingredients are directed specifically against women, such as crane brain (Ael. *NA* 1.44) and sparrow's gizzard, which, given secretly in a drink, promotes pleasure and eros in the maiden who swallows it (Cyran. 1.18.42–4). Evidently most of this popular-technical writing was composed for men, yet in literature and historical anecdote suspicions are regularly directed to women as food handlers who might add secret ingredients to affect men's eros.[42] Here, too, the network of competition and suspicion motivates the action. Aretaphila, married to the hateful tyrant of Cyrene, tried to poison him; detected in her food-tampering, she defends herself by claiming merely to be preparing antidotes to the drugs and devices of other women, so that her husband's affection (*εὔνοια*) will not be drawn away and her honorable position will continue to be secure (Plut. *De Mul. Vir.* 256A–C.)

Deianeira, in Sophocles' *Trachiniai,* is similarly motivated: she hates "bold women," presumably those who use erotic charms to gain lovers, but uses Nessos' blood to keep Heracles faithful to her (lines 575–87).[43]

These self-help procedures, in their very ordinariness, do not constitute a reliable system for managing the deepest and most disturbing problems of passion, particularly the deeply unwanted experience of falling helplessly and hopelessly in love. Let us turn then once more to Polemo's wedding to study the medical and psychological aspects of such invasive, antisocial eros and the covert, symbolizing activities used by lovers on unwilling subjects.

REMEDIUM AMORIS

Polemo's detection of the truth concealed in two lovers' hearts was couched in a specifically medical idiom.[44] In effect, he was diagnosing an illness on the basis of signs visible in the face and bearing of the bride and her lover. The belief runs deeply through ancient medicine, social practice, and literature that intense desire is a diseased state affecting the soul and the body, an illness that up to a point can be discerned and analyzed but that is remarkably difficult to treat. The pathology is fundamentally melancholic: "Those who possess a large quantity of hot black bile become frenzied or clever or erotic or easily moved to anger and desire."[45] Nowadays, free-associating on the world *love* would not elicit *frenzy* and *anger* as primary responses. Since the premises we bring to "love stories" are somewhat different from those of the ancient Mediterranean, we must always be on our guard against misreading such narratives. For instance, when Herodotos begins a dynastic tale by saying "Kandaules fell in love with his own wife" (1.8), his audience knew from the word eros itself, as well as from the ominous conjunction of eros with marriage, that desperate events were in the offing.

If falling in love is, in many contexts, much the same as falling ill,[46] this is particularly the case for those whose desire has been roused for someone they must not associate with or marry. Of course, it is true for young lovers in general: "What is sweeter for a human being than the desire for a woman, especially a young man's desire? . . . Yet griefs and toil follow close behind."[47] If the lover's desire is not reciprocated or is not sanctioned by his or her family, the situation is desperate indeed, often resulting in the victim's taking to bed, wasting away and, if untreated, meeting his end.[48] Suicide is a common end to stories of hopeless love, as in the old Greek folk song of Kalyke, who leaped from a cliff rather than live without the young man who rejected her.[49] Eros in such circumstances is felt to be the sort of constraint or external pressure that may make life simply unlivable. The moral justifications for suicide in classical philosophy refer generically to god-sent constraints, among which eros counted in popular opinion if not in Aristotle's.[50]

Since eros, for all its beauty, loosens the limbs and dominates one's better judgment (Hes. *Theog*. 120–22), it is crucial to be able to diagnose and treat the affliction.[51] Polemo's demonstration belongs to a long line of claims and stories showing the limits of ordinary medicine's ability to cope with the devastating and disruptive power of eros. The physician can detect it, but he cannot control it. A few such narratives (proving the rule) do manage to have happy endings. The famous physician Erisistratos not only correctly analyzed the cause of prince Antiochos' sickness as unfulfilled eros but detected the object of his love by feeling his pulse as various members of the household entered the sickroom.[52] (Galen believed the story and claimed to have done the same.)[53] But more typically, eros confounds social expectations and medical expertise. People who lived near the River Selemnos in Achaia had told Pausanias that bathing in its waters cured eros for men and women alike, and he comments wryly, "If there is any truth in the tale, the water of the Selemnos is more valuable than great wealth" (7.23.3). The continual, quasi-medical complaint is that there is no drug to cure that disease except the beloved in person.[54]

The core experience represented in erotic literature is that of powerful involuntary attraction, felt as an invasion and described in a pathology of physical and mental disturbance. There are many well-known examples from Sappho[55] to *Daphnis and Chloe*,[56] but let us rather look at a less-celebrated example, that of the lovely Sosipatra, from the pages of Eunapios' *Lives of the Philosophers*.[57] Trained from the age of five in religious and philosophical lore by two old Chaldean magi, Sosipatra acquired the powers of prescience and telepathy that were the aim of many iatrosophists in the fourth century A.D. Her kinsman Philometor, "conquered by her beauty and her speech," falls into a state of eros that "was constraining him and doing him violence." He begins to ply unspecified arts at which he is adept to make Sosipatra feel the same, and she begins to be aware of his attempts. She describes her *pathos* to her confidant Maximus: "When Philometor is present, he is just Philometor . . . ; but when I see him leaving, my heart inside me is stung and twists about at his departure." Philometor perseveres in his rites, and Maximus struggles against him, "learning by divinatory sacrifices what Power Philometor had summoned to help him and then invoking a more violent and forceful Spirit to dissolve the spell of the lesser one."[58] Sosipatra is freed from her unwelcome feelings, and Philometor stops his plotting.

What covert symbolizing activities did Philometor practice? Many low-tech devices are known from *Papyri Graecae Magicae,* such as words to whisper over a wine cup before giving it to someone to drink[59] or ingredients to add to the drink itself.[60] A sun scarab properly boiled in myrrh and with just a touch of vetch ("the constraining plant," *καταναγκη βοτάνη*[61]) enables you to compel a woman to follow you once you have touched her, presumably with a dab of the wonderful oil.[62] Eye contact is powerful even without pharmaceutical or spiritual assistance,[63] but the effect can be enhanced by saying the secret name of Aphrodite to yourself seven times while looking at her (*PGM* IV.1265–74) or by saying a formula and breathing deeply three times while you stare at her. If she smiles back it is a sign that the spell has worked![64]

But Philometor was not fooling around with amatory *jeux*. The compulsion he tried to project because he felt it himself belonged rather to the far more extensive and expressive set of rites known as *agōgai,* spells to lead or draw a person to one's house and bedroom. The compelling power, if it works ideally, knows no resistance. I will use two literary pictures of perfectly effective *agōgai*—albeit both contain a measure of authorial irony—to frame an account of the rites and procedures known from *Papyri Graecae Magicae*.

The first occurs in Apuleius' *Metamorphoses,* or *Golden Ass* (2.32 and 3.15–18).

Fotis, servant to the witch Pamphile, is directed to filch some hairs from a handsome Boeotian youth while he is sitting in the barbershop. The barber prevents her, so, to cover her failure, she brings her mistress some hairs of the same color taken from goatskins recently shorn and hanging up, inflated, to dry. Pamphile takes them at night up to her rooftop, where she has her laboratory stocked with herbs, engraved metal tablets, pieces of shipwreck wood, parts of human corpses, and animal bones. With fire and incantations and various liquids she knots and burns the hairs, which causes the bodies from which they came, not the Boeotian youth's but the inflated goatskins, to come bouncing along the street and beat at Pamphile's house door.

Before the recovery of the rituals in *Papyri Graecae Magicae* one might have thought that Apuleius' picture was so much fantasy. But everything in it, with the stunning exception of the untypical gender of the agent, belongs to the regular procedures for drawing a person helplessly out of her house and into one's bed. At *PGM* VII.462–66, for instance, a copper nail from a shipwreck is used to write characters on a tin plate, which is then bound with *ousia*—some real material from the body of the person being enchanted, typically hair[65]—and thrown into the sea. The inscription reads, "Make her, NN, love me." *PGM* CXVII (first century B.C.) requires "two strands of her hair." *PGM* XVI, XIXa, and LXXXIV—all prescription love spells—were actually found wrapped with hair.[66] The progressive power of a spell to force its victim from her house and along the streets to yours can be watched in the flickering of seven wicks made from the hawser of a wrecked ship (*PGM* VII.593–619): "If the first flame sputters, know that she has been seized by the daemon; the second, that she has come out; the third, that she is walking; the fourth, that she is arriving; the fifth, that she is at the gate; the sixth, that she is at the doorlatch; the seventh, that she has entered the house."

Many *agōgai,* like Pamphile's, employ fire; ἔμπυρον (*empuron,* "in the fire") is even used as a rubric.[67] The obvious symbolism of burning passion felt as internal heat and fire occurs also in the commands ("May the soul and heart of her, NN burn, and be on fire until she comes loving to me, NN," XXXVI.81–82) and in the act of placing the inscribed papyrus with *ousia* in the dry-heat room of a public bathhouse (XXXVI.75).[68] But in some ways a more revealing aspect of the spell worker's fire is not its heat but the fact that it is specifically lighted at night. For *agōgai* are fundamentally generated not by a belief in some thermal technology as such but from a dramatic scene of nocturnal isolation with well-defined psychological features and a consistent strategy of duplicitous projection. If we look not at the prescientific beliefs (such as the power of "sympathetic magic") that may be extrapolated from the procedures but rather at their rhetoric, drama, and social psychology, we will be able to reach a much more authentic understanding of their tenor and function in the lives of ancient lovers.

A Night Scene

When the setting for an *agōgē* is specified, the time is night, the place is ordinarily a high room or rooftop from which the agent may speak to and observe the moon or the planet Venus, and the equipment includes a lamp or fire and sundry materials. There are twelve secure examples of such rites in *Papyri Graecae Magicae* and

Papyri Demoticae Magicae: *PGM* IV.1496–1595 (= XXXVI.333–60), IV.1716–1870 (*Sword of Dardanos*), IV.2006–2125 (*Pitys's agōgē*), IV.2441–2621, IV.2708–84, IV.2891–2942, VII.862–918 (*Lunar spell of Claudianos*), and XII.376–96, *PDM* xiv.1070–77, *PGM* xxxvi.134–60 (a night rite at one's own house door), *PDM* lxi.112–7, and *PGM* LXI.1–38. Other *agōgai* involving lamps (*PGM* VII.593–619, LXII.1–24) are almost certainly to be performed as night scenes, even though darkness and stars and sleep are not mentioned. In order not to misperceive the psychological and social relations implied in this scene, it must be divided into two aspects: the ritual scene of the agent (typically a man about to go to bed) and the imagined scene of the victim (typically a woman asleep in her own house).

The Agent. "Keep the offering (a mixture of drowned field mouse, moon beetles, goat fat, baboon dung, two ibis eggs, etc.) in a lead box, and whenever you want to enact it, remove a little and build a coal fire and climb up onto a high house and offer it saying the following formula at moonrise" (*PGM* IV.2466–70). The agent stands facing the night sky, looking at the moon (or the planet Venus), addressing a long prayer to her and watching for the goddess's reactions: "If you see the star (Venus) glowing, it is a sign that she (the victim) has been hit; if you see it scattering sparks, she has begun to walk along the road; if it assumes an oblong shape like a lamp, she has arrived" (IV.2939–42).[69] Certain preliminary steps may be required during the preceding day, such as placing an inscribed ass's hide under a corpse at sunset (IV.2038–41) or burying a wax Osiris under her doorsill (*PDM* 61.116), but the dormant power of such preparations is only awakened to life in the dead of night.

The Victim. The person to be affected by an *agōgē* is usually sleeping in her own bed, and what the agent wishes for her is an increasingly powerful feeling of restlessness and inner torment so that she cannot sleep: "Take sweet sleep away from her, let her eyelids not touch and adhere to each other, let her be worn down with insomniac anxieties focused on me" (*PGM* IV.2735–39);[70] "Isis is twisting and turning on her holy bed. . . . Make NN, daughter of NN, have insomnia, feel flighty, be hungry and thirsty, get no sleep, and lust for me, NN, son of NN, with a gut-deep lust until she comes and makes her female genitals adhere to my male genitals. If she wants to sleep, put thorn-filled leather whips underneath her and impale her temples with wooden spikes" (XXXVI.142, 147–52). The anxiety wished upon her is variously elaborated in terms of physical and mental symptoms such as burning ("Burn her *psyche* with a sleepless fire," IV.2767), disorientation ("Make her dizzy, let her not know where she is," LXI.15–16), and frenzy ("Let her be terrified, seeing phantoms, sleepless with lust and affection for me," VII.888–89).

Between the agent and the victim, as depicted in these scenarios, there is a curious transference. The rite assigns a role of calm and masterful control to the performer and imagines the victim's scene as one of passionate inner torment. But if we think about the reality of the situation, the intended victim is in all likelihood sleeping peacefully, blissfully ignorant of what some love-stricken lunatic is doing on his roof; while the man himself, if he is fixated on this particular woman, is really suffering in that unfortunate and desperate state known as eros. The spells direct that the woman's mind be wholly occupied with thoughts of the lover: from the evidence of the ritual we can say, rather, that the lover himself is already powerfully preoccupied with thoughts of the victim.[71]

The experience of eros as a victimization by unwanted invasive forces requires powerful therapy. The method of behavior modification employed in these rites is to make the lover go through motions that are masterful and dominating, with a text that suppresses all reference to his felt anxiety and conjures up instead an image of his love object's experiencing the torments that he is actually feeling: "Let her mind be dominated by the powerful constraint of eros." (IV.2762–63) The texts, of course, are technical manuals for professional (or at least expert) use and they do not provide us with information about the clients who consulted the expert. But it fits better with what we have seen of the psychology and type scenes of Mediterranean eros to imagine that the typical client for such a rite was not a Don Juan who wanted to increase the sheer number of his conquests but rather some young male who needed it rather desperately. Philometor, for instance, resorted to such rites because he was first "conquered" by the beauty and speech of Sosipatra. His attempt to project eros onto her was a calculated response to his own miserable plight.

The second literary picture of an *agōgē,* from Lucian, features just such a client. Allowing, again, for a certain authorial irony affecting the tone of the narrative, it presents an altogether more credible and realistic context for the actual employment of professional expertise in a problem of lovesickness. Young Glaucias, who is eighteen and has just inherited his father's estate, has fallen in love with Chrysis, the wife of Demeas, and it throws him into utter helplessness so that he can no longer study philosophy. His teacher comes to his rescue by bringing in a Hyperborean mage who can summon daemons, call stale corpses to life, and send *erōtes* to get people. The mage expects four minas up front to buy materials and sixteen more if the rite works. He waits for the waxing moon and at midnight, in an area of the house open to the sky, first summons the shade of Glaucias' father (who reluctantly gives his blessing to the affair) and then makes Hecate come up from the underworld and the Moon come down from the sky. Finally he fashions a little statue of Eros from clay and tells it to go fetch Chrysis. It flies off and after a little while she knocks at the door and rushes in to embrace Glaucias "like a woman absolutely mad with lust." At cockcrow Hecate, the Moon, and Chrysis herself all return to their proper places.[72]

Under the Lucianic icing of irony there is a substantial and perfectly plausible rendition of the real-life concerns and motives that led people to use magic rites. Looking at the fuller social context of their performance, we can see that *agōgai* are structured as a system of displacements. The first displacement, presumably of therapeutic value in itself, is the intense imaging of the client's illness as a thing felt by someone else. It might be very healthy for a self-conceived victim of love to act out a scene of mastery and control and to see from the outside and at a psychic distance what those torments look like. An *agōgē,* too, is the kind of last-ditch therapy made necessary by a certain cultural conception of eros, and as such it is a therapy that not only proclaims its own extremity but even in a certain sense its own impossibility. For the implied message of the rite is the home truth enunciated above, that there is no cure for eros except the beloved herself.

The control exercised by the agent is in some part a control over his own desperation, as he summons chthonic powers to do terrible things: it puts him in a role opposed to that of the erotic victim he "actually" is. The spiritual authority assumed by the lover is a second kind of displacement, for he speaks with the backing of, and sometimes in the person of, a mighty Power: "for I have about me the power of the great god, whose name cannot be uttered by anyone except by me alone on account of his power. . . . Hear me because of the constraint, for I have named you on account of NN, daughter of NN, that she have affection for me and do whatever I want" (LXI.23–29). Within this displaced authority ("It is not I, the helpless lover, who command you but I, the god or friend of gods") we can also detect the real authority of the expert who has designed and administered the rite. Desperate lovers like Glaucias are helped by their friends and also by experts, whose wisdom about personal problems may be couched in language very different from that now current but that was surely effective in its way (a point not appreciated by Lucian).

A third, and particularly revealing, displacement is that which occurs in the diabolic strategy occasionally adopted to enlist the aid of the goddess. To inflict so awful a condition as eros on an unwitting human, the goddess must be persuaded that the intended victim deserves to be punished:

> Let all the cloudy darkness part asunder and let the light of the goddess Aktiophis shine forth for me and let her hear my holy voice. For I come announcing the slander (διαβολή, *diabolē*) of the foul, unholy woman, NN. For she has slandered your holy mysteries, making them known to mortals. It was NN who said this, it was not I who said, "I saw the greatest goddess descending from the celestial pole, walking on earth without sandals, carrying a sword, naming a disgusting name." It was NN who said, "I saw the goddess drinking blood." NN said it, I did not: AKTIOPHI ERESCHIGAL NEBOUTOSOUALETH PHORPHORBA SATRAPAMMON CHOIRIXIE SARKOBORA. Proceed to NN and take away her sleep and make her *psyche* burn; give her mental torment, sting her out of her mind, chase her from every place and every house and bring her thus to me, NN. (IV.2471–92, cf. XXXVI.138–44)

The procedure is remarkably duplicitous, and therein lies its resonance with the larger patterns of Mediterranean social relations and with the cultural configuration given to eros. The projection of the lover's διαβολή (*diabolē*), like the similar transfers of his own victimage onto another and of another's commanding power onto himself, can also guide us in interpreting the deep dissonance that exists between the literary creations and the material artifacts in this field. There are two contrasts. First, in literature lovesick clients are usually female and the ritual experts whose help they seek in learning how to counteract or fulfill eros are usually male,[73] whereas the prescription papyri and tablets are predominantly composed by (or on behalf of) men in pursuit of women.[74] The generic rites in manuals, too, regularly and unselfconsciously assume that the client will be a man aiming at a woman.[75] The second contrast is that poetry and novels reveal a fascination with the powerful crone, often in groups like Macbeth's weird sisters.[76] Yet gangs of ugly women raiding cemeteries and swooping down on handsome young men do not figure in the papyri or tablets. In real life the persons famous for their "magical" powers and knowledge are regularly men, not women, as in Apuleius' *Apologia* 90.[77]

Both contrasts make sense as part of a cultural habit on the part of men to deal with threats of eros by fictitious denial and transfer. When weakened by invading eros, men could seek help through a personal ceremony that reassigned the roles of victim and master and, in the more generalized forum of literature, through the construction of public images that relocated both the victimage (in young women—Simaetha in Theoc. 2 and others) and the wickeder forms of erotic depredation (in older women—Canidia in Apul. *Met.* and others).

Women, for all we know, might have resorted to the same ceremonies; in a few cases (note 74) we know that they did, and this is testimony to the cultural belief that women were potentially victims of Eros and agents of daemonic eros in the same way that men are. But we certainly do not find them resorting to these ceremonies in the tangible materials left to us in even nearly equal numbers to men.[78] At a guess I would say that insofar as women's conceptions of eros as a problem for family politics overlapped with men's, their activities were more likely to take the form of vigilance and direct intervention in their immediate neighborhood. Young women who might fall into lovesickness are considerably more watched and guarded and disciplined than their brothers and presumably had less easy access to the male experts with their books and to the money required for hiring them. The "old women who know incantations" (Theoc. *Id.* 2.91) have expert knowledge but tend not to leave much behind them in the way of papyrus, lead, and published poetry. Clement of Alexandria imagines that rich ladies, transported in their litters to public temples, associate with old women and mendicant priests who teach them whispered spells to gain lovers (*Paed.* 3.28.3).

It might be tempting to identify such old ladies, wrinkled and dressed in black, as a source for the witch fantasy in men's imagination, but they are at most a Rorschach blot onto which men projected facets of their own behavior. One, more paradoxical conclusion to be drawn from the confrontation of real *agōgai* with literary fantasies is that Horace's Canidia behaves in a masculine style—and not only because she is energetic in going after what she wants (*mascula libido, Epod.* 5.41). The conceptual or imaginative source of the witch fantasy in men's erotic rituals is revealed in the poetic διαβολή (*diabolē*) (*PGM* IV.2574–2601), where the agent accuses his victim of distinctly witchy behavior: She, NN, offered to the goddess an unholy brew—fat of dappled goat and blood and filth, gore of a dead virgin, heart of one untimely dead, οὐσία [*ousia*] of a dead dog, a human fetus.[79] He goes on in the same creative vein to paint a picture of a recognizable witch, an unholy, dangerous outcast from the goddess's true worshippers. Yet no secret is made to us of the fact that this is a lie, deliberately concocted as a strategy to discredit the victim and to enlist the goddess's dread powers on the side of the lover. Further, the lie is a version of the lover's own truth, for the macabre handling of charnel material that regularly figures in witch fantasies is all of a piece with the *agōgai,* which require a good deal of animal mutilation, contact with cemeteries, and converse with the violent dead.

A Dream of Passion

> *Credimus? an qui amant ipsi sibi somnia fingunt?*
>
> Do we believe? Or do lovers make up their own dreams?
>
> VERGIL, *Eclogues* 8.108

A final facet of the lover's therapeutic procedure has yet to be disclosed in order to appreciate the many-sidedness of these highly (and in some ways deliberately) misleading texts. He frequently employs a compelling go-between—an Eros,[80] an unquiet corpse,[81] the goddess herself,[82] or any messenger chosen by her.[83] The imagined scenario is that the daemonic assistant will literally fly to the victim's house, enter her bedroom, and torment her until she comes to the agent's house. The most physical version is "Drag (ἕλκε) her by the hair and by the feet" (VII.887).[84] The visitation of the compelling assistant is acted out in a puppet show at *PGM* IV.1852–59. After making a clay statuette of Eros to be your assistant, "go late at night to the house of her whom you wish, tap on her door with the Eros and say to it, 'Look, here is where she, NN, dwells; *stand above her* and say the words I have chosen, assuming the appearance of the god or daemon she worships.' "

But that last clause introduces a notion that, at least in our categories, is something completely different. The power who will stand over her head in the likeness of a revered deity is a dream. The classic descriptions of significant messages received in sleep represent the dream speaker hovering over the head of the dream receiver. The erotic assistant in *agōgai* accomplishes his mission in a psychic form indistinguishable from influenced dreaming.[85] Indeed, the same ritual is frequently employed for dream sending (ὀνειροπομπή) and for love drawing.[86] It is not a question of redeploying the same procedure for a different purpose; rather an *agōgē* is inherently a nocturnal drama set in or near the lover's own bedroom; the imagined narrative of what happens to the victim is a projection of the lover's own disturbed sleep and erotic dreaming:[87] "Put the leaf (inscribed with an *agōgē*) under your head while you sleep" (*PDM* xiv.1070–77); "(The assistant you summon) will stand by you in the night *in your dreams*" (*PGM* IV.2052–53).[88]

The interlacing of dreams and sex could be explored in some detail, but I will cite only a few intriguing texts. Dio Chrysostom interprets Paris as a man who fantasized about a perfectly beautiful woman in a daydream. The entire story of Helen was originally Paris's dream, based on his erotic desire, but then he had the status and wealth to carry it out in waking life (*Or.* 20.19–23). In Dictys of Crete's journal of the Trojan War (6.14–15), Odysseus was frightened by powerfully erotic dreams, which interpreters saw as a warning against incest with his own son. The omen is fulfilled when Telegonus and Odysseus kill each other. Apollonios' Medea is entranced at the sight of Jason: as he leaves, his image stays in her mind as if she were dreaming and it remains there as a second level of reality during her waking moments (*Argonautica* 3.442–58). Best of all, "when night came on again, Artaxerxes was on fire, and Eros reminded him how beautiful were Callirhoe's eyes, how fine her features. He praised her hair, her walk, her voice, how she had entered the courtroom, how she had stood and spoken and kept silent, how she had showed modesty and how she had wept. He kept awake most of the night and only fell asleep long enough to see Callirhoe in his dreams" (Chariton *Callirhoe* 6.7).

These literary elaborations are based on a widespread but little studied association of dreams and sex that is summed up in the verb (ἐξ)ονειρώττω and noun (ἐξ)ονειρωγμός or ὀνείρωξις. The verbal suffix–ωττω/ωσσω indicates a physical disturbance, usually of an unhealthy sort.[89] Observations concerning such dreams are plentiful in Hippocrates, Aristotle, Galen, and similar writers. The term applies not only to the dreams of men accompanied by seminal emissions but also to women's erotic dreams (Ar. *Gen. An.* 739a21–27, *Hist. An.* 10.6 [637a27–28] and 10.7 [638a5]). The most striking case of its use I know occurs in Celsus' charge that Mary Magdalene's encounter with the resurrected Jesus was only the ὀνειρωγμός of a sexually excited [πάροιστρος] woman.[90]

To give a rounded account of this erotic therapy one should also include (though there is not room to do it here) the rites of dream sending.[91] In Pseudo-Callisthenes' *Alexander* the Egyptian pharaoh-in-exile, Nectanebos, meets Olympias and agrees to help her with his knowledge.

> Nektanebos went forth from the palace and quickly picked and gathered a plant which he knew suitable for provoking dreams. And having rapidly done this, he made a female body of wax[92] and wrote on the figure Olympias' name. Then he made a bed of wax and put on it the statue he had made of Olympias. He lit a fire and poured thereon the broth of the plant, until the spirits appeared to Olympias; for he saw, from the signs there, Ammon united with her. And he rose and said, "My lady, you have conceived from me a boy child who shall be your avenger." And when Olympias awoke from her sleep, she was amazed at the learned diviner, and she said: "I saw the dream and the god that you told me about, and now I wish to be united with him."[93]

The same disturbing, malevolent imagery occurs both in *agōgai* and in dream-sending rites, such as that of Agathocles (*PGM* XII.107–21): "Take an all-black cat who died a violent death, write with myrrh on papyrus the following inscription with the dream you want to send, and place it in the mouth of the cat." With this we should compare two erotic *agōgai* in which papyrus inscriptions are to be placed in the mouth of dogs who died a violent death (XIXb, XXXVI.361–71) and one that requires that a skull fragment of a man who died violently be placed in the mouth of a wax dog (IV.1872–1926). The symbolic and gestural language is similar because the cultural configuration of classical and later Greek eros is shaped with elements of violence and dreaming that our culture keeps at arm's length.

There are many *agōgai* like these two that do not happen to contain explicit references to the "night scene." If we correctly place the typical enactment of such rites in the lamp-lit world of a lovesick man alone with his feelings and about to enter the powerful underworld of his own *psyche,* then *agōgai* again turn out to make sense as psychodramas in which intensely disturbing emotions are manipulated and treated. Taken together, they give us a uniquely vivid view of personal anxiety in a Mediterranean cultural setting, with their characteristic self-dramatization, suicidal intensity, and masking procedures.

THE TORMENTS OF *PSYCHE*

A tiny but deeply disquieting terra-cotta statuette in the Louvre shows a woman on her knees, hands behind her back, pierced with thirteen nails.[94] Instructions for such an artifact are given at IV.296ff., where a wax or clay image of a kneeling and bound woman is to be accompanied by a statuette of Ares standing over her and plunging a sword into her neck. Each body part on the female is to be inscribed with a magical phrase and pierced with one of thirteen copper needles while saying "I pierce such-and-such part of her, NN, in order that she have no one in mind but me, NN." The purpose expressed in the words is psychological—the lover aims to create in his victim a state of mental fixation on himself—but the imagery is physically violent, even sadistic. A contrast between psychological "bonding" and physical torment is even clearer in the spell known as the *Sword of Dardanos* (IV.1716–1870), in which a magnet is engraved with Psyche ridden by Aphrodite and burned from beneath by a torch-holding Eros on one side, and on the other side with Eros and Psyche embracing.

The problem can be posed in more general terms by noting that a good deal of covert erotic ceremonial employs the objects, methods, and language found in the procedures against enemies known as *κατάδεσμοι* (*katadesmoi*, or binding spells).[95] The statue instructions cited above are entitled *φιλτροκατάδεσμος* (*philtrokatadesmos*), which might seem to be a simple contradiction in terms—at least on a certain understanding of affection and desire and the desirability of mutual affection. The fundamental idea behind binding spells and *agōgai* alike is constraint. To the Louvre figurine compare the procedure at *PGM* V.304–69. An iron ring is wrapped inside a sheet of papyrus or lead; inscribed "Let his mind be bound (*καταδεθήτω*)"; pierced with the pointed stylus; tied round with knots while the agent says, "I bind (*καταδεσμεύω*) NN to such-and-such an action;" then placed either in an unused well or in the grave of one untimely dead. Or compare the colorful "I bind you, Theodotis daughter of Eus, by the tail of the snake and by the mouth of the crocodile and by the horns of the ram and by the venom of the asp and by the whiskers of the cat and by the penis of the god, that you may not be able to have intercourse ever with another man either frontally or anally or to fellate[96] or to take pleasure with another man except me alone Ammonion Hermitaris."[97]

Audollent rather indiscriminately bundles together incantations designed to curse love rivals, make couples divorce, make a pimp's business decline, or bring a desirable person to one's door.[98] It is only the last that interests us here. Lead texts, rolled up, frequently pierced with a nail, and often found in tombs contain commands to powerful spirits identical to those in papyrus texts—to take away the sleep of a named woman, to make her burn with love for a named man, to make her think constantly of him: *coge illam mecum coitus facere* (*DT* 230).[99] Operations for bringing about a divorce or enmity between friends naturally invoke Typhon,[100] but he also figures prominently in *agōgai,* both by name and symbolically in the use of ass's blood.[101] Like binding spells, erotic rites employ the violent and untimely dead as agents.[102] The first effect of some *agōgai* is to send the affected person to her bed feeling ill;[103] to this we may compare *PGM* CXXIV, a spell designed specifically to cause illness (*κατακλιτικόν*), that uses a wax doll pierced with bones and placed in a pot of water.[104]

The forces brought to bear for erotic constraint are in principle deadly: the *agōgē* demonstrated to Hadrian "evoked [one] in one hour, sickened and sent [one] to bed in two hours, killed [one] in seven."[105] Hence both binding spells and *agōgai* can employ animal mutilation to strengthen their point.

Here is a test for the reader. Which of the following procedures do you think is used for love and which for hate: (1) smear an inscribed lead tablet with bat's blood, cut open a frog and place the tablet in its stomach, stitch it up and hang it from a reed and (2) take out a bat's eyes and release it alive, put the eyes in a wax or dough figurine of a dog, pierce the eyes and bury it at a crossroads with an inscribed papyrus? The former (*PGM* XXXVI.231–55) requests the supreme angels that he, your enemy, NN, drip blood as the frog drips blood, the latter (IV.2943–66) asks Hecate that she, NN, lie awake with nothing on her mind except you.

The Louvre figurine raises thoughts of systemwide female victimage and male dominance. Is this what eros meant (or means) to men—women's bondage, pain, humiliation, submission? The answer is complex. In the previous section I sketched the pretensions of male control and the patterns of gender transfer used to hide men's vulnerability and erotic agency. The Louvre image, existing within a cultural system that assumed and demanded a woman's submission to the controlling male of her family, could certainly not escape that network of meanings. But the submission in question is a social protocol, not a sexual practice; and we should at least be cautious about assuming a perfect isomorphism between the public stance and the private posture. Ancient texts are not as chatty and revelatory about personal and sexual histories as modern writers, but what we can see of personal psychology in the *agōgai* (second section above), framed by the social constraints depicted in the first section, points to four other facets of the bondage imagery that should be given weight as well.

First, insofar as the operations are a wish that she, NN, come to feel eros as deeply and disturbingly as the operator himself feels it, the binding and piercing represent not a will to dominate but a replication in her of his own experience. The submission in this case is portrayed as a submission to erōs itself, to a painful state of being in which one is afflicted by "affection, desire, pain" (*PGM* XVI.5–6), though willing someone into that state is of course selfishly motivated. The problem here is one of translating emotions and gestures from one cultural system to another. We—at least some of us—lack the categories, and hence the experience to say "I bind you, Nilos, with great evils: you will love me with a divine eros." (*PGM* XV).[106]

Second, some of the torments—insomnia, loss of appetite, dizziness—are temporary phenomena, inducements to action rather than ends in themselves. That at least seems to be implied by the frequent stipulation that she feel those things "until she comes to me."[107] This instrumental view of erotic torments runs contrary to the previous one, which sees them as constitutive of eros; but it shows in the agent that same quality of radical self-centering and indifference to the needs of others, which somehow also remains innocent of malice. The aseptic distance and respect for persons that are fundamental to some modern social ideologies make it difficult for us to see the petulant bravado of these private rites on its own terms. In a modern context they would be *simply* malicious; in an ancient context they are that and other things as well. In order not to misinterpret the significance of rites of erotic compulsion, we have to grant that they grow organically and naturally in a culture that assumes a far higher degree of self-serving activity in every sphere of social relations on the part of every agent and counteragent, individually and in groups.

Third, the erotic rhetoric reckons little with days and months: its units of measurement are now and forever.[108] When Pamphile casts her eye on a handsome youth, she aims to bind him with "eternal bonds of profound love" (*amoris profundi pedicis aeternis alligat, Met.* 2.5). The agent in at least some of these cases seems to be aiming at the bondage of marriage or the equivalent, praying for continuous, life-long love: "all the time of her life" (*PGM* VII.913–14); "Let him continue loving me until he arrives in Hades" (XVI.24–25); *usque ad diem mortis suae* (*DT* 267). In one case this is explicit: Domitiana adjures the daemon to bring the tortured and sleepless Urbanus to her and ask her to return to his house "to be his mate" (*σύμβιον γενέσθαι*). Her prayer concludes, "Yoke them as mates in marriage and desire, for all the time of their lives; make him subject to her in desire as her slave, desiring no other woman or maiden" (*DT* 271). Bondage in this event means permanence and stability. From the last text quoted, it might also seem to mean dominance, but I take that to be a private metaphor for Domitiana's real aim, which is not that her husband be publicly known to be her devoted slave but that he be faithful to her.

Fourth and most important, we must look at the real social context of these covert operations. *Agōgai* are a kind of sneak attack waged in the normal warfare of Mediterranean social life. Plato's Diotima generalizes in calling Eros a crafty hunter and a bold plotter against the beautiful (*Symp.* 203D); in fact *agōgai* are aimed as a rule at women and maidens, who are constantly guarded and watched by their own families and by all the neighbors. The means of eliciting consent and independent action on the part of these "passive actors" in the dynamic game of interfamily competition is to rouse their sexual desire. Let us return a final time to the Samian bride observed by Polemo: "And I later learned that it had happened with her consent."[109] Covert erotic rites operate on exactly the same protocols as parental vigilance, namely, that young women are quite apt to have sexual feelings and minds of their own and might well act on them even though it results in social tragedy and a calamitous fall in the fortunes of a house. Thus, not only are such rites of use to lovelorn swains, they are very useful for the face-saving needs of families whose daughters have actually eluded parental control. If they can claim that some devil made her do it, family honor is not so deeply hurt as it would be by her voluntary wantonness. *Agōgai,* viewed from this angle, are a backhanded tribute to the potential power of female sexual autonomy, though their language is that of divine compulsion, not that of *Our Bodies, Ourselves*.

If the heat of this last analysis properly brings out the invisible writing on love spells—their implied social script—then we have the paradoxical result that these wishes for bondage are a discourse (of sorts) about women's desire; and they speak of women's desire as an experience that will seem not to be forced on them by the expressed choice of a parent or suitor but to well up from within. They are one of the few categories of ancient text that speak at all of female pleasure, and they do so persistently and with a wide range of expression: "May she come melting with eros and affection and intercourse, fully desiring intercourse with Apalos" (*PGM* XIXa.53); "May she accomplish her own sexuality (*τὰ ἀφροδισιακὰ ἑαυτῆς*)" (IV.404). "Bring her loving me with lust and longing and cherishing and intercourse and a manic eros" (*PGM* CI.30–31). The social implications of this autonomous desire are alluded to in the neighboring clauses that request forgetfulness of parents and relatives, husband and children.[110]

These four considerations do not—and are not intended to—dispel the anguish roused in us by the Louvre image. Rather they are baby steps on a methodological path that has not yet been widely followed in studying classical and later Greek culture, particularly the subjects of sex, gender, and magic. These terms and practices are bundles of complex, historically specific meanings that are socially constructed according to the interests of cultures and economies very different from our own and hence difficult to translate without losing not just their savor but their very soul. There is no magic phrase, such as ANTHROPOLOGY that will guarantee success to our hermeneutic project any more than earlier slogans worked to unlock all secrets: FISHER KING, MOTHER GODDESS, CLASS CONFLICT, STRUCTURALISM. But cultural and social anthropology does at least raise questions and provide comparisons that illuminate much of the ancient material, letting us see much more clearly just how familiar and how strange it really is.

Notes

1. Polemo, *De Physiognomonia Liber,* chap. 69, in R. Foerster *Scriptores Physiognomonici,* vol. 1 (Berlin, 1893). Works frequently cited in this essay: *PGM* = *Papyri Graecae Magicae,* ed. Karl Preisendanz, 2nd ed. rev. Albert Henrichs, 2 vols. (Stuttgart, 1973–74). *PGM* contains Greek papyrus texts numbered I–LXXXI, "Christian" papyrus spells numbered P1–24, ostraca numbered O1–5, and two wooden tablets numbered T1–2. The English translation of *PGM* is *The Greek Magical Papyri in Translation, Including the Demotic Spells,* vol. 1, *Texts,* ed. H. D. Betz (Chicago, 1986), abbreviated *GMPT,* which unfortunately does not contain Preisendanz' nonpapyrus and "Christian" material. As the editor makes clear on p. ix, *GMPT* is conceived within a Christian interpretive framework, which explains its sectarian omission of Preisendanz' "Christian" papyri. I suspect that the omission of the ostraca is similarly motivated, since two of them are "Christian" magic (O3–4). For demotic spells I use PDM plus the number. *DT* = A. Audollent, *Defixionum Tabellae* (Paris, 1904); D. Wortmann, "Neue magische Texte," *Bonner Jahrbücher* 169 (1969): 56–111.

Many friendly readers have helped me improve this essay, all of whom I thank—in particular, Froma I. Zeitlin, Michael Herzfeld, Amy Richlin, Ludwig Koenen, Marilyn Arthur, Robert Daniel, and Ann Hansen.

2. Current examples in D. E. H. Russell and N. Van de Ven, ed., *Crimes against Women: Proceedings of the International Tribunal* (Millbrae, Calif., 1976).

3. D. E. Aune, "Magic in Early Christianity," in *ANRW* vol. II, pt. 23, sec. 2, (Berlin, 1980) 1506–57; an exciting and magisterial treatment of the larger interpretive issues in C. R. Phillips III, "The Sociology of Religious Knowledge in the Roman Empire to A.D. 284" in *ANRW* vol. II, pt. 16, sec. 3, pp. 2677–2773, esp. 2711–32 on "magic."

4. Gerrit Huizer and Bruce Mannheim, eds. *The Politics of Anthropology* (The Hague, 1979); James Clifford and George E. Marcus, *Writing Culture: The Poetics and Politics of Ethnography* (Berkeley, 1986); specifically on Greece, Michael Herzfeld, *Anthropology through the Looking-glass: Critical Ethnography in the Margins of Europe* (Cambridge, 1987).

5. A meaningful ethnography would require at the minimum not only a social object well-defined in space and time but the possibility of interviews, observations, and counting, which our resources cannot provide. Lacking this, I have resorted to the now familiar, though still debatable, notion of a Mediterranean family of cultures whose broad lines of similarity have persisted across the *longue durée*. The features of this cultural family and in particular its erotic items may well be found elsewhere in the world: I make no claim about a specifically or uniquely "Mediterranean" character in their configuration. Similarly, though I often con-

trast "us" and "them" as cultural strangers, many of "their" notions and practices may have a familiar ring. That is to say, my working portrayal of a twentieth century "us" is as much in want of qualification as my global picture of an ancient Mediterranean "them". The heuristic value of this strategy will be greatest if the reader first accepts the temporary role and character of a "twentieth-century Western consumer-capitalist" and then resists it.

6. Alvin W. Gouldner, *Enter Plato: Classical Greece and the Origins of Social Theory* (New York, 1965), 45–55. The economic metaphor is natural to us but should not mislead: the value in question is precisely not one that can be quantified and exchanged but depends on personal assertion, vindication, and negotiation in the intimate forums of a small-scale society. Good examples of misreading the honor code as an exchange relation are found in P. Bourdieu, "The Sentiment of Honour in Kabyle Society," in *Honour and Shame: The Values of Mediterranean Society,* ed. J. G. Peristiany (Chicago, 1966), 191–241.

7. Clarence Maloney, ed., *The Evil Eye* (New York, 1976), especially R. Dionisopoulos-Mass, "The Evil Eye and Bewitchment in a Peasant Village" and Vivian Garrison and Conrad M. Arensberg, "The Evil Eye: Envy or Risk of Seizure? Paranoia or Patronal Dependency?"; Michael Herzfeld, "Meaning and Morality: A Semiotic Approach to Evil Eye Accusations in a Greek Village," *American Ethnologist* 8 (1980): 560–74; Anthony H. Galt, "The Evil Eye As Synthetic Image and Its Meanings on the Island of Pantelleria, Italy," *American Ethnologist* 9 (1982): 664–81. For ancient material, see O. Jahn, *Aberglaube des* bösen Blicks bei den Alten, SB Leipzig 7 (Leipzig 1855); A. Moreau "L'Oeil malefique dans l'oeuvre d'Eschyle" *REA* 78/79 (1976/77): 50–64.

8. J. G. Peristiany (see n. 6) shows that the values are not a uniform system throughout the area, a point made by M. Herzfeld, "Honour and Shame: Problems in the Comparative Analysis of Moral Systems," *Man* n.s. 15 (1980): 339–51; David Gilmore, ed., *Honor and Shame and the Unity of the Mediterranean* (Washington, D.C., 1987); K. J. Dover, *Greek Popular Morality in the Time of Plato and Aristotle* (Oxford, 1974), 95–102, 205–13, 226–42; Julian Pitt-Rivers, *The Fate of Shechem; or, The Politics of Sex* (Cambridge, 1977).

9. *Histoire de la sexualité,* vol. 2, Michel Foucault, *L'usage des plaisirs* (Paris, 1984); English translation by R. Hurley, *The Use of Pleasure* (New York, 1985).

10. Ernestine Friedl, *Vasilika: A Village in Modern Greece* (New York, 1962); John K. Campbell, *Honour, Family, and Patronage* (New York, 1964); Peter Walcot, *Greek Peasants, Ancient and Modern* (Manchester, 1970); Juliet du Boulay, *Portrait of a Greek Mountain Village* (Oxford, 1974); R. B. Hirschon, "Under One Roof: Marriage, Dowry, and Family Relations in Piraeus," in Michael Kenny and David I. Kertzer, eds., *Urban Life in Mediterranean Europe: Anthropological Perspectives* (Urbana, 1983); Jill Dubisch, ed., *Gender and Power in Rural Greece* (Princeton, 1985); and above all Michael Herzfeld, *The Poetics of Manhood: Contest and Identity in a Cretan Mountain Village* (Princeton, 1985).

11. J. Schneider, "Of Vigilance and Virgins: Honor, Shame, and Access to Resources in Mediterranean Societies," *Ethnology* 10 (1971):1–24.

The charge of adultery brought against Lycophron was supported by an allegation that he had followed the mule cart on which she was riding during the wedding procession and had openly begged her not to consummate the marriage. Since her brother, an Olympic wrestler, was there, "could I have been crazy enough to utter such shameless words about a free woman in the hearing of all present and not be afraid of being strangled to death on the spot? For who could have endured to hear such words about his own sister . . . and not have killed the man who uttered them?" (Hypereides, *In Defense of Lycophron* 6). "One of the men in Chios, . . . was taking a wife and, as the bride was being conducted to his home in a chariot, Hippoklos the king, a close friend of the bridegroom, mingling with the rest during the drinking and laughter, jumped up into the chariot, not intending any insult but merely being playful according to the common custom. The friends of the groom killed him" (Plut. *De Mul.*

Vir. 244E). See also B. M. Lavelle, "The Nature of Hipparchos' Insult to Harmodios," *AJP* 107 (1986):318–31.

12. In the Cretan mountain village of "Glendi," card games in the coffee house are an emblematic male contest in which the cards themselves are frequently given feminine names, successful players are spoken of as sexually charged ("He's hot," "He's ploughing straight" [= "He is copulating"]), and unlucky players are taunted as having had recent sexual contact, which polluted them. Michael Herzfeld (see n. 9), 152–62.

13. "While an understanding of stratified agrarian societies may provide a first-level explanation for 'virginity complexes,' any particular instance of the complex must be understood within its specific historical context. Women's chastity may be a primary idiom used by people in stratified agrarian societies for negotiating claims to unequal privileges, but it is not the only idiom" (J. Collier, "From Mary to Modern Woman: The Material Basis of Marianismo and Its Transformation in a Spanish Village," *American Ethnologist* 13 [1986]: 100–107, quotation from p. 102).

14. J. du Boulay, "Lies, Mockery, and Family Integrity," in *Mediterranean Family Structure,* ed. Jean G. Peristiany (Cambridge, 1976), 389–406; P. Walcot, "Odysseus and the Art of Lying," *Ancient Society* 8 (1977):1–19.

15. R. Dionisopoulos-Mass (see n. 7), 58–60. For another modern account of love magic see the novel *Love with a Few Hairs* by Mohammed Mrabet, translated from the Moghrebi and edited by Paul Bowles (New York, 1968)—a discovery for which I owe thanks to Josiah Ober.

16. See J. G. Frazer's discussion of the customs of disguising the bride in his edition of *Ovid's Fasti* (Cambridge, Mass., 1931), 410–11.

17. Themistocles, a less lovable character, must have used an amulet (*περίαμμα*), "attaching (*περιάψας*) some good thing to the city" (Xen. *Mem.* 2.6.10–13).

18. Sulla's success was summed up in the name granted to him by the Senate—Epaphroditos (Appian, *Civil Wars* 1.97). Other *χαριτήσια* (spells or amulets to acquire "charm" [*χάρις*]): *PGM* VII.215 (*Stele of Aphrodite;* a tin amulet); IV.2226–29 (a gold amulet for *φίλτρα*); VII.186–90 (right leg of a living gecko caught in a graveyard and worn as an amulet; brings victory as well as charm); XII.182–89 (a prayer to the lord who is the *χάρις* of the cosmos to grant unfailing free speech "and let every tongue and every voice listen to me"). A Coptic shard contains the prayer, "Fill (this engraved shard) for me with every wish, every love which is sweet, every peace, every delight at once!" (*Enchoria* 5 [1975]: 115–18, revised by G. Browne *ZPE* 22 [1976]: 90–91).

19. Cyranides 2.23.21–28. Similarly, the right eye of a seal wrapped in deerskin makes one successful and desirable (*ἀξιέραστος*, Cyranides 2.41.9–10, 4.67.14–16). The seal's tongue brings victory and its whiskers and heart are a *χαριστήριον μέγιστον* guaranteed to bring success. I cite the Cyranides by book, chapter and line numbers in the edition of Dimitris Kaimakis, *Die Kyraniden,* Beiträge zur klassischen Philologie 76 (Meisenheim am Glan, 1976).

20. Or, in some cases, one's friends: Antigone reminds her angry father that "other men too have terrible sons and a sharp anger (*θυμός*) against them, but assuaged by the spells (*ἐπῳδαί*) of friends their angry nature is dis-spelled (*ἐξεπᾴδονται*)." Soph. *OC* 1192–94.

21. "Give me praise and love (before NN, son of) NN today. . . . (But as for my enemies) the sun shall impede their hearts and blind their eyes" (*PDM* xiv.309–34).

22. = IV.467–68. Empedocles once saved his host Anchitos from being murdered by the son of a man who had been executed in a capital case brought by Anchitos. As the young man rushed forward with sword drawn and in a state of terrible anger (*θυμός*), Empedocles struck a soothing and restraining (*καταστaλτικός*) chord on his lyre and recited *Od.* 4.221. (Iambl. *Vit. Pyth.* 113 = Diels-Kranz *FVS* 31.A.15) For other uses of Homeric and Vergilian lines cf. Ricardus Heim, *Incantamenta Magica Graeca Latina,* Jahrbücher für classische Philologie Suppl. 19 (Leipzig, 1892), 514–20.

23. "Restraint" is the first of many points of contact between two fields usually kept separate in modern accounts—erotic rites and binding spells, e.g., "Hermes Restrainer, restrain Manes" (*DTA* 109).

24. O1 in *PGM*, vol. 2, p. 233.

25. E. N. O'Neil in *GMPT,* p. 146.

26. *χάρις / νίκη* (charisma and victory, *PGM* VIII.36), *πράξις / χάρις* (successful business and charm, VIII.62), *ἀλκή / μορφή* (power and beauty, VIII.30), *προσώπου εἶδος / ἀλκὴ ἁπάντων καὶ πασῶν* (a handsome face and power over all men and women, VIII.5–6). A similar juxtaposition of personal success, triumph over enemies, and erotic ambitions are promised for a ring at *PGM* XII.270–350: "Wearing it, whatever you may say to anyone, you will be believed, and you will be pleasing to everybody" (XII.278–79).

27. cf. XXXVI.283–94; PDM xiv.335–55, a fish oil to be applied both to phallus and to face before intercourse. PDM xiv contains eight other penis ointments plus an erotic fish oil just for the face (355–65).

28. Hans Licht, "Sexuelle Reizmittel und Verjüngungskuren in Altgriechenland," *Zeitschrift für sexuelle Zwischenstufen* 13 (1926/27):134–37; Theodor Hopfner *Das Sexualleben der Griechen und Römer,* vol. 1, pt. 1 (Prague, 1938; repr. New York, 1975), 273–305. The Latin for "drug," *venenum,* is apparently derived from *Venus* and thus fundamentally meant aphrodisiac. Cf. Afranius, frag. 380–81 Ribbeck.

29. "A vessel inquiry that a physician in the district of Oxyrhynchos gave me" (PDM xiv.528).

30. The original purpose of wine was to promote intercourse, according to Schol. Lucian (p. 280.3ff Rabe): *ὅτι ὁ Διόνυσος δοὺς τὸν οἶνον παροξυντικὸν φάρμακον τοῦτο πρὸς τὴν μίξιν παρέσχεν.*

31. VII.167–86. Max Wellman, *Die Physika des Bolos Demokritos und der Magier Anaxilaos aus Larissa,* Abhandlungen der preussischen Akademie der Wissenschaften, Phil.-hist. Klasse, (Berlin, 1928). Hippol. *Haer.* 4.28–42 interprets all magician's performances as tricks: Richard Ganschinietz, *Hippolytos' Capitel gegen die Magier,* Texte und Untersuchungen zur Geschichte der altchristlichen Literatur, vol. 39, pt. 2 (Leipzig, 1913).

32. Bulbs are prominent (*Deipnosophists* 1.5B, 2.63E–64B, 64E–F, 8.356E), evidently from their resemblance to testicles (especially those that grow in pairs, cf. Pliny, *HN* 26.95, Diosc. *Mat. Med.* 3.126). *Satyrion,* frequently cited as an aphrodisiac, has a "bulbous root like a fruit, ruddy with a white inside like an egg" (Diosc. *Mat. Med.* 3.128). Sometimes the aphrodisiac effectiveness resides not in the shape but in the significant name, as in the case of scallops (*κτένες*, also meaning "vaginas") and the sea creatures known as *fascina* and *spuria* (Apul. *Apol.* 35; Adam Abt, *Die Apologie des Apuleius von Madaura und die antike Zauberei,* RGVV 4.2, (Giessen, 1908; repr. Berlin, 1967), 223–24 = 149–50. The words also mean male and female genitals, respectively (sources cited in Abt).

33. What works for humans, works for animals: squill and deer's tail will stimulate reluctant bulls (*Geoponika* 27.5, Varro, *Rust.* 2.7.8) and "this same procedure works for humans too" (*Geoponica* 19.5.4). Red mullet is an antaphrodisiac for men, a contraceptive for women and birds (Ath. 7.325D). The root of all-heal (*Ferulago galbanifera*) is good for birthing, other gynecological problems, and flatulence in cattle.

34. Literally, "rootcutters," *ῥιζοτόμοι*—also the title of a play by Sophocles concerning Medea, cf. D. F. Sutton, *The Lost Sophocles* (Lanham, Md., 1984), 117–18.

35. G. E. R. Lloyd, *Science, Folklore, and Ideology* (Cambridge, 1983), 119–35.

36. E. Degani, "Hipponactea," *Helikon* 2 (1962):627–29.

37. Hence it is eaten regularly by temple attendants (Cyran. 1.5.13–14), preferable perhaps to the three cold baths per day described by Chairemon apud Porph. *De Abstin.* 4.6–8. Hemlock is consumed for the same reason by the hierophant at Eleusis (Hippol. *Haer.* 5.8). Also useful are red mullet (Terpsicles [a pseudonym if I ever heard one] apud. Ath. 7.325D),

associated with Artemis (Plato Comicus apud Ath. 7.325A); and water lily root (*nymphaia:* named for a nymph who was hopelessly in love with Heracles; see Pliny *HN* 25.75; cf. 26.94, Diosc. *Mat. Med.* 3.132).

38. Cyran. 1.5.15–18; rocket seed ground with pine cones in wine was a widely known aphrodisiac (*P.Lit. Lond* 171; cf. Diosc. *Mat. Med.* 2.140, *P.Lond.* 121[r]182–84).

39. The wearing of a crow's heart (male by man, female by woman) ensures affection (εὔνοια) forever—"an unsurpassable miracle" (Cyran. 1.2.14–19).

40. Pliny *HN* 26.95 and 96, a knowledge attributed to Thessalian women by Diosc. *Mat. Med.* 3.126.

41. Pliny *HN* 27.65, Diosc. *Mat. Med.* 3.126; cf. Diosc. *Euporista* 2.96.

42. Plut. *De Tuend. San. Praec.* 126A, *Conj. Praec.* 139A.

43. A woman accused of using aphrodisiacs to gain the love of Philip was brought before Olympias who, seeing her beauty and hearing her intelligent conversation, said "The accusations are baseless, you have aphrodisiacs in yourself." Plut. *Conj. Praec.* 141B–C; a similar story in a discussion of misogyny in Satyros' *Life of Euripides* (a dialogue in which at least one of the three speakers was a woman), *P. Oxy.* 1176, frag. 39, col. 14: "When he saw her stature and beauty, he said 'Hail, lady; the accusations are false, for you have drugs in your face and eyes.'" The trope is reversed at Lucian, *Dial. Meret.* 8: "His wife told everyone that I had driven him crazy with drugs, but the only drug involved was his own jealousy."

44. As in the detective stories of Hippocrates visiting Democritos, Diog. Laert. 9.42. For an ancient drawing of two hearts, or what Audollent saw as such, see the "valentine" at *DT* 264.

45. Ps.-Aristotle *Problems* 30.1.954a52, cf. 954a25: "when black bile is overheated, it produces cheerfulness accompanied by song, and frenzy, and the breaking forth of sores."

46. E. J. Kenney, "Doctus Lucretius," *Mnem.* 4th ser. 23 (1970):366–92, esp. 380–90; J. L. Lowes "The Loveres Maladye of Hereos," *Modern Philology* 11 (1913–14):490–547; M. Ciavolella, "La tradizione dell' *aegritudo amoris* nel *Decameron," Giornale storico della letteratura italiana* 147 (1970):496–517.

47. Antiphon Soph. *On Likemindedness, FVS* 87B49 D–K. "A sign of the onset of eros is not delight in the presence of the beloved, which is only normal, but rather the sting and pain felt in the beloved's absence" (Plut. *Quomodo Quis Suos in Virtute Sent. Prof.* 77B). "These signs—groans and tears and pallor—indicate nothing other than eros." Lucian, *Iupp. Trag.* 2.

48. Some famous literary cases: Euripides' Phaedra, Callimachos' Cydippe, Chariton's Callirhoe. Lesser-known sufferers are the son of Diogenes (Suda, s.v. *Diogenes*) and the rich young man who fell in love with a farmer's daughter (Ath. 12.554C–E, from Cerkidas' and Archelaos' *Iamboi*).

49. Ath. 14.619D–E (from Aristoxenos), traditionally attributed to Stesichoros (D. Page, *Poetae Melici Graeci,* 277). Dimoites, cursed by his wife, falls in love with a corpse washed up on the shore; since it is in an advanced state of decay, he buries it and kills himself on the grave (Parth. *Amar. Narr.* 31). When Antiope gently but firmly refused the suit of Soloeis, he leaped into a river and drowned himself (Plut. *Thes.* 26.2–5). Daphnis's mother is afraid he will commit suicide (Longus 3.26.3). Phidalios of Korinth (*FGrHist* 30 F 2) states the general principle: "It is natural for lovers to cling to the beloved and to die for her . . . ; for they are made savage by desire and do not use their minds to reason with." So common is the motif that Lucian makes fun of it: in a checklist of those entering Hades are "seven who slew themselves for eros" (*Catapl.* 6).

50. "Some constraint (ἀνάγκη) sent by god" (Pl. *Phd.* 62C); "constrained (ἀναγκασθείς) by an unbearably painful and inescapable misfortune or meeting with a hopeless and unlivable shame" (Pl. *Laws* 9.873C); "to die to escape poverty or eros or some anguish is the mark of a cowardly rather than manly person" (Arist. *Eth. Nic.* 3.7.1116a12–14).

51. Plutarch recommends that the old practice of public diagnosis of illness be applied also to emotional disorders, imagining the following comments from bystanders: "You're suffering from anger; stay away from x." "You're feeling jealous; do y." "You're in love; I was in love once but I recognized my mistake (*μετενόησα*)" (*De Lat. Viv.* 1128E).

52. Appian *Bell. Syr.* 59–61; Plut. *Demetr.* 38; Lucian *Syr. D.* 17–18; Val. Max. 5.7.3. E. Rohde, *Der griechische Roman,* 5th ed. (Hildesheim, 1974 [orig. pub. 1876]), 55–59; P. M. Frazer, "The Career of Erasistratus of Ceos," *Istituto Lombardo, Rendiconti, Classe di Lettere e Scienze Morali e Storiche* 103 (1969):518–37; D. W. Amundsen, "Romanticizing the Ancient Medical Profession" *Bull. Hist. Med.* 48 (1974):320–37. R. Asmus analyzes the physiognomonic tradition reflected in this widely known story: "Vergessene Physiognomika," *Philologus* 65 (1906):415–21. Pl. *Lysis* 204C reflects the topos: commenting on a young man's blush, Socrates says, "I have this god-given talent for instantly discerning who feels eros and who is its object."

53. The pulse of Justus's ailing wife quickened when the name of the dancer Pylades was mentioned: Gal. *Prognosis* 6 (XVI.630–34 Kühn = *CMG* V.8.1, ed. V. Nutton, pp. 100–1044); the same incident is referred to in his *Commentary on Hippokrates' Prognostics* I.8 (XVIIIB.40 Kühn = *CMG* V.9.2, ed. Diels, p. 206).

54. Pl. *Phdr.* 252A–B. Orderly education will moderate other desires, but as for eros, the source of a million evils for individuals and whole cities, "what herbal drug can you cut to liberate these people from so great a danger?" (*Laws* 8.836B); "O King, there is no other *φάρμακον* for eros but the beloved in person" (Chariton 6.3.7); a magus declares that he can command the moon and sun and sea and air, "but for eros alone I find no drug" (*PGM* XXXIV, (fragment of a novel). Theocritos declares that there is no cure except poetry (*Id.* 11.1–3); Callimachos accepts the point and adds starvation as another cure (*Epig.* 47).

55. Plutarch nicely observes that the physical symptoms described in Sappho 31 are exactly those of eros (*Dem.* 38). They are also unwelcome, involuntary, and perhaps the result of a spell. One of the earliest magical papyri (*PGM* CXXXII, Augustan period) is a collection of verse rites with echoes of the topoi of Greek lyric (throwing fruit, *μῆλα*, col. 1, lines 6 and 9) and possibly of Sappho in particular—"I run after, but he flees" (col. 2, line 12, and Compare Sappho 1.21).

56. Chloe tentatively diagnoses her own symptoms as a spring fever (Longus 1.13ff.).

57. In the Loeb edition of Philostratus and Eunapius, *Lives of the Sophists* (Cambridge, Mass., 1921), 398–417.

58. So the love spells of Canidia were countered, she imagines, by "the incantation of a more scientific witch," *veneficae scientioris carmine* (Hor. *Epod.* 5.71–72).

59. *PGM* VII.285–89, 619–27, 643–51. VII.969–71 (partly in code) is a written equivalent that obviates the awkwardness of long mumbling while the other is waiting for the drink.

60. *PGM* 13.319–20 (wasps caught in a spider web!); PDM xiv.376–94, 428–50, 636–69, 772–804.

61. Pliny *HN* 27.57; Diosc. *Mat. Med.* 4.131.

62. *PGM* VII.973–80; *PGM* CXIX. A related form of touch magic at Pl. *Meno* 80A, where Meno feels numb and helpless because of Socrates' wizardry, as if he has been touched by an electric eel. *PGM* CXXVII: a man whose loins have been touched with the brain of an electric eel will bend over and not be able to stand upright.

63. As Hippodamia says of the beautiful Pelops in Sophocles' *Oinomaos:* "Pelops has such a magician's implement to capture eros, some lightning in his eyes; it warms him, it scorches me entire" (frag. 474). "As fire burns those who touch it, so beautiful people ignite a subtle fire in those who see them even from a distance, so that they glow with eros" (Xen. *Cyr.* 5.1.16). See D. M. Halperin, "Plato and Erotic Reciprocity," *CA* 5 (1986):63, n.5.

64. *PGM* X.19–23. Surely, friendly eye contact from an admirer for the space of three

deep breaths is an unmistakable message in its own right: the formula serves more for self-confidence.

65. "Lead forth and bind fast Matrona, daughter of Tagenes, whose *οὐσία* you now have, the hairs of her head, that she may not have sex . . . with any other man except Theodoros, son of Techosis." Wortmann (see n. 1), p. 60, lines 19–23. *PDM* xiv.1075. "Put the hair of the woman in the leaf"; see K. Preisendanz, "*Ousia,*" *WS* 40 (1918):5–8. In other *agōgai, οὐσία* is attached to the head or neck of a kneeling doll (IV.302–3) or wrapped with a papyrus incantation and placed in a box (*PGM* XV) or put inside an inscription on ass's hide and placed with vetch in a dead dog's mouth (XXXVI.361–71).

66. See Wortmann, p. 69 and Faraone in this volume (chapter 1) p. 14 and n. 64.

67. *PGM* XXXVI.68–101; 102–34; 295–310; in the lunar calendar at VII.295 the moon in Aries is a propitious time for *ἔμπυρα* and *ἀγώγιμα*.

68. E. Kuhnert, "Feuerzauber," *RhM* 49 (1894):37–58.

69. Looking at the moon: *PGM* IV.2708; "Late at night, in the fifth hour, facing Selene (the Moon) in a pure room . . . ; when you see the goddess turning ruddy, know that she is already attracting her" (VII.874–75, 889–90; LXI.6); "before Isis in the evening when the moon has risen" (*PDM* lxi.118); "in the waning of the moon when the goddess is in her third day," (*PGM* XII.378–79; the same day is chosen for a necromantic ceremony in Heliodoros, *Aithiopica* 6.14.2. If a rooftop is not available, one may make do with the ground, *PGM* LXI.6.

70. Sleeplessness is a central aim, too, of erotic rituals that do not mention a time of performance, such as *PGM* IV.2943–66; VII.374–76, 376–84. These could be plausibly grouped with our "night scenes".

71. "Let her have only me in her mind (*κατὰ νοῦν*)" (*PGM* IV.1520, 2960–61); *Solum me in mente habeat* (*DT* 266.19); Maltomini (*Civiltà class. e cris.* 1 [1980]:376) supplements *PGM* CCXXII along these lines. A rare trace of the lover's obsession is found in Wortmann [see n. 1], 64, line 78: "Matrona . . . whom Theodoros has on his mind (*ἐν νόῳ*)."

72. Lucian *Philops*. 13–15. The narrator comments: "But I know that Chrysis; she is a lusty and forward lady, and I don't see why you had any need of a clay ambassador and a magus from the Hyperboreans and Selene herself when one could lead (*ἀγαγεῖν*) her all the way to the Hyperboreans for twenty drachmas."

73. Theoc. 2.161–62 (Assyrian stranger; the local old ladies who knew spells were unable to help, 2.91); Verg. *Ecl.* 8.95–99 (Moeris); Lucian *Dial. Meret.* 4.4–5 (Syrian herbalist, an old lady); Heliodoros *Aithiopika* 3.17; cf. 3.19 and 4.5 (an Egyptian priest, pretending to be a love wizard).

74. Men aiming at women: *PGM* XVIIa, XIXa, LXXXIV, CI, CVII, CVIII, CIX; ostracon O 2 (*PGM*, vol. 2, p. 233); *DT* 100, 227, 230, 231, 264–71, 304; two tablets for the same agent and victim (see below, n. 99); two tablets and one ostracon for the same agent and victim in Wortmann [see n. 1], 57–84; *BIFAO* 76 (1976):213–30; *ZPE* 24 (1977):89–90. Women aiming at men: *PGM* XV, XVI, XXXIX, LXVIII; *DT* 270, 271. Women aiming at women: *PGM* XXXII; *PSI* 28, cf. F. Maltomini, *Miscellanea Critica,* Papyrologica florentina 7 (1980), 176). Men aiming at men: *PGM* XXXIIa, LXVI; *JEA* 25 (1939):173–74.

75. Rare traces of other situations: *PGM* I.98 ("fetches women and men"); IV.2089–92 ("*οὐσία* of her/him . . . where she, or he, dwells . . . bring her to me")—both these instances might equally have in view male or female lovers working the spell on a male; XXXVIa.70 is unambiguous on this point ("brings men to women and women to men," but at XXXVI.73–74 the client is directed to take "*οὐσία* of whatever woman you want," thus reverting to the generic norm); XII.364–75 is a spell to create hatred between two men, with an alternate version for a man and a woman.

76. Hor. *Epod.* 5 and 17, *Sat.* 1.8; Apul. *Met.* 1.5–19, 2.5; Petron. *Sat.* 63; Luc. 6.413–830.

77. A. Abt, *Die Apologie des Apuleius von Madaura und die antike Zauberei,* RGVV 4.2 (Giessen, 1908; repr. Berlin, 1967), 318–329.

78. In real life (or what has some claim to be such) women are charged with administering φίλτρα in food or drink rather than with performing ἀγωγαί: Antiphon 1.9, 19; Arist. *Mag. Mor.* 1.16; Plut. *Conj. Praec.* 139a.

79. A second version at *PGM* IV.2643ff.; the two are written as verse and compared at *PGM*, vol. 2, pp. 255–57.

80. *PGM* IV.1840–59; cf. the Hyperborean mage cited above from Lucian *Philops.* 13, ἔρωτας ἐπιπέμπων. Galen asserts that eros is a strictly human passion, "unless of course one believes the tales that some people are led (ἄγεσθαι) to this passion by a tiny baby god holding burning torches" (*Comm. in Hippoc. Prognost.* I) vol. XVIII.2, p. 19 Kühn.

81. *PGM* IV.2031–32 (νεκυδαίμων); *IV*.2088 (καταχθόνιος δαίμων); cf. the related rite over a skull at 4.1928–2005; *PDM* xiv.1070 (a mummy); *PGM* XXXVI.139 (demons of the darkness).

82. *PGM* IV.2486 (Hecate); IV.2730–36 (Hecate accompanied by troops of the shrieking untimely dead); IV.2907–9 (Aphrodite).

83. VII.884–85: Selene is asked to send a holy angel or sacred assistant to serve the lover in the course of this night.

84. "Drag (ἕλκε) Matrona by her hair, her guts, her *psyche,* her heart" (Wortmann [see n. 1], 66. In earlier periods the ἴυγξ served to drag or draw unwilling persons to a lover's bed: Pind. *Pyth.* 4.214, *Nem.* 4.35, playfully at Xen. *Mem.* 3.11.18. Hephaistos is shown making one as a torture wheel for Ixion on an Attic vase from the circle of the Meidias Painter; see E. Simon in *Würzburger Jahrbücher für die Altertumswissenschaft,* new series 1 (1975):177ff. Eros plays with one; see S. G. Miller, *AJA* 90 (1986): plate 14, figs. 10, 12. *PGM* has one poetic reference to Ixion's wheel (IV.1905–6).

85. *PGM* IV.2500, 2735. Thus Cupid assumes the form of Perdica's mother and appears to him in a dream (*Aegritudo Perdicae* 77–83).

86. *PGM* IV.2443, VII.877; *PDM* xiv.1070; *PGM* 64.

87. Arist. *On Dreams* 2 has a very acute analysis of the parahypnotic fantasies of emotionally aroused people: the timorous man seems to see his enemy approaching, the amorous man his beloved (460b1–15).

88. The victim's dreaming of the agent is explicit in *PGM* XVII.15: ἐνυπνιαζομένην, ὀνειρωττοῦσαν. For the sexual implications of the latter term, see next paragraph.

89. R. Kühner, *Aus Führliche Grammatik der griechischen Sprache,* vol. 1, pt. 1; 3rd ed., rev. F. Blass (Hannover, 1890), sec. 328.9.

90. Origen *C. Celsum* 2.55.

91. See S. Eitrem's essay (chap. 3).

92. For wax figurines in erotic procedures, see pp. 178–81, 231.

93. *The Romance of Alexander the Great by Pseudo-Callisthenes,* trans. from the Armenian version by A. M. Wolohojian (New York, 1969), 27.

94. P. du Bourguet, "Ensemble magique de la periode romaine en Egypte," *La revue du Louvre* 25 (1975): 255–57; the text of the accompanying lead sheet is edited by S. Kambitzis, *BIFAO* 76 (1976): 213–30 (= *SEG* 26.1717).

95. Clearchos offers a variety of explanations for the popular notion that men whose wreath comes undone at a symposium are suffering from eros, one of which is that it is only the bound who come undone: "for men in love have been bound" (καταδέδενται γὰρ οἱ ἐρῶντες, Clearchos frag. 24 Wehrli [= Ath. 15.670C]). Other instances of unbound or slipping wreaths signifying love: Asclepiades *A. P.* 12.135 (no. 18 Gow-Page); Callimachos *Epigr.* 44; Ov. *Am.* 1.6.37–38.

96. λαικάζειν , cf. G. P. Shipp, "Linguistic Notes" *Antichthon* 11 (1977): 1–2; H. D. Jocelyn, "A Greek Indecency and Its Students: LAIKAZEIN," *PCPS* 26 (1980): 13–66.

97. V. Martin, "Une tablette magique de la Bibliothèque de Genève," *Genava* 6 (1928): 56–63. The text refers to itself as a φιλτροκαταδέσμος (line 8).

98. Under the heading *amatoriae* in the subject index (Index pp. 472–83) to *DT*.

99. *DT* 267–71; F. Boll, *Griechischer Liebeszauber aus Aegypten,* SB Heidelberg (Heidelberg, 1910, no. 2); Wortmann (see n. 1), 56–84; *ZPE* 24 (1977): 89–90.

100. "Give to NN fighting, war, and to NN contempt and hatred—such as Typhon and Osiris had." *PGM* XII.372–73, 449–52 (= PDM xii.62–75, with a drawing of an ass-headed figure labeled Sēth).

101. *PGM* VII.467–77, XXXIIa, XXXVI.69–101.

102. *PGM* IV.1390–1495, XIXa (a prescription spell to be placed in the mouth of a dead person, lines 15–16), XIXb (a generic spell to be placed with a corpse). The witchy versions of such rites in poetry traffic in materials gathered in graveyards and from those disturbing animals that belong simultaneously to two categories, such as frogs and snakes (Prop. 3.6.27–30). The similarity of necromancy and erotic rites is the organizing principle of L. Fahz, *De Poetarum Romanorum Doctrina Magica Quaestiones Selectae,* RGVV 2.3 Giessen, 1904).

103. Indicated by (κατα)κλίνειν: *PGM* IV.2076, 2442, 2624. "Caligula was unhealthy both in mind and body. . . . He was believed to have been drugged by his wife Caesonia with a love potion that sent him mad. He was most of all tormented by sleeplessness" (Suet. *Calig.* 50).

104. Dolls in erotic operations: Verg. *Ecl.* 8.75, 80 (wax and clay); Hor. *Sat.* 1.8.30–33, 43f. (wax and wool); Hor. *Epod.* 17.76 (wax); PDM lxi.112–27 (wax Osiris); *PGM* XCV. 1–6 (uncertain material and purpose). *PGM* CI, found in a pot evidently from a cemetery ("you daemons who lie here," line 2), was wrapped around what appears to be two crudely made wax figurines embracing, the male and female made of darker and lighter waxes respectively (Wortmann [see n. 1]). With all these should be compared the dolls employed to harm enemies: R. Wünsch *Philologus* 61 (1902):26–31; J. Trumpf, "Fluchtafel und Rachepuppe," *AM* 73 (1958):94–102; Faraone in this volume (chapter 1) p. 7, n. 31.

105. *PGM* IV.2449–51; "Be sure to open the door for the woman who is being led by the spell, otherwise she will die" (IV.2495). Cf. Theophr. *Hist. Pl.* 9.11.6 on the effect of different doses of strychnine. Pitys's *agōgē* "attracts, sickens, sends dreams, restrains, answers in dreams" (*PGM* IV.2076–78). Is XII.376–96 an *agōgē* or a recipe for death by insomnia? It compels a woman to lie awake, day and night, "until she consent" (ἕως συμφωνήσῃ, line 378); but the words command her to lie awake until she die (ἕως θανεῖ, line 396), and directions promise that she will die without sleep before seven days have passed. In recommending that lovers not use drugs and spells, Ovid argues both that they do not work and that they harm the psyche and the sanity of young women: *Nec data profuerint pallentia philtra puellis; philtra nocent animis vimque furoris habent* (*Ars Am.* 2.105–6). A humorous statement of the death brought by a love-potion (metaphorically for love itself): *Si semel amoris poculum accepit meri eaque intra pectus se penetravit potio, extemplo et ipsus periit et res et fides* [property and credit] (Plaut. *Truc.* 42–44).

106. Similarly in the Coptic *agōgē* published by P. C. Smither (*JEA* 25 [1939]: 173–74), the speaker conjures Iao, Sabaoth and Rous "that even as I take thee and place thee at the door and the pathway of P-hello the son of Maure, thou shalt take his heart and his mind and thou shalt master his whole body. When he stands, though shalt not allow him to stand; when he sits down, though shalt not allow him to sit down; when he sleeps, thou shalt not allow him to sleep; but let him seek after me from village to village, from city to city, from field to field, from country to country, until he comes to me and becomes subject under my feet—I Papapolo the son of Noah, his hand being full of every good thing, until I have fulfilled with him my heart's desire and the longing of my soul, with a good will and an indissoluble love."

107. *PGM* IV.1531, XII.490 (= *PDM* xii.155), XVIIa.16, XXXVI.82, *GMPT*, pp. 113, 149, 359; *DT* 230, 265A.

108. Two extraordinary cases specify lengths of time: F. Boll (see above, n. 99), 9–11: *ἐπὶ ε μῆνας*; *PGM CI.* 36–37 *ἐπὶ χρόνον μήνων δέκα*. Eitrem takes ε in the former text to be a five-month trial marriage (*P. Oslo* II, p. 33, n. 1). The latter text may refer to confirming a marriage by pregnancy, ten months being considered the normal period of pregnancy (J. Bergman, "Decem Illis Diebus," in *Ex Orbe Religionum: Studia Geo Widengren oblata,* Studies in the History of Religions—Supplement to *Numen* XXI [Leiden, 1972], I.332–46, esp. 340–41).

109. CF. M. Herzfeld, "Gender Pragmatics: Agency, Speech, and Bride-Theft in a Cretan Mountain Village," *Anthropology* 9 (1985):25–44.

110. *PGM* IV.2757–60, XV.4, XIXa.53, LXI.29–30 (= PDM lxi.173); *DT* 266, 268; cf. Sappho 16.

9

Magic and Mystery in the Greek Magical Papyri

Hans Dieter Betz

The topic of magic raises expectations and questions pertaining to *Religionswissenschaft* and *Religionsgeschichte*. One is reminded of the intense debates in the first thirty years of this century[1] centering upon the relationship between magic and religion on the one hand and between myth and ritual on the other.[2]

MAGIC AND MYSTERY

The topic of magic was usually treated as subordinate to the larger issues mentioned above. A key problem in this debate was the definition of magic in order to distinguish it from religion, which itself was in no way easier to define. Implied in this distinction was another question, that of the nature of ritual and ritual's relation to myth. In the course of these debates, the Greek mystery religions had become a focus of interest because there appeared to be sufficient evidence for arguing for or against various comprehensive theories concerning the fundamental problems of religion.[3] Since the mystery cults consisted primarily of rituals, of first concern was the relationship to their corresponding myths.[4] Later, similar questions were raised concerning Gnosticism. There was plenty of myth in Gnosticism; the question was whether there was any ritual and if so, what its role might be.[5] Most of these issues are still being discussed today in one way or another, due to the dearth of evidence and problems of methodology.

One will do well to realize that behind these scholarly problems have always stood the fundamental philosophical and historical problems concerning the origins of religion in general. Was magic the beginning stage and religion a later derivative,[6] or was it the other way around? Was the beginning marked by religion, while magic was a product of later degeneration?[7] Confusion, as well as unrealized presuppositions, created major difficulties for answering these questions.

Religion used to be treated, without further questioning, as something intrinsically positive, while magic was from the outset stained by negative connotations. Combined with anthropological theories of evolution,[8] sequences and hierarchies were established with some force of persuasion. Consequently, since humanity was thought to be "primitive" at the beginning of its development, an association with magic with this initial phase seemed logical. Depending upon the scholar's generosity, primitive mankind's irrationality could account for this involvement with magic, or else such an involvement could be explained as either primordial stupidity (*'Urdummheit'*) or mistaken science, that is, erroneous conclusions unfortunately drawn from correct observations. Consequently, religion could be presented as the opposite of magic, that is, as the result of later enlightenment and a compromise attempt to reconcile primitive irrationality with developing rationality in such a way that a rational overlay of "theology" covered up the irrationality of ritual and myth, which remained in place. Understood in this way, religion became a transitional phase between prescientific ignorance and what has been called the modern scientific worldview—which turned out to be, for the most part, agnosticism or atheism.

This view of religion is now outdated, to be sure, but it is still held by many in present society who believe it corresponds to scientific evidence. One should also realize that this viewpoint was the antithesis to another theory endorsed by ancient texts.[9] Conforming to cosmic theories of good origins and subsequent depravity, primitive religion was for the ancients religion in its purest form: the worship of the deity by an unspoiled human race created for this very purpose. Depravity set in when, because of human foolishness, or because of the error of polytheism, or even because of seduction by evil demons, there developed the worship of images, the offering of bloody sacrifices, the building of temples, and, of course, magic. Superstition and magic thus became synonyms, social evils to be rooted out by law and force in the name of true religion.[10]

Both these theories were based on ideologies that may have been persuasive in their day but that can hardly be corroborated by historical evidence. Contrary to their claims, these theories about religion are not based on archeological findings, anthropological evidence, or critical evaluation of texts but on preconceived ideas and speculations often indebted—unconsciously—to old myths and prejudices.[11]

Where are we left, then, as far as questions about magic and religion are concerned? Are these terms simply empty of content and ready to be interpreted any way one chooses? Is one person's religion simply another person's magic, and vice versa? An age such as ours, inclined to believe that realities are what the labels stuck on them say they are, tends to answer in the affirmative. Furthermore, sociologists seem to have convinced almost everyone that whatever human beings undertake, they do so in order to gain or to retain power over others. As a result, people who pursue the question of defining religion and magic tend to find in the definitions themselves nothing but another set of tools for the manipulation and control of social realities. If there is a difference between religion and magic at all, it does not really matter as long as they work in much the same way. The distinction itself is said to matter only to religious leaders because they implement and prove the old rule *Divide et impera*.

Alan Segal has recently summed what appears to be the current consensus: "Take the relationship between magic and religion: Religious leaders are often interested in strict distinctions between magic and religion so that the purity of religion can be maintained."[12] Consequently, social scientists, who by definition have no interest in religious leadership, assume a posture of freedom from any interest in power and are thus more competent "to isolate phenomena cross-culturally so that a consensus of methodology can be reached."[13] Where does this approach lead us? Again Alan Segal states the matter aptly:

> I will argue that no definition of magic can be universally applicable because "magic" cannot and should not be construed as a properly scientific term. Its meaning changes as the context in which it is used changes. No single definition of magic can be absolute, since all definitions of magic are relative to the culture and sub-culture under discussion. Furthermore, it is my contention that we have been misled by our own cultural assumptions into making too strict a distinction between magic and religion in the Hellenistic world. As we shall see, in some places the distinction between magic and religion will depend purely on the social context.[14]

There is much here that one can agree with. Yet caution is in order because the assessment coincides so beautifully and conveniently with our own social context. If we adopt such a perspective, we seem to be justified in assuming a posture of objectivity to meet the scientific requirement. The problem, however, is that this sort of objectivity represents nothing but a total relativism that must not be confused with the required scientific ethos. Omitted from such consideration is also the fact that the social sciences serve in many ways as an enormously powerful instrument of social control. Far from being disinterested in the outcome of the debate about definition, the social science approach to religion is, in fact, in competition with traditional churches and synagogues and insists that ongoing debates be fought on its own terms, thereby putting those religious institutions at a disadvantage from the outset.

Another problem with this posture of objectivity is that it permits precisely the bilevel commitment found among many academics. On one level, one can, as stated, perform in a purely objective, relativistic and even agnostic manner; while on another level, that of the so-called subcultures, one can subscribe to the terms as defined by the subculture one has chosen or has been born into. If this subculture is one of the religious communities and if that community does not make any public claims but remains strictly "private," holding on to convictions of merely "particular" traditions, a marriage of convenience is all but assured. The social context in which this bilevel commitment is commended and made possible constitutes our religious pluralism. Like most marriages of convenience, however, this one is built on very fragile premises. Divorce will be inevitable once the demands of truth, which have been put on ice, as it were, in the relationship, are allowed to play their rightful role.

Juxtaposing magic and religion as such may also be misleading. Such a juxtaposition tends to obscure the fact that magical ideas and practices pertain to areas outside of religion—assuming, of course, that religion is defined in the narrower sense as the worship of deity and not in the wider sense as an attitude toward reality per se. Sigmund Mowinckel in his *Religion und Kultus* rightly emphasizes that magic is a worldview: "Magical thinking and its practical application, called 'magic,' is not a kind of religion, but a worldview, that is, a particular way to understand things and their mutual connectedness. Magic is a *Weltanschauung* which in a certain way is analogous to the view of the universe as we today attempt to formulate it on the basis of the laws of causality and the interaction between cause and effect, demonstrated by physics, chemistry, biology, and psychology."[15]

Another potentially misleading "opposition" is that of magic and science. Both are in fact connected historically as well as in many other ways.[16] Until empiricism became dominant in the sixteenth and seventeenth centuries, magicians and Hermetists, as they called themselves, were leaders in the sciences, and many discoveries originated in their speculations and experimentations.[17] Even Isaac Newton was deeply involved with the magical arts of alchemy and Hermetism.[18]

Currently there is still another "opposition" causing confusion, that of magic and "Western civilization."[19] This "opposition", popular as it may be, is contradicted by the fact that within Western civilization magic is very much alive, as it has ever been.

I am not attempting to solve all these intricate problems. Rather, I am assuming that good reasons exist for the fact that no one definition appears acceptable to everyone at this time.[20] The primary reason seems to be that the many different complications have yet to be fully understood. In fact, at this moment research is under way in so many areas concerning magic that it would be premature to try to settle these issues before more of this research is completed. Every new study that one comes across introduces new materials and questions or exposes unjustifiable assumptions so far taken for granted. Our situation is to some extent ironic because our disagreements and uncertainties occur in a world in which we see ever more evidence of magical ideas and practices all around us. Those who believe they have control of the problem and the terminologies, not to mention the evidence, will have to ask themselves whether they are not the victims of self-delusion. Yet there are some points we can make even at this juncture. For one, no amount of evidence, analysis, and revision of conceptuality will automatically solve the problem of the relationship between magic and religion. Also, staying at the level of a social science–inspired relativism will inevitably mean that definitions of magic and religion will always turn out to be a matter of personal, subjective preference.

In order to understand what was meant by distinguishing between magic and religion, however, one will have to shift from social science to theology. The basic questions, as even pre-Christian antiquity well recognized, are theological in nature. Which words, attitudes, and practices are appropriate or inappropriate before the deity? Is forcing the gods into the service of human desires, wishes, or needs appropriate? Is the complete separation of moral issues from the practice of magic appropriate before the deity? Is it legitimate to place the greatest mysteries of divinity at the disposal of those who are willing to pay for it? Is the enormous finesse in deceiving gullible people and in exploiting their various conditions of misery religiously legitimate? These are theological questions, to be sure, but refusing to consider them amounts to deciding in advance. We will have to recognize that our dilemma of defining magic versus religion will simply remain unsolvable if we do not allow theological questions to play their role.

I will examine a body of texts and a set of issues that at first glance seem to have little to do with these theoretical problems. This examination, however, will show how in the Greek magical papyri and in the context of Greco-Egyptian syncretism the intricate relationships between magic and religion are played out.[21] A limited case study such as this demonstrates how careful one must be when one tries to understand the various facets of the theoretical problems outlined above.

MAGIC IN THE GREEK MAGICAL PAPYRI

> "The Greek magical papyri" is a name given by scholars to a body of papyri from Graeco-Roman Egypt containing a variety of magical spells and formulae, hymns, and rituals. The extant texts are mainly from the second century B.C. to the fifth century A.D. To be sure, this body of material represents only a small number of all the magical spells that once existed. Beyond these papyri we possess many other kinds of material: artifacts, symbols, and inscriptions on gemstones, on ostraka and clay bowls, and on tablets of gold, silver, lead, tin and so forth.

This opening paragraph from my introduction to *The Greek Magical Papyri in Translation, Including the Demotic Spells*[22] indicates that whatever magic may be, the magical papyri have plenty of it. Descriptive terms for magic and even definitions occur in the papyri.[23] "Holy magic" (*ἱερὰ μαγεία*) is a positive term, and the one who is initiated into the art is called "blessed initiate" (*ὦ μα[κάρι]ε μύστα*).[24] There are, however, different levels of cultural sophistication in the papyri, and it is in sections representing a higher cultural level that we find descriptive terms such as *μαγεία* (magic), *μαγικός* (magical), and *μάγος* (magician). These terms are used always with positive connotations so that for *Papyri Graecae Magicae* magic and religion are a single entity.

This does not mean, however, that for the magicians whose writings are included in the *Papyri Graecae Magicae* all magic is simply legitimate and acceptable. The magical handbooks that make up most of the material represent collections, that is, selections of those texts that were deemed by the collectors to constitute an authoritative tradition. By implication, other materials not judged worthy of tradition were discarded. Also at times the magician will admit his fear that his operations may be illegitimate and dangerous. He then assures the gods that he only does what they themselves have revealed and commanded him to do.[25] Just in case, he always has some protective charms (*φυλακτήρια*) at hand.[26] There is at least one passage where an inferior magic, using tools, is distinguished from a superior magic, employing magical words only.[27] On the whole, therefore, one can say that although magic is accepted and approved by the magicians of the *Papyri Graecae Magicae,* there are some indications that they were aware of its problematic aspects. Theologically, there was some concern about the legitimacy of "forcing the gods," a concern subdued in the *Papyri Graecae Magicae* but fully discussed by Neoplatonist philosophy and the Patristic Fathers.[28] For the most part, however, the concern is merely whether the magic "works." This is indicated by constantly recurring "advertising slogans" inserted by the redactors to commend the material, for example, "This really works" or "If it does not work, try this other spell."

Seen as a whole, the *Papyri Graecae Magicae* contain magical material of a syncretistic nature. The Greek and demotic bilingual texts, especially, demonstrate that "the corpus as a whole derives in a very large measure from earlier Egyptian religious and magical beliefs and practices."[29] The interrelationships between the Greek and the demotic spells are complicated and have not yet been determined with accuracy: "Some of the spells were written in Greek, others in Egyptian, all within the same texts and all for use by the same magician. Perhaps even more telling is the fact that even in the spells written in Greek, the religious or mythological background and the methodology to be followed to ensure success may be purely Egyptian in origin."[30] The linguistic phenomena of the bilinguals are truly fascinating:

> Most have passages in Greek as well as in Demotic, and most have words glossed into Old Coptic (Egyptian language written with the Greek alphabet [which, unlike Egyptian scripts, indicated vowels] supplemented by extra signs taken from the Demotic for sounds not found in Greek); some contain passages written in earlier Egyptian hieratic script or words written in a special "cypher" script, which would have been an effective secret code to a Greek reader but would have been deciphered fairly simply by an Egyptian.[31]

What is true of these texts, however, is not necessarily true of others that appear to be derived from a Greek religious background, while still others come from a Jewish provenance. Not all the Greek material can be shown to be simply imposed upon an earlier Egyptian, non-Greek foundation. There are also materials of older Greek origin that became secondarily integrated with Egyptian or Greco-Egyptian spells. The way in which the materials evolved to their present stage is clear, at least in broad outlines.

In accordance with their geographical place of origin, the materials in the *Papyri Graecae Magicae* continue the older Egyptian magic. Perhaps brought in by Greek settlers in Egypt, Greek magic began to exert its influence perhaps as early as the period of classical Greece[32] and certainly as part of the hellenistic religious syncretism. Jewish materials appear to derive from hellenistic syncretistic Judaism[33] rather than from Jewish religion of the time of the Old Testament. The few "Christian" elements are part of the hellenistic-Jewish syncretistic spells.[34] Christian magic rapidly expanded after the sixth century A.D., transforming and largely—but not totally—replacing the older material then classified as "pagan."[35]

MYSTERY CULTS AND THE GREEK MAGICAL PAPYRI

How, then, does the magical papyri material relate to the mystery cults? We must consider several points in this regard. For the *Papyri Graecae Magicae* there is no distinction between magic and mystery cult. Magic is simply called *μυστήριον* (mystery, *PGM* IV.723, 746; XII.331, 333) or *μυστήρια* (mysteries, IV.476; V.110), *μέγα μυστήριον* (great mystery, I.131; IV.794), *μυστήριον μέγιστον* (greatest mystery, (IV.2592), *μυστήριον τοῦ θεοῦ* (mystery of God, XIII.128, 685), *θεῖον μυστήριον* (divine mystery, XIXa.52), *μεγαλομυστήριον* (supermystery, XII.322), and *τὰ ἱερὰ μυστήρια* (the holy mysteries, IV.2477). The magicians call themselves *μύστης* ("initiate," I.127; IV.474, 744; cf. *συμμύστης*, fellow-initiate, IV.732; XII.94; *παῖς μυστοδόκος* initiate child, XX.8), *μυσταγωγός* (mystagogue, IV.172, 2254); an outsider is called *ἀμυστηρίαστος* (uninitiated, XII.56, 380, 428).

Clearly, this identity of magic and mystery expresses a hermeneutical tendency on the part of the magicians. The mystery cult terms, being of Greek origin, are found only in those papyrus sections that belong to the higher cultural level within the body of texts. These papyrus sections come along with Greek deities, hymn fragments, rituals, and bits of myth.[36] Obviously, in the eyes of the magicians who wrote and transmitted these texts, the mystery cult language and ritual provided religious legitimacy and cultural approval for all the other magical materials included in the spells as well. We must not misjudge this situation, however. The undeniably Greek origin of the mystery cult terminology in the *Papyri Graecae Magicae* does not decide the general question of whether the mystery cults as such were the result of peculiar developments in Greek religion or whether mystery cults existed in Egypt even prior to and apart from Greek influences.[37] Furthermore, the question whether the Eleusinian and perhaps other mysteries received an initial impetus from outside of Greece, perhaps Asia Minor, Crete, or Egypt, has still not been clarified.[38] While these questions must be answered separately, it is clear from the *Papyri Graecae Magicae* that the mystery cult material there derives from Greek sources and therefore was imported in some form from Greece into Egypt.

Another misunderstanding involves the assumption that the Greek mystery cults had originally nothing to do with magic and that only later syncretism could have combined them. Information about and relics from the Greek mystery cults definitely suggest that magic was a constituent element in the rituals of the mysteries. From Eleusis it may suffice to mention the fire ritual by which Demeter made the king's son (Demophon in the Homeric *Hymn,* Triptolemos in later tradition) immortal[39] and, furthermore, the purifications, processions, sacrifices, and the oath to keep it all secret.[40] The symbols (σύμβολα),[41] the formulas (συνθήματα),[42] and the quotations on the "Orphic" Gold Tablets[43] cannot be understood without the assumptions of magic. Consequently, one can assume that magic was a constitutive element of the mystery cults from their inception. This inherent relationship does not contradict, but rather helps to explain, the fact that in the *Papyri Graecae Magicae,* the Greek mystery cult traditions constitute a secondary interpretation imposed upon the older magic derived from different cultural origins.

To the historian of ancient religions, this secondary imposition of mystery cult terms, ideas, rituals, and traditions is not an unfamiliar phenomenon. The hellenistic period as a whole testifies, through countless pieces of evidence, to how the Greek mysteries expanded their influence. This expansion included the invasion and transformation of other cults into mystery cults, of which we have impressive examples. In his recent book, *Mithras,* Reinhold Merkelbach has shown convincingly that the transformation of the older Persian Mithra religion into the Mithras mysteries took place only in the hellenistic period and may have been the result of skillful planning and organization by religious "founder-figures" or "reformers."[44]

Hellenistic Judaism soon appropriated mystery cult terms and ideas. Especially, the term μυστήριον (mystery) appears to be irresistible. Hellenistic Jewish wisdom literature,[45] apocalypticism,[46] the Testaments of the Twelve Patriarchs,[47] Qumran,[48] and especially Philo of Alexandria[49] employ the terminology. For Philo, the Jewish religion is the true mystery religion,[50] with Moses serving as the great μυσταγωγός (mystagogue),[51] while the Greek mysteries are to be regarded as false mysteries: "Certainly you may see these hybrids of man and woman continually strutting about through the thick of the market, heading the processions at the feasts, appointed to serve as unholy ministers of holy things, leading the mysteries and initiations and celebrating the rites of Demeter" (*Spec. Leg.* III. 40). Philo points up the absurdity that "we often find that no person of good character is admitted to the mysteries, while robbers and pirates and associations of abominable and licentious women, when they offer money to those who conduct the initiatory rites, are sometimes accepted" (*Spec. Leg.* I.323; cf. also *Cherub.* 94).

Expansion of mystery cult terms and ideas is evidenced also by the early Christian literature.[52] Paul frequently employs *μυστήριον* (mystery) as a term designating the revelation of the transcendental realities of the divine world and of wisdom, prophecy, history, the afterlife and, by implication, the sacraments of baptism and the eucharist as well.[53] Ephesians extends the usage, calling the Gospel itself *μυστήριον* (mystery), something Paul himself did not do. The *agapē* relationship between the heavenly Christ and his church on earth is called *τὸ μυστήριον μέγα* (the great mystery).[54] In all probability, Ephesians received this language from Colossians, which more closely reflects Paul's usage.[55] I Timothy (3:9,16) speaks of *τὸ μυστήριον τῆς πίστεως* (the mystery of faith) and *τὸ τῆς εὐσεβείας μυστήριον* (the mystery of religion).

Another contributor of the terminology to the New Testament was apocalypticism. 2 Thessalonians 2:7 speaks of the Antichrist and his antimysteries as *τὸ μυστηύριον τῆς ἀνομίας* (the mystery of lawlessness),[56] while Revelation uses it to describe the seven stars in Christ's right hand and the seven golden lamps of the Christophany, the seven stars being the angels of the seven churches and the seven lamps being those churches (Rev. 1:20). The great eschatological event envisioned in 10:7 is called completion of *τὸ μυστήριον τοῦ θεοῦ* (the mystery of God).

Revelation also knows of antimysteries: the great whore has written her name, Babylon the Great, on her forehead, a *μυστήριον* (mystery, 17:5, 7). Mark's Gospel, when explaining the meaning of Jesus' parables (4:11), takes them to reveal *τὸ μυστήριον τῆς βασιλείας τοῦ θεοῦ* (the mystery of the Kingdom of God), a concept appropriated also by Matthew (13:11, *τὰ μυστήρια τῆς βασιλείας τῶν οὐρανῶν*, "the mysteries of the Kingdom of the heavens") and Luke (8:10), *τὰ μυστήρια τῆς βασιλείας τοῦ θεοῦ*, "the mysteries of the Kingdom of God"). Didache (11:11) and Ignatius of Antioch (*Eph.* 19:1; *Mag.* 9:1; *Trall.* 2:3) provide further evidence for the influence of the term in early Christianity. All this is, however, only a beginning of the enormous expansion of mystery cult terms and ideas in the patristic period, including Gnosticism, Manicheism, and Mandaism. If we look at the *Papyri Graecae Magicae* again with this perspective in mind, then, the nature of the syncretistic transformation of the magical material into mystery cult rituals becomes clear. It is important to realize, however, that such mystery cult transformation is limited to a few highly conspicuous texts of the *Papyri Graecae Magicae*.

The *Spell of Pnouthis* (*PGM* I.42–195) is a magical letter of the famous scribe, Pnouthis, to a certain man named Keryx, a name that by itself refers to the "herald" of a procession. The (pseudepigraphical) letter contains instructions for acquiring an "assistant daemon" (*πάρεδρος δαίμων*).[57] The whole ritual is to be treated as "the great mystery" and must be kept secret (I.130), and its recipient is addressed enthusiastically as "O [blessed] initiate of the sacred magic" (I.127). Most prominent are mystery cult terms and ideas in the famous *Mithras Liturgy* (IV.479–829),[58] which as a whole is called "mysteries handed down [not] for gain but for instruction" (IV.476), that is, for the initiation of a young adept into the magical arts. As the initial prayer states, the ritual was revealed by "the great god Helios Mithras" (IV.482) so that "I alone may ascend into heaven as an inquirer and behold the universe" (IV.484–85). This purpose is confirmed by the remarkable self-definitory statement in the prayer (IV.718–24): "O lord, while being born again, I am passing away; while grown and having grown, I am dying; while being born from a life-generating birth, I am passing on, released to death—as you have founded, as you have decreed, and have established the mystery."

While the whole ritual is called *μυστήριον* (mystery), its core appears to be an ointment to be prepared according to the recipe in IV.750–811. This ointment is called "the great mystery of the scarab" (*τὸ μέγα μυστήριον τοῦ κανθάρου*) and must be applied to the face of the initiate ("anointing his face with mystery," IV.746); the ointment ritual conveys immortality (IV.747, 771).

The fact that the *Mithras Liturgy* is a product of syncretism should not be used to deny that it has genuine connections with the mysteries of Mithras. The long-standing debate principally between Albrecht Dieterich and Franz Cumont appears to have been finally decided in favor of the former.[59] While Dieterich had always taken the *Mithras Liturgy* as belonging to the Mithras cult, Cumont and those who followed him argued that the name of Mithras was inserted into the text by the magician who copied the text but that it had nothing to do substantially with the rest of the material. Because of Cumont, scholars until now have been reluctant to assign the text to the Mithras cult. Even Merkelbach, in his book *Mithras* mentioned above, does not include a discussion of this text. His evaluation of the wide range of new archaeological evidence, however, has in fact prepared the ground for a reevaluation of the *Mithras Liturgy*. As he points out, the Mithras mysteries were a creation of the syncretism of the hellenistic era, not the continuation of the older Persian cult, as Cumont had believed. There is, furthermore, a wealth of material demonstrating magical elements.[60] Since the Mithras mysteries developed differently in different countries, there is really no reason to exclude the possibility of an Egyptian version. Whether the *Mithras Liturgy* is the product of just one magician's efforts or whether there were connections with a Mithraic cultic community cannot be determined on the basis of this one text alone; but even if the former holds true, the author of the *Mithras Liturgy* may still be a serious devotee of the god. The text itself refers only to the magical operator and one apprentice who is to be initiated (see esp. IV.484–85, 732–50).[61]

Other *Papyri Graecae Magicae* texts show a mystery cult terminology that reveals Jewish influences.[62] *PGM* V.96–172, entitled *Stele of Jeu the hieroglyphist in his letter,* includes a summons of the Headless One in the name of Moses: "I am Moses your prophet to whom you have transmitted your mysteries celebrated by Israel" (V.108–11).[63] The so-called *Eighth Book of Moses* refers to "the mystery of the god, which is [called?] 'Scarab' " (XIII.128; cf. IV.794). In another place, the same text orders, "Now begin to recite the stele and the mystery of the god," again referring to the scarab (XIII.685). In *PGM* XII.331 and 333 a magical ring with an engraved gemstone is called *μυστήριον* (mystery). Perhaps the name results from the image of the scarab engraved in the gemstone (XII. 275–76). While these texts reveal Egyptian, Greek, and Jewish religious ingredients, a love spell (XXXVI.295–311), in an invocation of the Jewish god, refers to the sex act as "the mystery rite of Aphrodite" (306).[64]

Finally there are two passages in the great magical papyrus of Paris that use the strange concept of religious slander.[65] Both belong to the "slander spell" (διαβολή) in *PGM* IV.2441–2621, about which Sam Eitrem published a penetrating study.[66] The invocation addresses the goddess Aktiophis[67] and then presents the following denunciation of the women sought by the spell:

> For I come announcing the slander of NN, a defiled and unholy woman, for she has slanderously brought your holy mysteries to the knowledge of men. She, NN, is the one, [not] I, who says, "I have seen the greatest goddess, after leaving the heavenly vault, on earth without sandals, sword in hand, and [speaking] a foul name." It is she, NN, who said, "I saw [the goddess] drinking blood. She, NN, said it, not I, AKTIŌPHIS ERESCHIGAL NEBOUTOSOUALĒTH PHORPHORBA SATRAPAMMŌN CHOIRIXIĒ, flesh eater. Go to her NN and take away her sleep and put a burning heat in her soul, punishment and frenzied passion in her thoughts, and banish her from every place and from every house, and attract her to me, NN." (IV.2474–90)[68]

Betrayal of the mysteries to the uninitiated was, of course, a notorious scandal in Athens; prominent figures such as Alcibiades, Andocides and their noble friends got themselves in trouble for doing this very thing.[69] A third coercive spell (IV.2474–2621) contains another slander reporting sacrilegious sacrifices, among them, "For you, a vulture and a mouse, your greatest myst'ry, goddess" (IV.2592).[70] Such sacrifice was, of course, sheer blasphemy, and the charge was designed to arouse the wrath of the goddess against the alleged perpetrator.

The interesting phenomenon that the divinely beneficial mystery can be turned into its opposite is paralleled in Judaism and Christianity, when in those contexts the Greek mysteries are declared to be a daemonic mimicry of the true mystery of God.[71] In Christianity, this juxtaposition begins in Paul (1 Cor. 10:18–22)[72] and finds clear expressions as well in *τὸ μυστήριον τῆς ἀνομίας* (the mystery of lawlessness, 2 Thess. 2:7) and the mystery of the great whore Babylon (Rev. 17:5, 7).

CONCLUSION

The passages discussed in the preceding pages show that the growing influence of Greek mystery cult terminology and ideas in the hellenistic era had a profound impact on the Greek magical papyri. Under this impact the earlier Egyptian magic was transformed, enriched, brought up to date, and thus legitimated. By presenting themselves as mystagogues, the magicians doubtless added to their prestige. This transformation shows that the older Egyptian magic, which at one time functioned as highly valued "religion," had now sunk to the lower level of mere "magic" as a result of the encounter with the Greek religious world, in particular the mystery cults. The mystagogue-magicians of the Greek mystery cults transformed the older magic into a new and higher "religion." For the mystagogue-magicians, the syncretistic amalgam was indeed "religion."

By the same token, however, the mystery cult materials that the magicians were able to obtain were also transformed. The magicians treated the mystery cult traditions in the same way as they treated other religious traditions: they appropriated, adopted, and subsumed. Taking advantage of the intrinsic relationship between magic and mystery, they took the tradition apart and reconstituted it as something new: magical spells and rituals of a seemingly greater appeal and force.

There is scarcely any doubt that the work of the author of the *Mithras Liturgy* would have horrified conscientious cult officials of the mysteries of Mithras. To them, the *Mithras Liturgy* would have meant dragging the great tradition down into the muddy waters of "magic." Their horror would have been more than a matter of subjective preference. They would have pointed to the absence in the *Mithras Liturgy* of such essentials as the moral ethos, the oaths, the fellowship, and the loyalty among members of the cult, not to speak of the concerns for the welfare of the imperial government and the world community. For them, in other words, the author of the *Mithras Liturgy* had betrayed the Mithraic "religion" for "magic."

Introducing the mysteries into the magical tradition in this way and making them into magic destroyed the internal coherence and integrity that the individual mystery cults had possessed. It was, of course, easier to have a magician bring the mysteries of Eleusis into the small towns of Egypt in order to let those who would never be able to travel to Eleusis become beneficiaries and partakers. Yet if one intended to become initiated into the Eleusinian mysteries in the "real" sense, one *had to* travel to Eleusis, and one had to absolve the long preparations, spend the necessary time to live through the rituals, and, most important, celebrate the Eleusinian festivals with awe, caution, and full preparation of the soul. Similarly, the mysteries of Mithras were not just for everyone. It took hard work, rigorous training, and serious trials and examinations before one could move from the lower to the higher grades of the Mithraic hierarchy. Being a Mithraist meant to be a member in a sacred covenant with the god Mithras, a covenant that involved one's entire life.

We find, then, real differences between magic and religion even within this body of highly syncretistic material. Certainly, even greater differences existed between the Greek magical papyri in their entirety and the official religions from which the traditions were derived. What determined the distinctions between, and the resultant definitions of, magic and religion were theological issues internal to the religious traditions and cults involved. These issues differed from one cult to another, but they were not arbitrary. In order to evaluate religious phenomena concerning the problems of magic and religion, one must have a high degree of sensitivity to the inner life and thought of the cults in question. What characterized the magicians of the Greek magical papyri was that they unashamedly lacked a full comprehension or appreciation of the inner integrity of the cults whose materials they appropriated. That is why they were right in calling themselves "magicians," and their art, "magic." They lacked what we would call "religion." They themselves no doubt believed that they possessed a "religion that worked," but what they in fact had produced was magic.

Notes

1. The literature on magic in general is immense. For surveys of the earlier debates, see Karl Beth, *Religion und Magie: Ein religionsgeschichtlicher Beitrag zur psychologischen Grundlegung der religiösen Prinzipienlehre* 2d ed. (Leipzig, 1927); Ludwig Deubner's "Magie und Religion," in his *Kleine Schriften zur klassischen Altertumskunde* (Königstein 1982), 275–98; several articles by Albrecht Dietrich, *Kleine Schriften* (Leipzig, 1911), especially "Der Untergang der antiken Religion," pp. 448–539; Carl Heinz Ratschow, *Magie und Religion* 2d ed. (Gütersloh, 1955); Geo Widengren, *Religionsphänomenologie* (New York, 1969), 1–19; John Middleton, "Theories of Magic," in *Encyclopedia of Religion* 9 (1987): 82–89; see also Betz, "Magic in Graeco-Roman Antiquity," in *Encylopedia of Religion* 9 (1987): 93–97.

2. See Widengren, *Religionsphänomenologie,* 150–257.

3. For the older discussion, see Sam. Eitrem, "Eleusinia—les mystères et l'agriculture," *SO* 20 (1940): 133–51; for the present state of research, see Walter Burkert, *"Homo Necans:" The Anthropology of Ancient Greek Sacrificial Ritual and Myth,* trans. Peter Bing (Berkeley, 1983); idem, *Greek Religion,* trans. John Raffan (Cambridge, Mass., 1985), esp. 276–304; idem, *Structure and History in Greek Mythology and Ritual,* Sather Classical Lectures 47 (Berkeley, 1979), 123–42.

4. For this point, see especially Burkert, *Structure and History* (see n. 3).

5. For discussion, see William C. Grese, *Corpus Hermeticum XIII and Early Christian Literature,* SCHNT 5 (Leiden, 1979), 40–43; Kurt Rudolph, *Gnosis: The Nature and History of Gnosticism,* trans. Robert McLachlan Wilson (San Francisco, 1983), 204–52.

6. I have discussed this problem in my earlier book, *Nachfolge und Nachahmung Jesu Christi im Neuen Testament,* BHTh 37 (Tübingen, 1967), 101–7.

7. This view was held by the great expert on magic, Alphonse A. Barb. See his "The Survival of Magic Arts," in *The Conflict between Paganism and Christianity in the Fourth Century,* ed. Arnaldo Momigliano (Oxford, 1963), 100–25; idem, "Mystery, Myth, and Magic," in *The Legacy of Egypt,* 2d ed., ed. J. R. Harris (Oxford, 1971), 138–69. Similarly, of course, Dieterich, *Kleine Schriften* (see n. 1), 512–13.

8. See on this problem Geo Widengren, "Evolutionism and the Problem of the Origin of Religion," *Ethnos* 10 (1945): 57–96. Widengren's conclusions were negative and put an end to the speculations: "The origin of religion is beyond scientific research; we can only get some idea of the oldest conceivable forms of religion. The attempt to find the origin of religion must be annulled as well as the old evolutionistic method" (p. 96).

9. The fact is that magic has been treated as decadent and pagan since the Old Testament. See, e.g., Exod 7:11, 8:18–19, 9:11, 22:18, Deut. 18:10–11; 1 Sam. 28; 2 Kgs. 9:22; 2 Chron. 33:6; Ps. 58:5; Jer. 27:9; Dan. 2:2; Mic. 5:12; Nah. 3:4; Mal. 3:5; also Wisd. of Sol. 12:4, 17:7; Jub. 48:9–10; 1 Enoch 64–65, 94; Test. XII, *Judah* 23; Ps.- Philo, *Lib. Ant. Bibl.* 34, 64. The New Testament continues the polemics: Acts 8:9–24, 13:6–12; Gal. 5:20; Rev. 9:21; 18:23, 21:8, 22:15, etc.

10. For the notion of superstition and related literatures, see Dieter Harmening, *Superstitio: Überlieferungs-und theoriengeschichtliche Untersuchungen zur kirchlich-theologischen Aberglaubensliteratur des Mittelalters* (Berlin, 1979).

11. Ground-breaking was the work by Marcel Mauss's "Esquisse d'une théorie générale de la magie," in her *Sociologie et Anthropologie* (Paris, 1950), 1–141; English edition: Mauss, *A General Theory of Magic,* trans. Robert Brain (New York, 1972). For the recent discussion, see Hans G. Kippenberg and Brigitte Luchesi, eds., *Magie: Die sozialwissenschaftliche Kontroverse über das Verstehen fremden Denkens* (Frankfurt, 1978); Leander Petzoldt, ed.,

Magie und Religion: Beiträge zu einer Theorie der Magie, WdF 337 (Darmstadt, 1978); Richard A. Horsely, "Further Reflections on Witchcraft and European Folk-religion," *HR* 19 (1979): 71–95.

12. Alan F. Segal, "Hellenistic Magic: Some Questions of Definition," in *Studies in Gnosticism and Hellenistic Religions,* ed. R. van den Broek and Maarten J. Vermaseren, EPRO 91 (Leiden, 1982), 349–75; the quotation is from p. 349.

13. Ibid., 349–50.

14. Ibid., 351.

15. Sigmund Mowinckel, *Religion und Kultus* (Göttingen, 1953), 15: "Das magische Denken und dessen Umsetzung in die Praxis, die 'Magie,' ist nicht eine Art Religion, sondern ein Weltbild, eine bestimmte Weise, die Dinge und deren gegenseitige Zusammenhänge aufzufassen; es ist eine 'Weltanschauung,' die etwa dem Weltbild entspricht, das wir heutzutage auf der Grundlage des Kausalgesetzes und der Zusammenhänge zwischen Ursache und Wirkung zu formulieren versuchen, so wie es die Physik und Chemie, die Biologie und die Psychologie herauszubringen bemüht sind." The problem with this description, however, is that it juxtaposes religion and *Weltanschauung* without clarifying their relationship; it also confuses *Weltanschauung* and *Weltbild,* two notions that may not be synonyms: *Weltanschauung* carries the connotation of unscientific propaganda, while *Weltbild* can be that of science or prescientific mentalities.

16. Recent studies have pointed out the connections between magic and the beginnings of science. See G. E. R. Lloyd, *Magic, Reason, and Experience: Studies in the Origins and Development of Greek Science* (Cambridge, 1979).

17. For a collection of important articles covering a wide area of issues, see Brian Vickers, ed., *Occult and Scientific Mentalities in the Renaissance* (Cambridge, 1984).

18. See, e.g., Betty Jo Dobbs, *The Foundations of Newton's Alchemy* (Cambridge, 1975); Peter J. French, *John Dee: The World of an Elizabethan Magus* (London, 1972); furthermore, the papers in *Magia naturalis und die Entstehung der modernen Naturwissenschaften, Symposion der Leibniz-Gesellschaft, Hannover, 14. und 15. November 1975,* Studia Leibnitiana 7 (Wiesbaden, 1978); Antoine Faivre and Rolf C. Zimmerman, eds., *Epochen der Naturmystik: Hermetische Tradition im wissenschaftlichen Fortschritt* (Berlin, 1979); Wolf-Dieter Müller-Jahncke, *Astrologisch-magische Theorie und Praxis in der Heilkunde der frühen Neuzeit,* Sudhoffs Archiv 25 (Wiesbaden, 1985); Walter Pagel, *Religion and Neoplatonism in Renaissance Medicine,* ed. Marianne Winder (London, 1985).

19. This view has been advocated by Murray and Rosalie Wax, "The Magical World View," *JSSR* 1 (1962): 179–88; also by the same authors, "The Notion of Magic," *Current Anthropology* 4 (1963): 495–513.

20. The difficulties for even a working definition can be seen from the one attempted by Kurt Goldammer, "Magie," *Historisches Wörterbuch der Philosophie* 5 (1980): 631: " 'M.[agie]' meint Zauber, abergläubische Handlung, Geheimritual und ist eine durch die Resultate ethnologischer und religionsgeschichtlicher Forschung sehr komplex gewordene, teils wertende, teils wertungsindifferente Bezeichnung für vorwissenschaftliches und 'ausserrationales' zweckhaftes Handeln des Menschen auf der Grundlage bestimmter Kausalvorstellungen, für eine damit zusammenhängende Weltanschauung, ferner für niedere Religionsformen oder für Religionsderivate und -surrogate, die durch derartiges Verhalten geprägt sind. Die irrationale Komponente und theoretische Grundlage der M. und das oft dahinterstehende metaphysische System sowie ein mit ihr nicht selten verbundenes kompliziertes Ritual schlagen jedenfalls eine Brücke zur Religion."

21. For a very useful collection of magical texts from antiquity in translation, see Georg Luck, *"Arcana Mundi": Magic and Occult in the Greek and Roman Worlds* (Baltimore, 1985).

22. (Chicago, 1986), xli [=GMPT].

23. Hans Dieter Betz, "The Formation of Authoritative Tradition in the Greek Magical Papyri," in *Jewish and Christian Self-Definition,* vol. 3, *Self-Definition in Greco-Roman World,* ed. Ben F. Meyer and E. P. Sanders (Philadelphia, 1982), 161–70, esp. 163–64.

24. *PGM* I.127. The Greek texts are quoted by papyrus numbers and lines according to the edition of Karl Preisendanz, ed., *Papyri Graecae Magicae: Die griechischen Zauberpapyri,* 2 vols., 2d ed. rev. Albert Henrichs (Stuttgart, 1973–74).

25. For references, see Betz, "Formation" (see n. 23), 165–66.

26. See Betz, *GMPT,* 338, s. v. *phylactery.*

27. *PGM* IV.2081–87: "Most of the magicians, who carried their instruments with them, even put them aside and used him [i.e., the daimon] as an assistant. And they accomplished the preceding things with all dispatch. For [the spell] is free of excessive verbiage, immediately carrying out as it does the preceding things with all ease." The translation is by Edward N. O'Neil in *GMPT,* 74.

28. For a detailed survey, see Francis C. R. Thee, *Julius Africanus and the Early Christian View of Magic,* HUTh 19 (Tübingen, 1984), 316–448.

29. Janet H. Johnson, "Introduction to the Demotic Magical Papyri," in *GMPT*, lv.

30. Ibid.

31. Ibid.

32. Greek inscriptions and influences on Egyptian art begin in the sixth century B.C. See Olivier Masson, with contributions by Geoffrey Thorndike Martin and Richard Vaughan Nicholls, *Carian Inscriptions from North Saqqara and Buhen* (London, 1978).

33. The most conspicuous specimens are the *Stele of Jeu the hieroglyphist, PGM* V.96–172; the prayer in V.459–89—see on this passage the commentary by Marc Philonenko, "Une prière magique au dieu créateur (*PGM* 5, 459–89)," *CRAI* (1985), 433–52; the love charm in *PGM* VII.593–619; the *Eighth Book of Moses, PGM* XIII.1–1077; the *Prayer of Jacob, PGM* XXIIb.1–26 (cf. also James H. Charlesworth, ed., *The Old Testament Pseudepigrapha,* vol. 2 [Garden City, N.Y., 1985], 715–23); an invocation to "the god of Abraham, Isaac, and Jacob," *PGM* XXXV.1–42.

34. See *PGM* IV.1233, 3019; XII.192, 391–92; XIII.289; XLIV.18.

35. Evidence from Christian magic is widely dispersed. For collections of some of the materials, see Preisendanz, *Papyri Graecae Magicae,* vol. 2, 209–32; Angelicus M. Kropp, *Ausgewählte koptische Zaubertexte,* 3 vols. (Brussels, 1930–31); idem, *Der Lobpreis des Erzengels Michael* (*vormals P. Heidelberg Inv. Nr. 1686*) (Brussels, 1966). For surveys and bibliography, see also David E. Aune, "Magic in Early Christianity," in *ANRW* vol. II, pt. 23, sec. 2 (Berlin, 1980), 1507–57.

36. In an earlier paper, I have tried to identify liturgical fragments from the mysteries of the Idaean Dactyls: Betz, "Fragments from a Catabasis Ritual in a Greek Magical Papyrus," *HR* 19 (1980): 287–95; also idem, "The Delphic Maxim 'Know Yourself' in the Greek Magical Papyri," *HR* 21 (1981): 156–71. On the liturgical fragments in *PGM* LXX see D. Jordan, "A Love Charm with Verses," *ZPE* 72 (1988): 245–59.

37. See on this question, with further bibliography, John G. Griffiths, "Mysterien," *LdÄ* 4 (1982): 276–77; Joris F. Borghouts, "Magie," *LdÄ* 3 (1980). 1137–51; Hartwig Altenmüller, "Magische Literatur," *LdÄ3* (1980):1151–62.

38. Burkert's, "From Telepinus to Thelpusa: In Search of Demeter" (in his *Structure and History* [see n. 3] 123–42), shows that the nature and variety of earlier influences may be rather complicated.

39. Homeric *Hymn to Demeter* 226–91. For the interpretation, see N. J. Richardson, *The Homeric Hymn to Demeter* (Oxford, 1979), 228–56; Burkert, *"Homo Necans"* (see n. 3), 280–81.

40. Ibid. 248, 251–53, 256, 293. In *PGM* the order is frequently given to keep the magic secret; see, e.g., I.41, 146; IV.75, 255–56, 745, 850–51, 923, 1115, 1251, 1870, 2512, etc.

41. See Burkert, "*Homo Necans* (see n. 3), 41, n. 29; 269, n. 19. Cf. *σύμβολα μυστικά* at *PGM* III.701; IV. 945. For further references, see Betz, (see n. 36), 291.

42. See Burkert, "Homo Necans (see n. 3), 265–74: "Myesis and Synthema". For *PGM* see Dieterich, *Eine Mithrasliturgie,* ed. Otto Weinreich (Darmstadt, 1966), 213–18, 256–58; Betz, "Fragments" (see n. 36), 292–93; idem, "Delphic Maxim" (see n. 36), 165–70.

43. The basic edition is by Günther Zuntz, *Persephone* (Oxford, 1971), 277–393. Since then, more gold tablets have been discovered; see the references in Burkert, *Greek Religion* (see n. 3), 293, n. 1; Martin West, *The Orphic Poems* (Oxford, 1983), 22–23, 25–26, 171.

44. Reinhold Merkelbach, *Mithras* (Königstein, 1984), 64–70, 75–77, especially concerning the royal cult at Commagene.

45. For the terminology of *μυστήριον*, *μύστης* (mystery, initiate), see Septuagiut Tobit 12:7, 11; Judith 2:2; Wisd. 2:22, 6:22, 8:4, 12:6, 14:15 and 23; Sir. 3:19, 22:22, 27:16–17 and 21; Dan. 2:18–19, 27–30, and 47.

46. See especially 1 Enoch 16:3; 3 Enoch 11:1; Test. Sol., passim. The *Hymn of Orpheus* tells of an initiation and apotheosis of Moses, who is celebrated as the great mystagogue of Israel. *Joseph and Aseneth* contains an elaborate conversion ritual, real or fictitious, that imitates mystery cult initiation ceremonies. For all these texts in English translation and with introductions and notes, see J. H. Charlesworth, ed., *The Old Testament Pseudepigrapha,* vols. 1 and 2 (Garden City, N.Y., 1983 and 1985).

47. Test. XII, *Judah* 12–16 calls Judah's wine-drunkenness his being "in mystery" (*ἐν μυστηρίῳ*), i.e., possessed by Dionysos.

48. The term *raz* (mystery) frequently occurs in the Qumran texts; see Eric Vogt, " 'Mysteria' in Textibus Qumran," *Bib* 37 (1956): 247–57; Raymond E. Brown, "The Semitic Background of the New Testament *mystērion,*" *Bib* 39 (1958): 426–48; *Bib* 40 (1959): 70–87. For the Iranian background, see also Geo Widengren, *Iranisch-semitische Kulturbegegnung in parthischer Zeit* (Köln, 1960), 55, 100.

49. Philo of Alexandria employs the full range of mystery cult terminology. His writings are as yet an unexhausted source for terms and ideas derived from the Greek mysteries. See the study of Joseph Pascher, Η ΒΑΣΙΛΙΚΗ ΟΔΟΣ: *Der Königsweg zu Wiedergeburt and Vergottung bei Philon von Alexandreia,* Studien zur Geschichte und Kultur des Altertums vol.17, pts. 3–4 (Paderborn, 1931). See also Betz, *Nachfolge* (see n. 6), 130–36.

50. Philo's mystery cult terminology is mixed with traditions of middle-Platonic philosophy. See especially *Sacr.* 62; *Leg. All.* 3.27, 71, 100; *Quod Deus* 61; *Vita Cont.* 27. Remarkable are also the addresses beginning with *ὦ μύσται* (O initiates); see *Cherub.* 48; *Fuga* 85; *Spec. Leg.* 1. 320.

51. See, e.g., *Virt.* 185; *Vita Mos.* 1.71; *Cherub* 49; *Spec. Leg.* 1.319.

52. See Günther Bornkamm, "*μυστήριον*, *μύω*," *TDNT* 4 (1967): 802–28.

53. Rom. 11:25, 16:25; 1 Cor. 2:1 and 7, 4:1, 13:2, 14:2, 15:51.

54. Eph. 1:9; 3:3, 4, and 9; 5:32; 6:19.

55. Col. 1:26 and 27; 2:2; 4:3.

56. This expression is unique in the New Testament, but cf. 1 Enoch 16:3; Test. XII, *Judah* 12–16; Test. of Sol., passim; Sibyll. Or. 8:56 and 58; IQM 14:9; IQH 5:36; Joseph. *BJ* 1.470; Matt. 24:12; Did. 16:4; Hermas, *Mand.* 8:3; Barn. 14:5, 15:7, 18:2.

57. For the translation and notes, see Edward N. O'Neil, in *GMPT* 4–8 and the glossary, 332–33, s.v. assistant daimon (*paredros*).

58. The translation is that of Marvin W. Meyer, in *GMPT,* 48–54; for commentary, see Dieterich, *Eine Mithrasliturgie* (see n. 42), which is still indispensable; furthermore, Roger Beck, "Mithraism since Franz Cumont," *ANRW* Vol. II, pt. 17, sec. 4 (1984): 2002–2115.

59. See, for this debate, Dieterich, *Eine Mithrasliturgie,* (see n. 42), 29–30, 230–32, 234–40, 250; also Marvin W. Meyer, *The "Mithras Liturgy"* (Missoula, Mont., 1976) vii–viii; Hans-Josef Klauck, *Herrenmahl und hellenistischer Kult: Eine religionsgeschichtliche Untersuchung zum ersten Korintherbrief,* (Münster, 1982), 156–58; 301, n. 97; 339, n. 31; Maarten J. Vermaseren, "La soteriologie dans les Papyri Graecae Magicae," in *La soteriologia dei culti orientali nell' impero romano,* ed. U. Bianchi and M. J. Vermaseren, EPRO 93 (Leiden, 1982), 17–30, esp. 25.

60. See also Reinhold Merkelbach, *Weihegrade und Seelenlehre der Mithras-mysterien* (Opladen, 1982); idem, *Mithras* (see n. 44), 77–146.

61. Phenomenologically, this situation corresponds to the phenotype of the magician, working as an individual craftsman with, perhaps, one apprentice. There appears to be no trace in *PGM* of connections with larger religious communities or institutions. See Walter Burkert, "Craft versus Sect: The Problem of Orphics and Pythagoreans," in *Jewish and Christian Self-Definition,* vol. 3 (see n. 23), 1–22.

62. The origins and precise nature of Jewish influences in *PGM* constitute an unresolved problem. See Ludwig Blau, *Das altjüdische Zauberwesen* (Strasbourg, 1898); Michael Morgan, *"Sepher Ha-Razim" : The Book of Mysteries* (Chico, Calif., 1983).

63. For Moses as a mystagogue of Israel, see the *Hymn of Orpheus* mentioned above (n. 46) and the passages from Philo noted above (nn. 49–51). On the whole subject, see John G. Gager, *Moses in Greco-Roman Paganism,* SBL Monograph Series 16 (Nashville, Tenn., 1972), 134–61; furthermore, Raphael Patai, "Biblical Figures As Alchemists," *HUCA* 54 (1983): 195–229, esp. 213–29 ("Moses the Alchemist").

64. For the mysteries of Aphrodite, see Martin P. Nilsson, *Geschichte der griechischen Religion,* vol.1, 3d ed., (Munich, 1967), 524.

65. For literature and references, see Betz *GMPT,* 83, n. 314.

66. Sam. Eitrem, "Die rituelle ΔΙΑΒΟΛΗ," *SO* 2 (1924): 43–61.

67. Probably an epithet of Selene; see, *GMPT* Betz, 322, s.v. *Aktiophis*.

68. Translated by Edward N. O'Neil, in *GMPT,* 83.

69. See, on the mystery scandals, Burkert, *Greek Religion* (see n. 3), 316–17, also 296, 299; idem, *"Homo Necans"* (see n. 3), 252–53; Nilsson, *Geschichte der griechischen Religion,* vol. 2, 2d ed. (Munich 1961), 90–91; George E. Mylonas, *Eleusis and the Eleusinian Mysteries* (Princeton, 1961), 298–99.

70. Translated by Edward N. O'Neil, in *GMPT,* 85.

71. See Justin Martyr, *Apol.* 1.66.4; also 1.54, 56, and 62.1–2; *Dial.* 70.5; furthermore, *Apol.* 1.57–58, 2.5–6 and 8. On the whole subject, see Klauck, *Herrenmahl* (see n. 59), 139–41.

72. See Klauck, *Herrenmahl* (see n. 59), 264–72.

10

Nullum Crimen sine Lege: Socioreligious Sanctions on Magic

C. R. Phillips III

But I know it when I see it.
POTTER STEWART

Like the late Justice Stewart on obscenity, most classicists confidently point to what they consider magic, on a modern viewpoint, in Greco-Roman literary texts, curse tablets, papyri, and astrology.[1] Many are wont to identify what they deem magical elements in the "developed" forms of Greco-Roman religion, albeit often lamenting the persistence of such allegedly primitive relics of irrationality.[2] Again, scattered ancient evidence for the repression of certain religious activities, some of which the ancients actually called magic, has encouraged a tacit working assumption that the ancients were as anxious to penalize such activities as, say, the witch-hunters of early modern Europe: according to Ammianus Marcellinus (359 A.D.), "Anyone who wore round his neck a charm against the quartan ague or some other complaint, or was accused by his ill-wishers of visiting a grave in the evening, was found guilty and executed as a sorcerer or as an inquirer into the horrors of men's tombs and the empty phantoms of the spirits which haunt them" (19.12.14).[3]

Indeed, as Christina Larner put it "Witchcraft is the labelling theorist's dream."[4] But in the case of the Greco-Roman world, traditional classical scholarship has produced something more like the Homeric baleful dream (*oulos oneiros, Il.* 2.6). It has accepted uncritically the nineteenth-century notion that magic is either "bad" religion or "bad" science—that "magic" represents a "primitive" worldview that has not evolved.[5] Sometimes it measures ancient religious phenomena against modern notions of religion and science. That which does not measure up becomes categorized as magic. Sometimes, and more charitably, it notes the different worldview of Greco-Roman religion, and finds "magic" in those ancient religious phenomena that do not seem in accord with the "developed" forms of that religion.[6] It does not attend to the possibility of polemic when charges of unsanctioned religious activities appear in ancient texts but rather accepts those charges as empirically valid.[7] In brief, it neglects to consider whether what it deems an unsanctioned religious activity represented the same thing to the ancients and, if it did, for what reasons; there is the tacit assumption that the ancients, like moderns, had universal standards. Regardless of the precise interpretational strategy or strategies employed, unsanctioned religious activity appears omnipresent. And that activity gets categorized as magic on the Christianizing view that magic is the antithesis to religion; as such, it ought to have been ruthlessly eradicated. But few have followed the logic of their position to examine how often and how severely unsanctioned religious activity was repressed in classical antiquity. The results are unsettling. It will appear that the ancients were far more tolerant of unsanctioned religious activity than modern scholars have assumed. And when they chose to repress it, the reasons for the repression lie more in specific circumstances surrounding a given activity rather than in a general societal revulsion.

LEGAL DEFINITIONS AND SANCTIONS

The evidence for legal actions against unsanctioned religious activity in classical antiquity appears sparse when considered against the dimensions of time, population size, and geography; even the relative increase in legislation under the Christian emperors does not significantly alter the picture. Sociology of law would traditionally argue that since a legal system supposedly mirrors what a society considers sufficiently important for its well-being either to encourage or repress, one must conclude that unsanctioned religious activity did not arouse sufficient concern to mobilize the legislative process to vigorous action. This view, however, now appears overly simplistic. First, although legal specialists of a socioeconomic elite can certainly influence legislation and sometimes enact it, their interests need not be identical with those of the entire elite, still less with those of society at large. A fourth-century A.D. senator enjoying the *otium* of his provincial estates will have very different priorities from, say, another senator in daily contact with the imperial court. Second, however the legislation arises, it may not necessarily be the most empirically efficacious way to repress, for instance, unsanctioned religious activity. Vested interests and political considerations may require legislative compromise, resulting in statutes betokening that compromise more than ideology. Thus, it may indeed be that the relative infrequency of ancient legislation against unsanctioned religious activity reflects general societal disinterest in repression and hence less fear of that activity than hitherto assumed. But it may equally well be that the infrequency implies only disinterest or lack of ability to reach a viable consensus among the legal establishment. These larger considerations, albeit of fundamental importance, lie outside my scope. Rather, I propose to foster discussion of such considerations through examination of ancient and modern views of the object of the legislation, usually called "magic."[8]

The problem lies with the use of modern definitions of "magic" to categorize ancient religious systems. As I will demonstrate, those definitions utilize combinations of Judeo-Christian and modern scientific models for, respectively, religion and science to identify phenomena that do not conform as magical—that is, "bad"—religion or science in the modern sense. Thus on a modern view a whole host of phenomena become magic. But since ancient religion and science did not offer universally accepted definitions, ancient law could not look there for guidance. Rather, the ancient legal systems could apply various labels to unsanctioned religious activity. Magic was one such label, but neither it nor any other label for unsanctioned religious activity appear consistently attached to a given set of phenomena. Moderns have abstracted *magic* to cover all ancient religious phenomena that do not conform to their notions of "true" religion and science, regardless of how the ancients viewed those phenomena. Thus the term *unsanctioned religious activities* appears here advisedly, to avoid the value-laden modern overtones in *magic*. Now the aforementioned imbalance disappears, since unlike more recent times, the ancients did not consider all unsanctioned religious activity necessarily to be criminal. I can claim that your religious activity is unsanctioned simply because I do not like it or because it does not conform to my particular definitions of legitimate religion. Further, I can assert that since you violate sacred norms (as I understand them), you probably also violate secular ones, since the sacred legitimates the secular, by definition. I may not have evidence for your secular transgressions, but I can nevertheless assert they must exist in the form of your theological miscreance. And if I can persuade those with coercive power of the correctness of my views, you are in trouble.[9]

Long ago, Hopfner observed that there was no action against magic per se in Greek law.[10] Rather, the legal action for impiety (*asebeia*) could encompass specific deviant actions. Consider concocting *pharmaka*. These could sometimes be deemed magical potions, and thus their concoction was actionable under *asebeia;* but the extant accounts do not name magic.[11] A given potion, for example, might be viewed in purely naturalistic terms, and it might equally well be viewed as part of a religious ritual, this latter case leaving the way open for the suspicion of unsanctioned religious activity.[12] Again, *asebeia* could encompass prosecution of philosophers or a charge of profanation but did not necessarily entail accusation of magic. For example, Diopeithes' law of 432 or 431 B.C. and the prosecution of Anaxagoras involved the pre-Socratic philosopher's astrophysical observations and may well have been a covert way to attack Pericles as well.[13] Alcibiades' profanation of the mysteries (415 B.C.) came under *asebeia,* and once again the political turbulence surrounding the event should be noted.[14] Finally, the trial of Socrates (399 B.C.) involved the threefold accusation of refusing to recognize the state gods, introducing new divinities, and corrupting the youth.[15] Theft of sacred objects could sometimes involve *asebeia,* while at other times special theft actions appeared. It would appear, then, a variety of activities—be they pharmaceutical, philosophical, pious, or political—could, on occasion, fall under *asebeia.*[16]

Any society will have norms for what is considered socially acceptable behavior; ancient societies traditionally legitimated those norms with reference to divine sanction. Thus, someone transgressing a given secular norm could readily be conceived as violating the divine "rightness" of the universe—and hence practicing improper methods of relating to the spiritual world as well. But ancient systems of religious and scientific "knowledge" did not have universally accepted definitions that would have influenced legal thinking. Ancient pagan religion never defined "orthodoxy" and "heresy" in the Christian sense of the words. Nor could it, since its polytheism, ever-receptive to new divinities, made it impossible to postulate a canonical divinity or divinities with the implication that worship of others constituted "wrong" religious thinking. A cult could have its ordinances, of course, but those rules sought to define the internal functioning of the cult rather than to debar the adherents from other religious associations.[17] Actions for impiety are not buttressed by a consistent, all-encompassing ideology. What *could* constitute unsanctioned religious activity? When one gets beyond, for instance, theft of sacred objects or transgression of ritual ordinances, the possibilities become vague and hence almost limitless.[18] It remains striking that the three prominent instances of Anaxagoras, Alcibiades, and Socrates all have very implicit secular political agenda as well. In a roundabout way, a charge of impiety could be a very potent form of social control, since one's secular actions could always be taken as contrary to the divinely sanctioned norms for mortal behavior.

Again, ancient science could not offer universally accepted models that could, say, differentiate between the naturalistic and religious uses of *pharmaka,*—the religions use providing the possibility of legal action.[19] This is precisely the opposite of early modern Europe, where theologians and scientists could pool their knowledge systems to elaborate "true" tests of whether someone was a witch.[20] Where there exists no one socially accepted set of knowledge systems to buttress a particular definition, as in antiquity, a whole host of local and personal standards will reign. Thus Plato (*Leg*. 933a–b) lamented on spells, charms, and enchantments, "It is not easy to know the truth about these and similar practices, and even if one were to find out, it would be difficult to convince others; and it is just not worth the effort to try to persuade people whose heads are full of mutual suspicion."[21]

Those who were in a position to take coercive action on unsanctioned religious activities, namely the socioeconomic elite, would be constantly reminded of those activities as represented in literature. Circe's charms and those that healed Odysseus' wound, the bewitching power of song that Sappho first noted and that received fullest elaboration in Gorgias' *Helen,* Heracles' robe dipped in Nessos' blood all provided mythic exempla of the wide variation in unsanctioned religious activity.[22] On the one hand, those *exempla* reflect contemporary religious "knowledge" projected onto the mythical past. On the other hand, those reifications influence the ongoing social construction of religious "knowledge." As a result of this dialectic, when, say, a Circe and Odysseus can both traffic in such activities, it becomes hard not to accept such wide variation as part of the socioreligious fabric. Ancient literary works represent productions by the socioeconomic elite for the socioeconomic elite, namely those with the time, money, education, and hence inclination to peruse the texts.[23] Unsanctioned religious activities, in literature and daily life, were givens. The elite would be less-than-likely to take action on actual instances of the phenomena except where those instances entailed danger to the social order as the elite conceived it. And, as indicated, the interests of the elite members in a position to influence legislation need not coincide with elite interests at large. The sheer bulk of preserved "magical" papyri and curse tablets argues against any large-scale repressions that have escaped historical notice.[24] Repression also depends on communication. Given the ancient socioeconomic elite's profound contempt for the lower orders, it would not usually trouble systematically to ferret out instances of "magic," whether in urban slums or countryside.[25] As Ammianus Marcellinus later put it (28.1.15), "not everything which has happened among the lower classes is worth my while to recount."

The Roman material until the fourth century A.D appears functionally equivalent. Specific unsanctioned religious activities could be subject to legal sanctions, but there appears no omnibus definition. For example, the famous fragment (8.8) from the fifth-century B.C. Twelve Tables ("The Person Who Has Enchanted the Crops," *qui fruges excantassit*) punishes an attack on private property (crops) rather than the means to that attack. Moreover, the Roman legal system was reactive rather than inquisitorial—that is, it operated in response to an individual plaintiff's complaint and did not trouble itself to seek out violations of statutes. Thus the law in question could only take effect if a plaintiff could persuade authorities that the attack had come from unsanctioned religious activity.[26] Again, the *Lex Cornelia de Sicariis et Veneficis* (81 B.C.) parallels the vagueness of Greek legislation on *pharmaka*. *Venenum* could be poison pure and simple, as in the case of Locusta, the poisoner of Claudius (Tac. *Ann.* 12. 66, 13.15), or it could be defined as unsanctioned religious activity. The lack of universally accepted definitions of unsanctioned religious activity and the parallel lack of a general theory of naturalistic causation made poison and unsanctioned religious activity appear similar since both apparently invoked hidden forces and thus frustrated normal forensic proofs.[27] More generally, someone who murdered by poison could be conceived as outside the secular norms of society, and anyone outside those norms of society would consequently be likely to engage in a variety of unsanctioned activities secular and sacred. Thus in 26 A.D. Claudia Pulchra was charged with unchastity, adultery, poisoning (*veneficia*), and production of curse tablets (Tac. *Ann.* 4.52).[28]

As for astrology, most classicists have labeled its ancient occurrences as examples of magic, influenced, no doubt, by modern naturalistic concerns. But astrology flourished; repressive action, such as Augustus' ban on certain kinds of consultations, can be seen on closer analysis to have been practical rather than based on theological notions that astrology always constituted objectionable activity (11 A.D.): "The seers were forbidden to prophesy to any person alone or to prophesy regarding death even if others should be present. Yet so far was Augustus from caring about such matters in his own case that he set forth to all in an edict the aspect of the stars at the time of his own birth."[29] Indeed, given the intense interest in, and regular practice of, astrology, how practical would it have been to impose a kind of authoritarian search-and-destroy procedure? Historians have noted the ten expulsions of the astrologers from Rome between 33 B.C. and 93 A.D. Most remark on the frequency; I would emphasize, on the contrary, the relative infrequency. If astrology was really the continual menace the repressions might imply, one might expect to see more measures more frequently and more vigorously. It may be that a few repressions escaped the sources; for example, the extant books of Tacitus'*Annals* preserve only parts of the reigns of Tiberius, Claudius, and Nero, and none of Caligula's. But the infrequency of repression of astrologers even in those extant narratives argues against the notion of a host of repressions that have dropped out of the sources. Clearly, someone had to come to the attention of the authorities in a particular set of circumstances—as in the expulsions, the famous trial of Apuleius for magic, or the Severan ordinance against divination.[30]

What about late antiquity? The pages of Ammianus Marcellinus, supplemented by Libanius, apparently present a picture of omnipresent unsanctioned religious activity. But other views seem more plausible. First, Ammianus chose to write more about those activities. Second, the later emperors chose, for political reasons, to view more unsanctioned activities as repressible than had the earlier emperors.[31] Given Christianity's doctrinal interest in defining what it considered "orthodoxy" and "heresy" in the light of biblical norms, it now became possible to offer theologically legitimated definitions of a hitherto-undefined phenomenon. Of course, Christian "orthodoxy" in the modern sense did not exist. Rather, there were numerous groups each claiming that its particular doctrines represented a faithful continuation of the apostolic tradition. But each group could, and did, define "magic" in light of its interpretation of what it meant to be a "true" Christian. Competing Christian groups could be dismissed as heretical and hence magical, as could Jews also, and even the mere practice of certain rituals became actionable from a theoretical point of view, as the laws of the Theodosian Code demonstrate. Now that proper religion could be defined, so could improper religion.[32]

Whether one looks at the Greek or Roman material, a striking picture emerges of the great frequency of what might be called unsanctioned religious activity and the infrequency of repression.[33] There were laws that could repress specific activities, but no general ban. "But I know it when I see it" aptly summarizes the legal situation, and it was a long step from that to actually taking legal action.[34]

Indeed, legal systems tend to produce a variety of responses when confronted with a phenomenon that seems contrary to state interests and yet not amenable to rigid definitions. Moreover, those in control of the legal system can and do change their minds about even tentative definitions. Consider, for example, the problems of defining "obscenity" in modern U.S. constitutional law. Here, as in antiquity, definitions from religion and science do not help, due to, respectively, the doctrine of the separation of church and state and the lack of attention paid to obscenity by, for instance, psychology, in legal connections. For example, Justice Brennan, writing for a Supreme Court majority, claimed that "sex and obscenity are not synonymous" and went on to posit that "contemporary community standards" of what constituted "prurient interest" should suffice to define obscenity.[35] In a later case, although a majority reaffirmed those principles, Justice Brennan, writing in dissent, now objected to terms such as "prurient interest", claiming that "The meaning of these concepts necessarily varies with the experience, outlook, and even idiosyncrasies of the person defining them."[36] In another case of the same year the majority actually attempted to define obscenity, which led Justice Douglas, in dissent, to remark, "We deal with highly emotional, not rational, questions."[37] Earlier, a majority had avoided definition altogether, claiming that the First Amendment entailed a right to privacy in viewing what one wished.[38] Thus a functionalist definition entirely avoids theory, a larger sociological claim that does not imply necessary equation of ancient and modern legal systems in other capacities.

The larger question of legal authority also has relevance. Whatever be the case for day-to-day legitimations, the ultimate legitimations rely on ancestral traditions such as *patrios nomos* or *mos maiorum* or the U.S. Constitution. In all cases, reference is made to a presumed intent *in illo tempore* of "founding fathers." But such concepts do not clarify intentions and thus leave the way clear for those accepted as guardians of the traditions to legitimate their ordinances in terms of those very phrases. Given the status that society had bestowed upon them as heirs to the traditions, their interpretations of those traditions would ipso facto have the stamp of correctness. And uncertainty tends to prevail even about the identifiable archaic lawgivers, as in the case of attributing laws—and possibly democratic intentions—to Solon.[39] Even where there exists more material than mere phrases, uncertainty on intent prevails. Again turning to a modern U.S. example, in the famous 1973 abortion decision the majority expressed some uncertainty as to whether the right to abortion was founded in the First, Fourth, Fifth, Ninth, or Fourteenth Amendment, before ultimately settling on the last-named.[40]

Legal systems have problems with categories of presumed criminal behavior that lack definitions from religious or scientific criteria. They may at best appeal to specific actionable circumstances. It is now time to examine the lack of definition of "magic" in the religious and scientific systems of antiquity, and how that lack has influenced modern views.

RELIGIOUS DEFINITIONS

If legal definitions of unsanctioned religious phenomena in antiquity were ambiguous at best, the religious ones were not even that. The lack of a body of sacred ordinances and theological elaborations accepted by society differentiates pagan antiquity from the later Christian centuries. Astrology did not necessarily constitute magic but could be so considered, depending on the circumstances. Again, the philosophical views of theurgy seem practically to have legitimated unsanctioned religious phenomena through reference to the philosophical doctrines of the practitioner.[41] Desire to know the future or levitate or heal was not necessarily criminal; but who tried it, in what ways, for what reasons, and at what times could make it so.[42] New cults could enter, unidentified *numina* could flourish, and philosophical systems could disagree on the nature and function of divinities.[43]

But here modern definitional problems arise. Most scholarship on Greco-Roman religion has proceeded under the influence of nineteenth-century Christianity and anthropology. Both agreed that human religious thought had evolved from an alleged original state of irrational, savage, magical practices.[44] They differed about the end to which it had evolved. The former tradition could simply dismiss the Greco-Roman material as non-Christian and hence magical by definition, although perhaps making exceptions for what seemed glimmerings towards "right" thinking in, for instance, the Platonic and Aristotelian traditions. Humankind, in its view, evolved to Christianity. For the anthropological tradition several complicating factors arose. Most investigators had a none-too-subtle anticlerical agenda, the heritage of the Enlightenment.[45] Religion represented an evolution from magic but had further evolved to science, whose rightness as an explanatory system seemed confirmed not only by the biological evidence for evolution but also by the century's remarkable feats of applied science in the form of technology. Despite a difference on the ideal end of societal evolution, both groups agreed that magic was a society's original thought system. This view received confirmation from the wealth of ethnographic data from allegedly "primitive" societies, "primitive" being measured against the standards of the nineteenth-century European gentry.[46] What seemed magical in the ancient texts could now convincingly be labeled as such, since contemporary "savage" societies seemed to be doing exactly the same thing. Modern science further confirmed the correctness of the labels. Thus everything nonrational became "magic," the forerunner, for anthropologists, of religion and for clerics, of science; that is, magic was either bad science or bad religion. In either case, its presence demonstrated that primitive humans could not think clearly. Modern material and ancient material circularly explained each other. Rodney Needham has summarized this outlook: "As for the savages, they were characteristically sunk in ignorance and pathetic incapacities of reason, their imaginations imbued with magical prejudices and their rational faculties stunted for lack of occasion, at best, to elaborate the critical concepts of empirical cosmogony."[47]

Many anthropologists had close connections with classicists, thus unwittingly ensuring that the anthropological evolutionary scheme magic–religion–science would influence classical scholarship. The anticlerical Frazer could observe on African anthropology, "So in spite of the deplorable ravages of Christianity and civilisation among the people, there is hope of putting on record a good deal of their old life," while the more devout R. R. Marett, classicist by training and anthropologist by trade, remarked, "If, however, he is to emerge from savagery . . . he must submit to be interpenetrated by those 'civilising ideas,' of which the highest moral religion of the world is perhaps the fittest and most natural vehicle."[48] But regardless of personal theology, all identified magic with what they deemed primitive religious practices, often bringing those considerations to bear upon classical texts. Thus Frazer remarked on the Roman *Lemuria,* a private family ritual of appeasing ghosts listed in the Roman religious calendar (May 9, 11, and 13), "But it is too much to expect that superstition should always be rigorously logical." Apparently he had a problem with ghosts and a ritual in the dead of night that involved, among other things, tossing beans over the shoulder.[49] Thus ancient religion supposedly could be correctly (i.e., scientifically) interpreted for the first time, a curious amalgam of alleged "magic" and some inklings of "higher" religion. It was all too easy to identify a given ancient religious usage as magical if it did not conform to religion or science as the nineteenth century understood them.

It may be, then, that the entire corpus of what is considered magic in classical antiquity needs rethinking, since, as a rule, the label has been attached for reasons that say much about the nineteenth century and little about how the ancients viewed the phenomena. If this be done, the instances of "magic" shrink to the comparatively small number that the ancients labeled as such. As for unsanctioned religious activity, it seems more responsible not to force it into the category of magic when the ancients did not do so. But perhaps this is premature: Might not modern science be able to offer definitions freed from the anachronistic taint of religious models?

SCIENTIFIC DEFINITIONS

Many would consider modern science to be a series of demonstrably "true" laws capable of explaining, sooner or later, all natural phenomena. Those phenomena for which a less-developed society does not offer naturalistic explanations and for which modern science does, have often been dismissed as "primitive religion" at best and, more often than not, "magic"; that is, if *we* can offer a naturalistic model and *they* cannot, *we* are thinking correctly—after all, science "works"—and *they* are not.[50] And yet considerations from the history of science suggest an inherent implausibility in claiming absolute truth for science. Rather, it appears a case of evolving and competing "paradigms." Scientific laws change, often drastically, as one era's true explanation becomes, for a later era, an example of muddled thinking.[51] Thus the previous attempts to identify alleged magical phenomena in antiquity by reference to the supposedly unimpeachable truth of contemporary science must fail, since science is just as subjective as anything else. In other words, it lacks a transcendental critique of its "truth"; thus its knowledge claims become relative.[52]

Nor should ancient science be pilloried for lacking modern science's alleged capability for naturalistic explanations. The distinction is unfair, since all science is subjective. Moreover, instead of the axioms of modern "normal" science in Kuhn's definition, ancient science possessed, rather, a series of competing, often contradictory axioms with a vocabulary problem as well: "Greek knowledge . . . was so much within language, so exposed to its seductions, that its fight against the *dunamis tōn onomatōn* never led to the evolution of the ideal of a pure sign language, whose purpose would be to overcome entirely the power of language, as is the case with modern science and its orientation towards the domination of the existent."[53] Although Gadamer is perhaps overly seduced by the presumed "truth" of modern science, his observations for antiquity have force. Greek mathematics was more concerned with dialectic and demonstration than with the possibilities of wedding its observations to those of natural phenomena.[54] Interest in nature could proceed along naturalistic or religious parameters and often both at once, as in the case of Democritus' interest in miraculous resuscitations.[55] Ludwig Edelstein put it well: "Epicurus' atoms were the elements of rational philosophy, not the working hypothesis of empirical science."[56]

Thus the lack of a unified empirico-deductive system based on mathematical models would prevent common agreement on what constituted the natural or unnatural.[57] Although even in the modern era appeals to "science" as a standard of proof are questionable, it is even more questionable to impute the same possibility to what passed for science in classical antiquity. Ancient science was in no position to render judgements on what was magical and what was not. Asclepius offered viable medical cures to incubators. Was this magic, unsanctioned religion, sanctioned religion, or science? Opinions could, and did, vary.[58] Moderns incline to label it something like "magico-religious but with some cures of real scientific value." Such labels say much about the twentieth century, little about antiquity. Much better to take a given ancient phenomenon, or class of phenomena, and see who affixed what labels and explanations. As in religion, so in science: the label depended on one's social status and particular versions of both pagan polytheism and the various competing philosophical doctrines.

CONCLUSION

Lack of universally accepted religious and scientific norms in classical antiquity precluded ancient law from comprehensively legislating against unsanctioned religious activity. What the ancients called magic formed a small subset of that activity. But, regardless of nomenclature, clearly there *did* exist phenomena that fell outside the purview of either state religion or private worship of state-recognized divinity. How to determine which unsanctioned religious activity might be repressible in a particular context? Peter Brown has aptly demonstrated that in the Roman Empire there was no question of the existence of persons with the ability to influence natural phenomena in ways that did not coincide with traditional religious usages. The question lay, rather, in the presumed source of the power and its utilization. There was no problem if one were aligned with what those with coercive power considered the "right" sources of power and if one used that power for the "right" ends. "Wrong" sources of power entailed "wrong" ends, and in practical terms, suspicion of the one caused suspicion of the other;[59] that is, the ends could either justify or incriminate the means.

Even a disreputable activity need not have involved total repression, as in the famous Roman case of the Bacchanalia (186 B.C.); the evidence suggests that specific actions inimical to social interests were involved. Once those actions were eradicated, the cult was allowed "to those who claim it necessary that they observe the rites."[60] Again, many considered Christianity in the Roman empire as disreputable, possibly with abominable social practices. And yet no unified body of legislation existed; rather, specific actions arose that depended on an inflamed populace and an official willing to prosecute within the extraordinarily flexible Roman system of *coercitio*.[61]

In short, neither the legal, religious, or scientific systems had an interest in precisely defining unsanctioned religious activities.[62] Perhaps moderns should not either. Rather, one should look at the particular cases where a particular activity is so labeled and prosecuted. Certainly the ancients were capable of religious repression, as in the case of Tiberius and the affair of Mundus, or Diocletian's "great" persecution. In the former case (19 A.D.) particular charges of immorality associated with the cult of Isis came to Tiberius' attention. Although the emperor crucified the priests and destroyed the temple, worship of Isis did not disappear in Italy. In Diocletian's persecution (303–11 A.D.) the edicts largely involved confiscation of property, were sporadically enforced, and were frequently evaded through the assistance of local pagans.[63] The absence of many such examples suggests that the multiple definitional structures of ancient law, religion, and science made it impossible to offer an omnibus definition in the modern sense.[64] Only modern Christian theology has attempted to do so. Modern law, devoid of overt theological content, has not succeeded in ethical areas such as obscenity, and one presumes it would do no better were it ever called upon to define magic. Modern scholars have unwittingly been enthralled not only by developments in nineteenth-century classical scholarship but also by false analogies from examples of witchcraft in early modern Europe. This will not do, since it says more about modern outlooks than ancient ones. Instead of looking for legal repression of ancient magic, we could more accurately—and hence profitably—look for legal repression of unsanctioned religious activity, some but by no means all of which might be magical in the ancient view.

Notes

1. I wish to thank John Matthews, Geoffrey de Ste. Croix, Alan Watson, and John Winkler for assistance; Watson in particular for demonstrating fundamental problems in

interpreting the relationship of law and society—cf. his provocative article referenced at n. 8 below. These acknowledgements do not imply a *nil obstat*. Many of this article's claims appear with fuller discussion and documentation in my "The Sociology of Religious Knowledge in the Roman Empire to A.D. 284," *ANRW* vol. 2 pt. 16 sec.3, 2677–2773, esp. 2711–32 on magic; cf. J. Z. Smith, "Towards Interpreting Demonic Powers in Hellenistic and Roman Antiquity," *ANRW* vol. 2 pt. 16 sec. 1,425–39. Reference and discussion of Stewart's axiom at n. 34 below.

2. For example, Martin P. Nilsson, *Geschichte der griechischen Religion,* 3rd. ed. (Munich, 1967), 1:110–32 (note the heading "Zauberriten im Kult"); Kurt Latte, *Rōmische Religionsgeschichte* (Munich, 1960), 107, which concludes the chapter on early (i.e., agrarian) religion: "Neben unzweifelhaft magischen Zeremonien gibt es andere die die Beziehungen zu den Mächten regeln sollen." Nineteenth-century anthropology and its doctrine of "survivals" are largely responsible for this tendency; cf. the second section below and, for "survivals," E. J. Sharpe, *Comparative Religion* (New York, 1975), 49–51.

3. Trans. Walter Hamilton (Harmondsworth, 1986). Ammianus describes the activities of the notorious Paulus Tartareus in the secluded town (19.12.8) of Scythopolis in Palestine; the seclusion of the town implies that certainly Ammianus, and possibly Paul, felt the prosecutions might be less successful in a more public forum. For Paul, see n. 25; cf. the next section for discussion of the significance of the Christian context.

4. Christina Larner, *Witchcraft and Religion: The Politics of Popular Belief,* ed. Alan Macfarlane (Oxford, 1984), 29. Cf. 30: "There are only gradations of recognition and labelling, and the strength of the labelling probably related to the level of interest in or panic about witchcraft more closely than to local standards of performance in cursing."

5. Cf. Frazer's view of the Roman festival of the *Lemuria*, (see n. 49) and H. H. Scullard, *Festivals and Ceremonies of the Roman Republic* (London, 1981), 15: "Some primitive ideas obviously survive into later times, but on the whole the Romans freed themselves from the cruder manifestations of magic and taboo."

6. Thus R. MacMullen (*Enemies of the Roman Order* [Cambridge Mass., 1966]) devotes chap. 3 to "Magicians" and chap. 4 to "Astrologers, Diviners, and Prophets." The implied antithesis to the "order" of allegedly developed, "official" religion is clear. Cf. his *Paganism in the Roman Empire* (New Haven, 1981), 83: "Magic without doctrine; devils without priests; prayers unintelligible; worship homeless; and ignominious realms of rule over a single house, a single field, cow, racehorse, gladiator, rival in love or adversary to one's career or party—all, together, constituted the broad underpart of the world above this one, the part with which mortals felt themselves to be most directly in contact."

7. But note the salutary observation in A. F. Segal, "Hellenistic Magic: Some Questions of Definition," in *Studies in Gnosticism and Hellenistic Religion Presented to Gilles Quispel*, ed. R. van den Broek and M. J. Vermaseren (Leiden, 1981), 368: "If someone called himself a magician, that was one thing. But more often, people who appeared to have divine favor or who exercised supernatural power could be charged with the crime of magic by their detractors. There were no objective criteria separating the miracle worker from the magician. So, it was often necessary for an adept to prove himself a miracle worker and not a magician."

8. In general, A. Watson, "Legal Change: Sources of Law and Legal Culture," *University of Pennsylvania Law Review* 131 (1982/83):1121–57, esp. 1121–25, 1151–57; cf. 1124: the fact "that law generally operates to protect the *status quo,* and hence to protect those having power, does not in itself mean that the rules are the best that could be achieved by the power elite." Again (1154), "Failure to appreciate the power and the autonomy of legal culture may lead scholars into interesting and illuminating errors." On *otium,* J. Matthews, *Western Aristocracies and Imperial Court, A.D. 364–425* (Oxford, 1975), 1–12. For general sociological considerations, Peter Berger, *The Sacred Canopy* (New York, 1969), 84–95; Peter Berger

and Thomas Luckmann, *The Social Construction of Reality* (New York, 1967), 72–79; Randall Collins, *Conflict Sociology* (New York, 1975), 364–80.

9. For this phenomenological view, see the fourth section and Phillips, "Sociology," 2689–97; cf. my "*Quae Per Squalidas Transiere Personas:* Ste.Croix's Historical Revolution," *Helios* n.s. 11 (1984): 54–56.

10. Hopfner, ("Mageia," *RE* 14 (1930):384, lines 37–38), claiming at 385.2–4 that *asebeia* included, e.g. the making or using of *pharmaka*. In general, *RAC* s.v. *Asebiesprozesse*. Nilsson (see n. 2) is disappointingly brief (pp. 791–92).

11. Thus Hopfner's list (see n. 10) at *RE* 14. 384.41–58.

12. Thus John Scarborough's essay (chap. 5, pp. 138–74 and cf. Lloyd (see n. 57). The famous Teian inscription (Dittenberger, *SIG*, 3d ed., vol. 1, no. 37) punishes makers of *pharmaka dēlētēria,* precisely what could be used when the sorts of circumstances I have detailed led to religious considerations. The Roman legal position on *venenum* (see n. 27) encourages this conclusion. Thucydides' observations on the plague at Athens (430 B.C.) are instructive. He reports the allegation that the Peloponnesians introduced *pharmaka* into the city's water supply (2.48.2) and follows with naturalistic interpretations and observations (2.48.3–2.52). But he also notes the populace took a more spiritual view: 2.53.4–2.54 and, in general, 1.23.3.

13. Plut. *Per.* 32, Diog. Laert. 2.12–15, Diod. Sic. 12.39, and "Diopeithes (no. 8)," *RE* 5 (1905):1046–47; Thuc. 2.65.3–4 observes the ambivalence of the *demos* about Pericles. The pre-Socratic philosophers produced a variety of competing philosophical and religious systems. If a given system struck an observer as "wrong," the way would lie open to various charges of sacred and secular transgressions because the system contradicted the observer's own system and hence became unsanctioned religious activity—precisely the case with Socrates.

14. Thuc. 6.27–29 (note the label at 6.27.2) and A. W. Gomme, A. Andrewes, and K. J. Dover, *A Historical Commentary on Thucydides, Volume IV.* (Oxford, 1970), 264–89. The troubled times following the failure of the Sicilian expedition led to a general distrust of many religious specialists, thus Thuc. 8.1.1: "They turned against the public speakers who had been in favor of the expedition . . . and also became angry with the prophets and soothsayers and all who at the time had, by various methods of divination, encouraged them to believe that they would conquer Sicily." (trans. R. Warner; Harmondsworth, 1954) See also Plut. *Nic.* 23.

15. Xen. *Mem.* 1.1.1; cf. Diog. Laert. 2.40, Pl. *Apol.* 24b, Xen. *Apol.* 10. Plato identifies the basis of the action as *asebeia* at *Epist.* 7.325c; I am not persuaded by arguments against the letter's authenticity. Some could claim Socrates' alleged astronomical interests were involved, an apparent parallel to Anaxagoras (Pl. *Apol.* 18b), but misperceptions of Socrates' activities in relation to other philosophers and sophists may be involved, thus K. J. Dover, *Aristophanes: Clouds* (Oxford, 1968), xxxii–lii. Of course, the charge of introducing new divinities could theoretically imply the "wrong" and possibly magical sorts of divinities, but such a charge would be notoriously hard to prove—and perhaps even hard to conceive—in a polytheistic system; cf. n. 60 on the Roman Bacchanalia. For other Greek philosophers, *RE* 2 (1896):1529 lines 58–63.

16. On the problems of *hierosulia* and *klopē* of sacred property, D. Cohen, *Theft in Athenian Law* (Munich, 1983), 93–103. Cf. his convenient list of evidence at 105–107, with examples involving *asebeia* (F. Sokolowski, *Lois sacrées. Supp.* [Paris, 1962], no. 117; Latte [see n. 18], 84, no. 12) and discussion at 110–11, 114–15.

17. Phillips, "Sociology," 2733–52.

18. Kurt Latte, *Heiliges Recht* (Tübingen, 1920); A. D. Nock, "A Cult Ordinance in Verse," *HSCP* 63 (1958): 415–21 and idem, *JBL* 60 (1941):88–95.

19. Thus the third section below.

20. J. P. Demos, *Entertaining Satan* (New York, 1982), 153–210; K. Thomas, *Religion and the Decline of Magic* (New York, 1971), chap. 14.

21. Trans. Trevor Saunders (Harmondsworth 1970).

22. Circe: Hom. *Od.* 10.213 (cf. Verg. *Aen.* 7.19 *saeua potentibus herbis,* glossed at 7.753–55); Odysseus: Hom. *Od.* 19.457–58; Sapph. 57 L-P; Gorgias 82B11.8 D-K; Deianara: Soph. *Trach.* 584 (*philtrois*), 585 (*thelktroisi*), 685 (*pharmakon*), 710 (*ethelge*) and cf. Charles Segal, *Tragedy and Civilization* (Cambridge, Mass., 1981), remarking on the "dark, mythic world of the Centaur and the Hydra" (p. 88). In general, C. P. Segal, "Eros and Incantation: Sappho and Oral Poetry," *Arethusa* 7 (1974):139–60, esp. 142–45. Cf. Pl. *Resp.* 364b–c, *Symp.* 202e–203a, *Gorg.* 484a with E. R. Dodds, *Plato: Gorgias* (Oxford, 1959), ad loc.

23. Despite the powerful assault on the notion of limited elite literacy by Bernard Knox in P. E. Easterling and Bernard Knox, eds., *The Cambridge History of Classical Literature,* vol. I (Cambridge, 1985), 1–16, I continue to remain persuaded by the considerations in G. E. M. de Ste. Croix, *The Class Struggle in the Ancient Greek World* (London, 1981), 539, n. 4; cf. Phillips, "Sociology," 2705–07 with n. 82

24. The other essays in this volume provide details; also cf. C. A. Faraone, "Aeschylus' ὕμνος δέσμιος (*Eum.* 306) and Attic Judicial Curse Tablets," *JHS* 105 (1985):150–54. Whether those papyri should, in fact, be called "magical" should be questioned in light of the considerations I have raised.

25. This idea permeates Ste. Croix (see n. 23). As always, the evidence for late antiquity shows a change, as in Ammianus Marcellinus' account of Paulus Tartareus (14.5.6–9, 15.3.4, 15.6.1, and especially 19.12; note here Paul busied himself in the countryside: see n. 3).

26. Cited in Plin. *Nat.* 28.17; cf. E. Fraenkel, review of *Zauberei und Recht in Roms Frühzeit,* by F. Beckmann, *Gnomon* 1 (1925):185–200. I owe the observation on property rights to G. E. M. de Ste. Croix (see n. 61), 11. On the notorious brevity of the Twelve Tables, which the fragment aptly represents, H. F. Jolowicz and Barry Nicholas, *Historical Introduction to the Study of Roman Law,* 3d ed. (Cambridge, 1972), 108–13. For property rights in the code, Alan Watson, *Rome of the XII Tables* (Princeton, 1975), 157–65. Obviously the concern to protect property rights would benefit the socioeconomic elite most; thus W. Eder, "The Political Significance of the Codification of Law in Archaic Societies: An Unconventional Hypothesis," in *Social Struggles in Archaic Rome*, ed. Kurt A. Raaflaub (Berkeley, 1986), 262–300. Humbler folk could learn of their rights via the Twelve Tables, but since that compilation did not discuss how to exercise those rights through the law of actions, the knowledge would be of minimal value; cf. Watson, *Rome,* 185–86. Of course, property rights can influence modern legislation on difficult moral categories; cf. my remarks below on Supreme Court obscenity decisions with N. Dorsen and J. Gora, "The Burger Court and the Freedom of Speech," in *The Burger Court*, ed. V. Blasi (New Haven, 1983), 34–41.

27. For the provisions of the law, E. Massonneau, *La Magie dans l'antiquité romaine* (Paris, 1934), 159–68 with *Dig.* 48.8, which does not mention magic per se; cf. Peter Garnsey, *Social Status and Legal Privilege in the Roman Empire* (Oxford, 1970), 109–10. On the association of magic with *venenum* in this law, see Paul. *Sent.* 5.23.19, with 15–18 showing other magic charges that could be subsumed under the heading; the topos is a common one in literature. In general, J. Scheid, ed., *Le Délit religieux dans la cité antique* (Paris, 1981).

28. In general, Berger, (see n. 8) chap. 2; cf. the critiques of Berger listed in Phillips, "Sociology," 2694, n. 47 and 2772, n. 314. Various collocations, including political disloyalty: Tac. *Ann.* 2.27, 4.52, 6.29, 12.22, 12.59, 12.64–65, 16.14, 16.30–31 wth F. H. Cramer, *Astrology in Roman Law and Politics* (Philadelphia, 1954), 254–67. Garnsey (see n.

27) observes (p. 110, n. 4), "Here, consulting magicians is an additional charge to add a touch of the sinister to the accused's activities, and to guarantee his condemnation and death." For later eras, Matthews, *Western Aristocracies* (see n. 8), 56–58 and Thomas (see n. 20), 343–45; 445, n. 2; 568–69; cf. *Theodosian Code* 9.38.1 (322 A.D.): on the birth of a child to Crispus and Helena, Constantine pardoned "all criminals except poisoners (*veneficos*), homicides, and adulterers."

29. Dio Cassius 56.25.5, trans. E. Cary (London, 1924); cf. the *senatusconsultum* of 17 A.D., the mere existence of which implies a general ineffectiveness of Augustus' rules: Ulpian apud *Mosaic. Rom. Leg. Coll.* 15.2, praef. 1–2, Paul. *Sent.* 5.21.3. For a convenient collection of passages implying the continued presence of astrologers, F. R. D. Goodyear, *The Annals of Tacitus,* vol. II (Cambridge, 1981), 266, n. 2. In general, W. and H. G. Gundel, *Astrologumena* (Wiesbaden, 1966).

30. Expulsions: Cramer (see n. 28), 233–48, Latte (see n. 2) 328–29; R. MacMullen, *Enemies of the Roman Order* (Cambridge, Mass., 1966), 132–34. J. Rea, "A New Version of P. Yale Inv. 299," *ZPE* 27 (1977): 151–56.

31. The evidence is conveniently collected in A. A. Barb, "The Survival of Magic Arts," in *The Conflict between Paganism and Christianity in the Fourth Century,* ed. A. Momigliano (Oxford, 1963), 100–25, although in a theoretically unsophisticated manner; cf. Phillips, "Sociology," 2711. On the politics, Peter Brown's "Sorcery, Demons and the Rise of Christianity: From Late Antiquity into the Middle Ages," in his *Religion and Society in the Age of Saint Augustine* (New York, 1972), 119–46

32. Here I follow the fundamental reinterpretation of orthodoxy and heresy in Walter Bauer, *Rechtgläubigkeit und Ketzerei im ältesten Christentum* (Tübingen, 1934); Bauer's work has occasioned immense discussion: Phillips, "Sociology," 2733, n. 166 and, in general, ibid. 2733–52. For the legislation, ibid., 2718–19 with nn. 116, 117. Polemic: ibid., 2737 n. 181 and, for magic, 2742, n. 194.

33. Thus despite the Christianization of the empire, the rustics remained more staunchly pagan, probably a function of the sheer difficulty of bringing Christianity to the difficult-of-access countryside. Diana Bowder, *Paganism and Pagan Revival: Constantius II to Julian,* (D. Phil. diss., Oxford University, 1976).

34. Justice Potter Stewart concurring in *Jacobellis* v. *Ohio,* 378 U.S. 184 (1964), at p. 197. Cf. *Harvard Law Review* 78 (1964/65):207–211, with reference to attempted lower court definitions at p. 208, n. 10.

35. *Roth* v. *United States,* 354 U.S. 476 (1957), at pp. 487, 489. Note that Brennan began to change his mind in *Jacobellis* v. *Ohio* (see n. 34), at p. 191. In general, C. Peter Magrath, "The Obscenity Cases: Grapes of Roth," in *The Supreme Court Review 1966*, ed. Philip Kurland (Chicago, 1966), 7–77.

36. *Paris Adult Theatre* v. *Slaton,* District Attorney 413 U.S. 49 (1973), at p. 84; and cf. the more general considerations in Justice Douglas's dissent at pp. 70–72. In general, *Harvard Law Review* 87 (1973/74):160–75, esp. 166–69 on definitional problems.

37. *Miller* v. *California,* 413 U.S. 15 (1973); definitions at pp. 24–26, Douglas at p. 46 (cf. his previous remarks at pp. 37–40).

38. *Stanley* v. *Georgia,* 394 U.S. 557 (1969), at pp 564–68.

39. *Patrios nomos:* Thuc. 2.34.1 with M. Ostwald, *Nomos and the Beginnings of the Athenian Democracy* (Oxford, 1969), 34–43, 175–76. Cf. the case of the late fourth-century B.C. Atthidographer Philochorus, described as both seer (*mantis kai hieroskopos*) and expounder (*exēgētēs*) of the *patrios nomos:* Jacoby, *FGH* 328 T1, T2 with notes at 256–60, esp. 259 with nn. 30–33 on political activities. Of course, laws could be attributed to founding fathers such as Draco, Solon, or Romulus. But that attribution tended to be for self-serving ends; thus the oligarchs at Athens in 411 B.C. attributed a Council of 400 to Solon to serve their political

ends. Given the difficulties of preserving "old" laws and the imprecise knowledge their contemporaries had about Solon, it would have been impossible persuasively to object. Thus C. Hignett, *A History of the Athenian Constitution* (Oxford, 1952), 1–30 (preservation of laws), 92–96, 273–74 (Council of 400). Sometimes, of course, the legitimacy of a group to pronounce could be questioned, thus the famous U.S. case of *Marbury* v. *Madison* (1803).

40. *Roe* v. *Wade* (1973) 410 U.S. 113, at pp. 152–53, 156–59; cf. Justice Rehnquist's dissent at 172–77 rejecting the interpretation of the Fourteenth Amendment.

41. E. R. Dodds, "Theurgy and its Relationship to Neoplatonism," *JRS* 37 (1947): 55–69.

42. Cf. O. Böcher, *Christus Exorcista* (Stuttgart, 1972) and M. Smith, *Jesus the Magician* (New York, 1978). Phillips, "Sociology," 2724, n. 139 (healing), 2727, n. 144 (levitation).

43. The case for ancient rationalizing views of the gods in the Homeric poems is instructive: relativism (Xenophanes) and allegory (Theagenes of Rhegium; cf. R. Pfeiffer, *History of Classical Scholarship* [Oxford, 1968], 9–11).

44. The major texts are reprinted in Tess Cosslett, ed., *Science and Religion in the Nineteenth Century* (Cambridge, 1984). J. W. Burrow, *Evolution and Society* (Cambridge, 1966); James R. Moore, *The Post-Darwinian Controversies* (Cambridge, 1979); Sharpe (see n. 2); George Stocking, *Race, Culture, and Evolution* (Chicago, 1982).

45. F. E. Manuel, *The Eighteenth Century Confronts the Gods* (New York, 1967).

46. Talal Asad, ed., *Anthropology & the Colonial Encounter* (New York, 1973); Katherine George, "The Civilized World Looks at Primitive Africa 1400–1800," *Isis* 49 (1958): 62–72; Robert J. Hind, " 'We Have No Colonies'—Similarities within the British Imperial Experience," *Comparative Studies in Society and History* 26 (1984): 3–35; Rodney Needham, *Belief, Language, and Experience* (Oxford, 1972), 152–246; David Pailin, *Attitudes to Other Religions: Comparative Religion in Seventeenth- and Eighteenth-Century Britain* (Manchester, 1984). Contrast the different perspectives on Borneo: Charles Hose and William McDougall, *The Pagan Tribes of Borneo,* (London 1912), vols. 1–2 and Hans Schärer, *Die Gottesidee der Ngadju Dajak in Süd-Borneo* (Leiden, 1946) with English translation by R. Needham, (The Hague, 1963); the former sees primitive irrationality; the latter, sophisticated cosmogony. The best introduction to nineteenth-century anthropology is now George Stocking, *Victorian Anthropology* (New York, 1987), although regrettably not detailing the dialectic with classical studies.

47. Schärer and Needham (see n. 46), 179.

48. In general, E. E. Evans-Pritchard's "Religion and the Anthropologists," in his *Essays in Social Anthropology* (New York, 1962), 29–45. The quotations come from unpublished material in the Balfour Library of the Pitt-Rivers Museum, Oxford University: Frazer to Tylor, December 4, 1896, accession F-8, pp. 3–4 on Uganda, interestingly observing (p. 3) that the compiler of the reports, a Mr. Pilkington, took "good honours in Classics" at Cambridge; "Animism and Savage Morality" (lecture), 1899, accession 13, p. 33(34). On other anthropologists, Robert Ackerman, "From Philology to Anthropology—The Case of J. G. Frazer," in *Philologie und Hermeneutik im 19. Jahrhundert,* ed. Mayotte Bollack et al. (Göttingen, 1983), 2:423–47; idem, "Sir James G. Frazer and A. E. Housman: A Relationship in Letters," *GRBS* 15 (1974): 339–64. Robert A. Jones, "Robertson Smith and James Frazer on Religion: Two Traditions in British Social Anthropology," in *Functionalism Historicized: Essays on British Social Anthropology,* ed. George W. Stocking, Jr. (Madison, Wis., 1984), 31–58; Jonathan Z. Smith, "When the Bough Breaks," in his *Map Is Not Territory* (Leiden, 1978), 208–39 (originally *History of Religions* 12 [1972/73]:342–71.)

49. Frazer, in his commentary on Ovid's *Fasti* (London, 1929) bk. 5 p. 46; cf. his comparative material at pp. 40–44 (e.g., 40: "a type of festival which has been observed by many races in many parts of the world" and, on the use of beans, "not the ripe fruit of philosophic thought but the crude fancies of primitive superstition"). More recently, H. H.

Scullard, *Festivals* (see n. 5) observed, in concluding his discussion of the festival (p. 119), "There was a strong streak of superstition even in many educated Romans."

50. Miraculous healings in antiquity and their relation to modern voodoo provide a classic demonstration: Phillips, "Sociology," 2724–26 with n. 142 on voodoo. For more general consideration of such ethnocentric problems, James Axtell, "Forked Tongues: Moral Judgements in Indian History," *Perspectives: American Historical Association Newsletter* 25, no 2 (Feb. 1987) 10–13.

51. Thus Thomas Kuhn, *The Structure of Scientific Revolutions,* 2d ed. (Chicago, 1970) has occasioned an enormous discussion. Cf. Barry Barnes, *T. S. Kuhn and Social Science* (New York, 1982) and Phillips, "Sociology," 2681–82. On the use of Kuhn in larger contexts, Michael A. Holly, *Panofsky and the Foundations of Art History* (Ithaca, 1984), 140, 176; cf. S. Bann's review, *History and Theory* 25 (1986):199–205.

52. Phillips, "Sociology," 2681–97, with 2684–85 on the philosophical problems.

53. H.-G. Gadamer, *Truth and Method* (New York, 1975), 413; cf. the rhetorical considerations for scientific writing in *Menander Rhetor,* ed. D. A. Russell and N. G. Wilson (Oxford, 1981), 12–14 (=336.24–337.32 Spengel).

54. Phillips, "Sociology," 2700–2701 with n. 66.

55. Democr. 68B Oc 1a D-K with discussions in Phillips, "Sociology," n. 69.

56. Ludwig Edelstein, "Recent Trends in the Interpretation of Ancient Science," *JHI* 13 (1952):594–95.

57. For Kuhn applied to ancient science, G. E. R. Lloyd, *Science, Folklore and Ideology* (Cambridge, 1983), 117–18.

58. Phillips, "Sociology," 2701 with n. 71.

59. *The Making of Late Antiquity* (Cambridge, Mass., 1978), chap. 1.

60. *CIL* 2d ed., vol. 1, p. 581, lines 3–4. The necessity would come from divine behest; in general, A. D. Nock, *Conversion* (Oxford, 1933), chap. 4. Cf. E. Fraenkel, "Senatusconsultum de Bacchanalibus," *Hermes* 67 (1932):369–96; A. H. McDonald, "Rome and the Italian Confederation (200–186 B.C.)," *JRS* 34 (1944):26–31. See the important general considerations in J. North, "Conservatism and Change in Roman Religion," *PBSR* N.S. 30 (1976):1–12. Etruscan evidence of the sixth and fifth centuries B.C. implies the early presence of the cult in Italy and, perhaps, later Roman respect for it: R. Bloch, "Recherches sur la religion romaine du VIe siècle et du début du Ve siècle av. J.-C.," in his *Recherches sur les religions de l'antiquité classique*, ed. R. Bloch (Paris, 1980), 370–73.

61. Despite the enormous body of material on this subject, the masterpiece remains G. E. M. de Ste. Croix, "Why Were the Early Christians Persecuted?" *Past and Present* 26 (1963): 6–38. The recent book by S. Benko, *Pagan Rome and the Early Christians* (Bloomington, Ind., 1984), due to a multitude of typographical, factual, and interpretational errors, may profitably be avoided. On abominations, Phillips, "Sociology," 2750–51 with n. 231.

62. I omit here the case of holy men, which coincides with my other observations on unsanctioned religious activity. Cf. Phillips, "Sociology," 2752–64.

63. Mundus: Joseph, *AJ* 18.65–84. Diocletian: evidence in A. H. M. Jones, *The Later Roman Empire 284–602* (Norman, Okla., 1964), 2:1079 and bibliography in *CAH* 12.789–95; cf. T. D. Barnes, *Constantine and Eusebius* (Cambridge, Mass., 1981), 15–27, G. E. M. de Ste. Croix, "Aspects of the 'Great' Persecution," *HTR* 47 (1954):75–113 and idem (see n. 61), passim.

64. Claims that an omnibus definition existed at a particular time usually founder on the definitional problems this essay has attempted to identify. For example, A. N. Sherwin-White, *The Letters of Pliny* (Oxford, 1966, 1968) makes such a claim for the legislation of A.D. 17 (see n. 29; I do not accept his dating to 16 A.D., despite Cramer [see n. 28], 237–38). Sherwin-White claims (p. 785) that the Ulpian passage constitutes "a kind of 'general law'"

that made it a capital offense "to practice magical or prophetic arts." He then characterizes both as "black arts." This is precisely the terminological confusion I have argued against at n. 6 above. Moreover, the ancient text does not support him; put differently, Ulpian seems clearer on the distinction than his modern interpreters. In section 1, Ulpian names the object of the law: *mathematicis Chaldaeis ariolis et ceteris, qui simile inceptum fecerunt. Magi* do appear in section 2, but that section discusses the course of previous legislation against unsanctioned religious activity: *apud ueteres dicebatur*. At most, then, one can claim that Ulpian saw the astrologers and magi as *examples* of unsanctioned religious activity; nothing implies an *equation* on his part. Sherwin-White (p. 785 n. 3) and Cramer (p. 238) cite Dio Cassius (57.15.8), and Tacitus (*Ann.* 2.32) as proof that the legislation linked astrology and magi. Although the passages do make the link (Dio with *goetas*), they only provide evidence for the authors' interpretations of the statute. Finally, Sherwin-White raises the question of whether knowledge alone, or practice, was actionable. This fundamental question would require a separate article; readers will find orientation in the exchange between him and Ste. Croix on *contumacia:* Sherwin-White, *Letters,* 787, Ste. Croix (see n. 61), and M. Finley, ed. *Studies in Ancient Society* (London, 1974), 250–62.

Selected Bibliography of Greek Magic and Religion

Abt, A. *Die Apologie des Apuleius von Madaura und die antike Zauberei*. Giessen, 1908.
Audollent, A. *Defixionum Tabellae*. Paris, 1904.
Aune, D. "Magic in Early Christianity." In *ANRW* vol. 2, pt. 23, sec. 2 1507–57. Berlin, 1980.
Ausfield, C. "De Graecorum Precationibus Quaestiones." *Neue Jahrbücher für Classische Philologie Suppl*. 28 (1903):502–47.
Barb, A. A. "St. Zacharias the Prophet and Martyr: A Study in Charms and Incantations." *JWCI* 11 (1948):35–67.
———. "The Survival of the Magic Arts." In *The Conflict between Paganism and Christianity in the Fourth Century,* ed. A. Momigliano, 100–25. Oxford, 1963.
———. "Three Elusive Amulets." *JWCI* 27 (1964):1–22.
———. "Antaura the Mermaid and the Devil's Grandmother." *JWCI* 29 (1966):1–23.
———. "Mystery, Myth, and Magic." In *The Legacy of Egypt,* 2d ed., ed. J. R. Harris, 138–69. Oxford, 1971.
———. "Magica Varia." *Syria* 49 (1972):343–70.
Basanoff, V. *Evocatio*. Bibliothéque de l'école hautes études, sciences religieuse 56. Paris, 1947.
Bell, H. I., A. D. Nock, and H. Thompson. *Magical Texts from a Bilingual Papyrus in the British Museum*. Oxford, 1933.
Betz, H. D. "Fragments from a Catabasis Ritual in a Greek Magical Papyrus." *HTR* 19 (1980):287–95.
———. "The Delphic Maxim 'Know Yourself' in the Greek Magical Papyri." *HR* 21 (1981): 156–71.
———. "The Formation of Authoritative Tradition in the Greek Magical Papyri." In *Self-Definition in the Greco-Roman World,* Jewish and Christian Self-Definition 3, ed. B. F. Meyers and E. P. Sanders, 161–70. Philadelphia, 1982.
———. ed. *The Greek Magical Papyri in Translation, Including the Demotic Spells*. Vol. 1, *Texts*. Chicago, 1986. [*GMPT*]
Bidez, J., and F. Cumont, *Les mages hellénisés*. 2 vols. Paris, 1938.
Björck, G. *Der Fluch des Christen Sabinus*. Papyrus Upsalensis 8. Uppsala, 1938.
Blau, L. *Das altjüdische Zauberwesen*. Strassburg, 1898.
Boll, F. *Griechischer Liebeszauber aus Aegypten*. SBHeidelberg no. 2. Heidelberg, 1910.
Bonner, C. "Liturgical Fragments on Gnostic Amulets." *HTR* 25 (1932):362–67.
———. "Witchcraft in the Lecture Room of Libanius." *TAPA* 66 (1932):34–44.
———. "Magical Amulets." *HTR* 39 (1946):25–39.
———. *Studies in Magical Amulets Chiefly Graeco-Egyptian*. University of Michigan Studies, Humanistic Series 49. Ann Arbor, 1950.
———. "Amulets Chiefly in the British Museum." *Hesperia* 20 (1951):301–45.
———. "A Miscellany of Engraved Stones." *Hesperia* 23 (1954):138–57.
Boyancé, P. "Théurgie et téletique néoplatoniciennes." *RHR* 147 (1955):189–209.
Brashear, W. "Ein Berliner Zauberpapyrus." *ZPE* 33 (1979):261–78.
Bravo, B. "Une tablette magique d'Olbia pontique, les morts, les héros et les démons." In *Poikilia: Études offertes à Jean-Pierre Vernant*. Recherches d'histoire et de sciences sociales 26, 185–218, Paris, 1987.
Brown, P. "Sorcery, Demons, and the Rise of Christianity." In idem, *Religion and Society in the Age of Saint Augustine,* 119–46. London, 1972.
Budge, E. A. W. *Amulets and Talismans*. New York, 1978.
Burkert, W. "ΓΟΗΣ: Zum Griechischen 'Schamanismus'." *RhM* 105 (1962):36–55.

———. *Lore and Science in Ancient Pythagoreanism*. Trans. E. L. Minar, Jr. Cambridge, Mass., 1972.
———. *Structure and History in Greek Mythology and Ritual*. Berkeley, 1979.
———. "Craft vs. Sect: The Problem of Orphics and Pythagoreans." In *Self-Definition in the Greco-Roman World*, Jewish and Christian Self-Definition 3, ed. B. F. Meyers and E. P. Sanders, 1–22. Philadelphia, 1982.
———. *"Homo Necans": The Anthropology of Ancient Greek Sacrificial Ritual and Myth*. Trans. P. Bing. Berkeley, 1983.
———. "Itinerant Diviners and Magicians: A Neglected Area of Cultural Contact." In *The Greek Renaissance of the Eighth Century B.C.: Tradition and Innovation*, ed. R. Hägg, Skrifter Utgivna av Svenska Institutet Athen. 30, 111–19. Stockholm, 1983.
———. *Die orientalisierende Epoche in der griechischen Religion und Literature*. SBHeidelberg, no. 1. Heidelberg, 1984.
———. *Greek Religion*. Trans. J. Raffan. Cambridge, Mass., 1985.
Calder, W. M. "The Great Defixio from Selinus." *Philologus* 107 (1963):163–72.
Cramer, F. H. "Expulsion of the Astrologers from Ancient Rome." *Classica et Mediaevalia* 12 (1951): 10–36.
———. *Astrology in Roman Law and Politics*. Philadelphia, 1954.
Crawley, A. E. "Cursing and Blessing." In *Encyclopedia of Religion and Ethics* 4:367–74. New York, 1911.
Crum, W. E. "Magical Texts in Coptic." *JEA* 20 (1934):51–53 and 194–200.
Cumont, F. *Astrology and Religion among the Greeks and Romans*. New York, 1912.
———. "Il sole vindice dei delitti ed il simbolo delle mani alzate." *Mem. Pont. Acc.* ser. iii (1923): 65–80.
———. *L'Égypt des astrologues*. Brussels, 1937.
Daniel, R. "Two Love Charms." *ZPE* 19 (1975):249–64.
———. "Some ΦΥΛΑΚΤΗΡΙΑ." *ZPE* 25 (1977):145–54.
Dedo, R. *De Antiquorum Superstitione Amatoria*. Gryphia 1904.
Delatte, A. *Anecdota Atheniensa*. Paris, 1927.
———. *Herbarius, Recherches sur le cérémonial usité chez les anciens pour la cuillette des simples et des plantes magiques*. 3d ed., Mém. Ac. Roy. Belg. vol. 54, pt. 4. Brussels, 1961.
Delatte, A., and Ph. Derchain. *Les intailles magiques gréco-égytiennes de la Bibliothèque Nationale*. Paris, 1964.
Detienne, M. *The Gardens of Adonis: Spices in Greek Mythology*. Trans. J. Lloyd. London, 1977.
Deubner, L. "Charms and Amulets (Greek)." In *Encyclopedia of Religion and Ethics* 3:433–39. New York, 1911.
———. "Magie und Religion." In *Kleine Schriften zur klassischen Altertumskunde*. Königstein, 1982, 275–98.
Dieterich, A. "Der Untergang der antiken Religion." In *Kleine Schriften*, 275–98. Leipzig, 1911.
———. *Abraxas: Studien zur Religionsgeschichte des spätern Altertums*. 2d ed. Leipzig, 1925.
———. *Eine Mithrasliturgie*. 3d ed. Darmstadt, 1966.
Dodds, E. R. *The Greeks and the Irrational*. Berkeley, 1951.
Dornseiff, F. *Das Alphabet in Mystik und Magie*. Stoicheia 7. Leipzig, 1975.
Dugas, Ch. "Figurines d'envoûtement trouvées à Délos." *BCH* 39 (1915):413–23.
Eckstein, F., and J. H. Waszink. "Amulett." *RAC* 1 (1950):397–411.
Edelstein, L. "Greek Medicine in its Relation to Religion and Magic." In *Ancient Medicine: Selected Papers of Ludwig Edelstein*, ed. O. and C. L. Tempkin, 205–46. Baltimore, 1967. Originally *Bulletin of the History of Medicine* 5 (1937):201–46.
Egger, R. *Römische Antike und frühes Christentum*. Vol.1. Klagenfurt, 1962–63.
Eitrem, S. *Opferritus und Voropfer der Griechen und Römer*. Kristiana, 1915.
———. "Die rituelle διαβολή." *SO* 2 (1924):43–61.
———. *Papyri Osloenses*. Fasc. 1, *Magical Papyri*. Oslo, 1925.
———. "Die *σύστασις* und der Lichtzauber in der Magie." *SO* 8 (1929):49–51.
———. "Aus 'Papyrologie und Religionsgeschichte': Die Magische Papyri." In *Papyri und Altertumswissionshaft*, Müchener Beiträge zur Papyrusforschung und antiken Rechtsgeschichte 19, ed. W. Otto and L. Wenger, 243–63. Munich, 1934.
———. "Die magischen Gemmen und ihre Weihe." *SO* 19 (1939):57–85.

———. "La théurgie chez les Neoplatoniciens et dans les papyrus magiques." *SO* 22 (1942):49–79.

———. "Apollo in der Magie." In *Orakel und Mysterien am Ausgang der Antike,* Albae Vigiliae 5, 47–52. Zürich, 1947.

Fabrini, P., and F. Maltomini. "Formulario magico." In *Papiri letterari greci,* ed. A. Carlini et al. Pisa, 1978, text no. 34.

Fahz, L. *De Poetarum Romanorum Doctrina Magica Quaestiones Selectae.* RGVV vol. 2, pt. 3. Geissen, 1904.

Faraone, C. A. "Aeschylus' *ὕμνος δέσμιος* (*Eum.* 306) and Attic Judicial Curse Tablets." *JHS* 105 (1985):150–54.

———. "Hephaestus the Magician and the Near Eastern Parallels for the Gold and Silver Dogs of Alcinous (*Od.* 7.91–94)." *GRBS* 28 (1987):257–80.

———. "Hermes but No Marrow: Another Look at a Puzzling Magical Spell." *ZPE* 72 (1988):279–86.

———. "An Accusation of Magic in Classical Athens (Ar. *Wasps* 946–48)," *TAPA* 119 (1989):149–61.

Faraone, C. A., and R. Kotansky. "An Inscribed Gold Phylactery in Stamford, Connecticut." *ZPE* 75 (1988):257–66.

Festugière, A.-J. "Le valeur religieuse des papyrus magiques." In *idem, L'idéal religieux des grecs et l'évangile,* 281–328. Paris, 1932.

———. *La révélation d'Hermès Trismégiste.* 4 vols. Paris, 1950.

———. "Amulettes magiques a propos d'un ouvrage recent." *CPh* 46 (1951):81–92.

Fowden, G. *The Egyptian Hermes: A Historical Approach to the Late Pagan Mind.* Cambridge, 1986.

Fox, W. S. "Submerged *tabellae defixionum.*" *AJP* 33 (1912):301–10.

Frazer, J. G. *The Golden Bough.* 3d ed. London, 1911–36.

———. *Ovid's Fasti.* Cambridge, Mass., 1931.

Friedrich, H. V. *Thessalos von Tralles: Griechisch und lateinisch.* Meisenheim an Glan, 1968.

Gager, J. G. *Moses in Greco-Roman Paganism.* SBL Monograph Series 16. Nashville, 1972.

Ganschinietz, R. *Hippolytos' Capital gegen die Magier.* Texte und Untersuchungen zur Geschichte der altchristlichen Literature vol. 39, pt. 2. Leipzig, 1913.

———. "Katochos." *RE* vol. 10, pt. 2 (1919):2526–34.

Giangrande, G. "Hermes and the Marrow: A Papyrus Love Spell." *Ancient Society* 9 (1978):101–116. Also in *Scripta Minora Alexandrina,* vol. 2, 573–88. Amsterdam, 1981.

Graf, F. *Eleusis und die orphische Dichtung Athens in vorhellenistischen Zeit.* RGVV 33, Berlin, 1974.

———. *Nordionische Kulte.* Bib. Hel. Rom. 21. Rome, 1985.

Griffith, F. L., and H. Thompson, eds. *The Demotic Magical Papyrus of London and Leiden.* 3 vols. London, 1904.

———, eds. *The Leyden Papyrus: An Egyptian Magical Book.* New York, 1974.

Guarducci, M. *Epigraphia greca IV: Epigrafi sacre pagane e cristiane.* Rome, 1978.

Gundel, H. G., and W. Gundel. *Astrolegumena: Die astrologische Literatur in der Antike und ihre Geschichte.* Sudhoffs Archiv 6. Weisbaden, 1966.

Heim, R. *Incantamenta Magica Graeca Latina.* Jahrbücher für classische Philologie Suppl. 19. Leipzig, 1892:463–576.

Heitsch, E. "Zu den Zauberhymnen." *Philologus* 103 (1959):215–36.

———. "Drei Helioshymnen." *Hermes* 88 (1960):150–58.

———. *Die griechischen Dichterfragmente der römischen Kaiserzeit.* 2 vols. Abhandlungen der Akademie der Wissenschaften in Göttingen 49 and 58. Göttingen, 1961–64.

Herrman, P. *Ergebnisse einer Reise in Nordstlydian.* Denkschr Wien 80 (Vienna 1962):1–63.

———. *Tituli Asiae Minoris.* Vol. 5.1. Vienna, 1981. [*TAM*].

Hopfner, Th. "Mageia." *RE* vol. 14, pt. 1 (1928):301–93.

———. "Theurgie." *RE* 6A.1 (1936):258–70.

———. *Griechische-ägyptischer Offenbarungszauber.* 2 vols., Studien zu Paläographie und Papyruskunde 21 und 23. (Leipzig 1921–24 [facsimile]; repr. Amsterdam, 1974 and [in part] 1983).

Huvelin, M. "Les tablettes magiques et le droit romain." In *Annales internationales d'histoire: Congrès de Paris,* 47–58. Paris, 1901.

Jahn, O. *Über den Aberglauben den bösen Blicks bei den Alten.*" SB Leipzig 7 (Leipzig, 1855): 28–110.

Jevons, F. B. "Graeco-Italian Magic." In *Anthropology and the Classics,* 93–120. Oxford, 1908.

Johnston, J. "Introduction to the Demotic Magical Papyri." In *The Greek Magical Papyri in Translation,* vol. 1, ed. H. D. Betz, lv–lviii Chicago, 1986.

Jordan, D. R. "Two Inscribed Lead Tablets from a Well in the Athenian Kerameikos." *AM* 95 (1980): 225–39.

———. "New Archaeological Evidence for the Practice of Magic in Classical Athens." In *Praktika of the 12th International Congress of Classical Archaeology*. Vol. 4, 273–77. Athens, 1988.

———. "*Defixiones* from a Well near the Southwest Corner of the Athenian Agora." *Hesperia* 54 (1985):205–55.

———. "A Survey of Greek *Defixiones* Not Included in the Special Corpora." *GRBS* 26 (1985):151–97. [*SGD*]

Kagarow, E. G. *Griechische Fluchtafeln*. Eos Suppl. 4. Leopoli, 1929.

Kaimakis, D. *Die Kyraniden*. Beiträge zur klassichen Philologie 76. Meisenheim am Glan, 1976.

Koenen, L. "Der brennende Horusknabe: Zu einem Zauberspruch des Philinna-Papyrus." *Chron. d'Égypte* 37 (1962):167–74.

Kotansky, R. "Two Amulets in the Getty Museum: A Gold Amulet for Aurelia's Epilepsy and an Inscribed Magical Stone for Fever, 'Chills,' and Headache." *J. P. Getty Museum Journal* 8 (1980):181–87.

———. "A Silver Phylactery for Pain." *J. P. Getty Museum Journal* 11 (1983):169–78.

Köves-Zulauf, T. *Reden und Schweigen*. Munich, 1972.

Kroll, W. *De Oraculis Chaldaicis*. Breslauer Philologische Abhandlungen vol. 7, pt. 1 (Breslau 1894).

Kropp, A. M. *Ausgewähte Koptische Zaubertexte*. 3 vols. Brussels, 1930.

Kuhnert, E. "Feuerzauber." *RhM* 49 (1894):37–58.

Laín Entralgo, E. *The Therapy of the Word in Classical Antiquity*. Trans. L. J. Rather and J. M. Sharp. New Haven, 1970.

Latte, K. *Heiliges Recht: Untersuchungen zur Geschichte der sakralen Rechtformen in Griechenland*. Tübingen, 1920.

———. *Römische Religionsgeschichte*. Munich, 1960.

Lewy, H. *Chaldean Oracles and Theurgy: Mysticism, Magic, and Platonism in the Late Roman Empire*. Recherches d'archéologie, de philologie, et d'histoire 13. Cairo, 1956.

Lloyd, G. E. R. *Magic, Reason, and Experience: Studies in the Origin and Development of Greek Science*. Cambridge, 1979.

———. *Science, Folklore, and Ideology: Studies in the Life Sciences in Ancient Greece*. Cambridge, 1983.

Lowe, J. E. *Magic in Greek and Latin Literature*. Oxford, 1929.

Luck, G. *Hexen und Zauberei in der römischen Dichtung*. Zurich, 1962.

———. *Arcana Mundi: Magic and the Occult in the Greek and Roman Worlds*. Baltimore, 1985.

Maas, P. "The Philinna Papyrus." *JHS* 62 (1942):33–38.

McCown, C. C. "The Ephesia Grammata in Popular Belief." *TAPA* 54 (1923):128–40.

MacMullen, R. *Enemies of the Roman Order*. Cambridge, Mass., 1967.

———. *Paganism in the Roman Empire*. New Haven, 1981.

Magia: Studi di storia delle religioni in memoria di R. Garosi. Rome, 1976.

Maloney, C., ed. *The Evil Eye*. New York, 1976.

Maltomini, F. "I papiri greci." *SCO* 29 (1979):55–124.

———. Review of D. F. Moke, *Eroticism in the Greek Magical Papyri,* by F. Maltomini. *Aegyptus* 59 (1979):273–84.

Massonneau, E. *La magie dans l'antiquité romaine: La magie dans la litterature et les auteurs romaines*. Paris, 1934.

Mauss, M. *A General Theory of Magic*. Trans. R. Brain. New York, 1972.

Mély, F. de. *Les lapidaires de l'antiquité et du moyen âge*. Paris, 1898.

Merkelbach, R. *Mithras* (Königstein, 1984).

Meyer, M. V. *The "Mithras Liturgy."* Missoula, Mont., 1976.

Miller, A. P. *Studies in Sicilian Epigraphy: An Opisthographic Lead Tablet*. Ph.D. diss. University of North Carolina at Chapel Hill, 1973.

Moke, D. F. *Eroticism in the Greek Magical Papyri: Selected Studies*. Diss. Minneapolis, 1975.

Moraux, P. *Une imprécation funéraire récemment découverte à Néocésarée*. Paris, 1959.

———. *Une défixion judicaire au Musée d'Istanbul*. Mém. Ac. Roy. Belg., vol. 54, pt. 2. Brussels, 1960.

Moreau, A. "L'oeil malefique dans l'oeuvre d'Eschyle." *REA* 78–79 (1976–77):50–64.

Morgan, M., trans. *Sepher Ha-Razim: The Book of Mysteries*. Chico, 1983.
Moutarde, R. *Le glaive de Dardanos: Objects et inscriptions magiques de Syria*. Mélanges de l'Université Saint-Joseph, vol. 15, pt. 3. Beirut, 1930.
Niggermeyer, J. H. *Beschwörungsformeln aus dem "Buch der Geheimnisse,"* Judaistische Texte und Studien 3. Hildesheim, 1975.
Nilsson, M. P. "Die Religion in den griechischen Zauberpapyri." *Bull. Soc. Roy. Lund* (1947) 48:59–93. Also in *Opuscula Selecta* 3:129–66. Stockholm, 1960.
———. *Geschichte der griechische Religion*. 3rd ed. Vol. 1. Munich, 1967. 2d ed. vol. 2. Munich, 1961.
Nock, A. D. "The Greek Magical Papyri." In *Essays in Religion and the Ancient World,* vol. 1, ed. Z. Stewart, 176–88. Cambridge, Mass., 1972.
———, and Festugière, A.-J., eds. and trans. *Hermès Trismégiste: Corpus Hermeticum*. 4 vols. Paris, 1946–54.
Norden, E. *Agnostos Theos*. 4th ed. Leipzig, 1923.
Parker, R. *Miasma: Pollution and Purification in Early Greek Religion*. Oxford, 1983.
Parrot, A. *Malédictions et violations de tombes,* Paris, 1939.
Parsons, P. Review of *The Greek Magical Papyri in Translation,* ed. H. D. Betz. *TLS,* Nov. 21, 1986, p. 1316.
Perdrizet, P. "Amulette grecque trouvée en Syrie." *REG* 41 (1928):73–82.
Petzoldt, L., ed. *Magie und Religion: Beiträge zu einer Theorie der Magie*. Wege der Forschung 337. Darmstadt, 1978.
Pfister, F. *Die Reliquienkult in Altertum*. Giessen, 1909.
———. "Epode." *RE* Suppl. 4 (1924):323–44.
———. "Pflanzenaberglaube." *RE* 19 (1938):1446–56.
Phillips, C. R. III, "The Sociology of Religious Knowledge in the Roman Empire." *ANRW* vol. 2, pt. 16, sec. 3, 2677–2773. Berlin, 1986.
Philonenko, M. "Une prière magique au dieu créateur (*PGM* V 459–89)." *CRAI* (1985):433–52.
Places, E. des. "Les oracles chaldaïques." *ANRW* 2.17.4, 2299–2335. (Berlin, 1984.)
Preisendanz, K. "Ousia." *WS* 40 (1918):5–8.
———. "Die griechischen und lateinischen Zaubertafeln." *APF* 9 (1928):119–54 and 11 (1933):153–64.
———. "Fluchtafel (Defixion)." *RAC* 8 (1972):1–29.
———, ed. *Papyri Graecae Magicae: Die griechischen Zauberpapyri*. 2nd ed. rev. A. Henrichs. 2 vols. Stuttgart, 1973–74. [*PGM*]
Prentice, W. K. "Magical Formulas on Lintels in the Christian Period in Syria." *AJA* 10 (1906):137–50.
Reitzenstein, R. *Poimandres*. Leipzig, 1904.
———. *Die hellenistischen Mysterienreligionen*. 3d ed. Leipzig, 1927.
Robert, L. "Échec au mal." *Hellenica* 13 (1965):265–71.
———. "Malédictions funéraires grecques." *CRAI* 1978:241–89.
———. "Amulettes grecques." *Journal des Savants* (1981):3–44.
Robinson, D. M. "A Magical Text from Beroea in Macedonia." In *Classical and Mediaeval Studies in Honor of Edward Kennard Rand,* ed. L. W. Jones. New York, 1938.
Rohde, E. *Psyche: The Cult of Immortality among the Greeks*. Trans. W. B. Hollis. London, 1925.
Scarborough, J. "The Drug Lore of Asclepiades of Bithynia." *Pharmacy in History* 17 (1975):43–57.
———. "Theophrastus on Herbals and Herbal Remedies." *Journal of the History of Biology* 11 (1978):353–85.
———. "Early Byzantine Pharmacology." *Symposium on Byzantine Medicine,* Dumbarton Oaks Papers 38, ed J. Scarborough (Washington, 1985), 213–32.
Scheid, J., ed., *Le délit religieux dans la cité antique*. Paris, 1981.
Schmiedeberg, O. *Über die Pharmaka in der Ilias und Odyssee*. Strassburg, 1918.
Schultz, W. "*Ephesia* und *Delphika Grammata*." *Philologus* 68 (1909):210–28.
Schwartz, F. M. and J. H. "Engraved Gems in the Collection of the American Numismatic Society." Pt. 1, "Ancient Magical Amulets," *ANSMusNotes* 24 (1979):149–95.
Schwartz, J. "Papyri Graecae Magicae und magische Gemmen." In *Die orientalischen Religionen im Römerreich,* EPRO 93, ed. M. J. Vermasseren, 485–509. Leiden, 1981.
Segal, A. F. "Hellenistic Magic: Some Questions of Definition." In *Studies in Gnosticism and Hellenistic*

Religion Presented to Gilles Quispel, EPRO 91, ed. R. van den Broek and M. J. Vermasseren, 349–75. Leiden, 1981.
Seligmann, S. *Die magischen Heil- und Schutzmittel aus der unbelebten Natur mit besonderer Berücksichtigung der Mittel gegen den bösen Blick: Ein Geschichte des Amulettwesens*. Stuttgart, 1927.
Siebourg, M. "Ein gnostisches Goldamulett aus Gellup." *BJ* 103 (1898):123–53.
———. "Ein griechisch-christisches Goldamulett gegen Augenkrankenheit." *BJ* 118 (1909):158–75.
Smith, J. Z. *Map Is Not a Territory: Studies in the History of Religions*. Leiden, 1978.
———. "Towards Interpreting Demonic Powers in Hellenistic and Roman Antiquity." *ANRW* 2.16.1, 425–39. Berlin, 1978.
———. *Imagining Religion*. Chicago, 1982.
Smith, M. *Jesus the Magician*. New York, 1978.
———. "Relations between Magical Papyri and Magical Gems." *Papyrologica Bruxellensia* 18 (1979): 129–36.
———. "The Hymn to the Moon, *PGM* IV 2242–2355." *Proceedings of the XVI Int. Congr. of Papyrology*, 643–54. Chico, 1981.
Solin, H. *Eine neue Fluchtafel aus Ostia*. Comm. Hum. Litt. vol. 42, pt. 3. Helsinki 1968.
Speyer, W. "Fluch." *RAC* 7 (1969):1160–1288.
Steinleitner, F. S. *Die Beicht im Zusammenhange mit der sakralen Rechtspflege in der Antike*. Munich, 1913.
Strubbe, J. "Vervloekingen tegen greafschenners." *Lampas* 16 (1983):248–74.
Tambiah, S. J. "The Magical Power of Words." *Man* 3 (1968):175–208.
———. "Form and Meaning of Magical Acts: A Point of View." In *Modes of Thought*, ed. R. Horton and R. Finnegan, 199–229. London, 1973.
Tavenner, E. *Studies in Magic from Latin Literature*. New York, 1916.
Thee, F. C. R. *Julius Africanus and the Early Christian View of Magic*. HUTh. 19. Tübingen, 1984.
Trumpf, J. "Fluchtafel und Rachepuppe." *AM* 73 (1958):94–102.
Tupet, A.-M. *La magie dans la poésie latine*. Paris, 1976.
Turner, E. G. "The Marrow of Hermes." In *Images of Man in Ancient and Medieval Thought: Studia Gerado Verbeke ab Amicis et Collegis Dicata*, 169–73. Louvain, 1976.
Vallois, R. "ARAI." *BCH* 38 (1914):250–71.
Versnel, H. S., ed. *Faith, Hope and Worship: Aspects of Religious Mentality in the Ancient World*. Studies in Greek and Roman Religion 2, Leiden, 1981.
———. "Religious Mentality in Ancient Prayer." In *Faith, Hope and Worship*, ed. Versnel, 1–64. Leiden, 1981.
———. " 'May he not be able to sacrifice . . . ': Concerning a Curious Formula in Greek and Latin Curses." *ZPE* 58 (1985):247–69.
Weinreich, O. *Antike Heilungswunder*. RGVV 8.1 Geissen, 1909.
———. "De Dis Ignotis." *ARW* 18 (1915):8–15.
Wessely, K. *Ephesia Grammata aus Papyrusrollen, Inschriften, Gemmen*. Vienna, 1886.
Wilhelm, A. "Über die Zeit einiger attischer Fluchtafeln." *ÖJ* 7 (1907):105–26.
Wolters, J. "Faden und Knoten als Amulett." *ARW* 8 (1905):1–22.
Worrel, W. H. "A Coptic Wizard's Horde." *American Journal of Semitic Languages and Literature* 46 (1930):239–62.
———. "Coptic Magical and Medical Texts." *Orientalia* n.s. 4 (1935):1–37 and 184–94.
Wortmann, D. "Die weisse Wolf." *Philologus* 107 (1963):157–61.
———. "Neue magische Texte." *BJ* 169 (1969):56–111.
———. "Neue magische Gemmen." *BJ* 175 (1975):63–82.
Wünsch, R. *Defixionum Tabellae Atticae*. In *IG* vol. 3, pt. 3. Berlin, 1897. [*DTA*]
———. *Sethianische Verfluchtungstafeln aus Rom*. Leipzig, 1898.
———. "Neue Fluchtafeln." *RhM* 55 (1900):62–85 and 232–71.
———. "Eine antike Rachepuppe." *Philologus* 61 (1902):26–31.
———. "Deisidaimoniaka." *ARW* 12 (1909):1–45.
———. "Amuletum." *Glotta* 2 (1911):219–30.
———. "Charms and Amulets (Roman)." In *Encyclopedia of Religion and Ethics* 3:460–65. New York, 1911.

Zeigler, K. *De Precationum apud Graecos Formis Quaestiones*. Breslau, 1905.
Ziebarth, E. "Der Fluch im griechischen Recht." *Hermes* 30 (1895):57–70.
———. "Neue attische Fluchtafeln." *GöttNachr*. 1899:105–35.
———. *Neue Verfluchungstafeln aus Attika, Boiotien, und Euboia*. SB Berlin 33 (Berlin 1934): 1022–50.
Zingerlie, J. "Heiliges Recht." *JOAI* 23 (1926):5–72.
Zuntz, G. "Once More: The So-called 'Edict of Philopator on the Dionysiac Mysteries' (*BGU* 1211)." *Hermes* 91 (1963):228–39.
———. *Persephone: Three Essays on Religion and Thought in Magna Graecia*. Oxford, 1971.

Index of Greek Words

ἀγωγή, 14, 214–43
ἀδικέω, 65–68, 97n.41
αἰδοῖα, 96n.31
ἀλεξητήριον, 127n.31, 146, 148, 149
ἀλεξιφάρμακον, 111, 146
ἄλιμον, 167n.62
ἄλυτος, 44–45, 79
ἁμαρτία, 36
ἁμαρτωλός, 34, 36, 48n.12, 64
ἀμυστηρίαστος, 249
ἀναβαίνω, 99n.63
ἀναθεματίζω, 65
ἀνάθεμα (=ἀνάθημα), 95n.25
ἀνατίθημι, 72, 74–75, 77, 79, 80
ἀναχωρέω, 119
ἀποχωρέω, 119
ἀνιερίζω, 73–74, 79, 80
ἀνιερόω, 72, 79, 80
ἀντίδικος, 16
ἀπογράφω, 5, 16, 24n.24
ἀπο(σ)τρέπω, 120
ἀρετή, 100n.76
ἀρχιτεκτονικός, 141
ἀσέβεια, 56n.106, 262
ἀσεβής, 36
ΑΣΚΙ ΚΑΤΑΣΚΙ, 111–12, 121, 137n.106
ἄωρος, 22n.6, 43, 61

βάσκανος ὀφθαλμός, 119
βιαιοθάνατος, 22n.6
βλήχων, 145

γράφω, 24n.24

δαίμων, 182, 251
δάφνη, 146
δεῖνα (ὁ or ἡ), 190, 209n.20, 212n.69
δεῦρο, 189
δέχομαι, 31n.85
δημιουργός, 140–41
διαβολή, 227, 228
διατάσσομαι, 53n.81
διδάσκαλος, 27n.49
διοργιάζω, 67, 96n.33
δύναμις, 100n.76, 147, 148, 149, 150, 151

ἐγγράφω, 5, 16, 24n.24
εἶδος, 220
ἐκδικέω, 65–68, 71–72
ἐκτίματρα, 78
ἕλκω, 229, 241n.84
ἔμπυρον, 224
ἐναγής, 36
ἐνέχομαι, 34
(ἐν)ορκόω, 49n.22, 54n.92
ἔνοχος, 34
ἔντευξις, 69, 97n.41
ἐξαγορεύω, 80
ἐξαίρομαι, 66
ἐξενπλάριον, 92
(ἐξ)ομολογέω, 75, 80
(ἐξ)ονειρωγμός, 229, 230
(ἐξ)ονειρώττω, 229
ἐπάναγκος, 194
ἐπαφροδισία, 219, 220
ἐπιδέω, 19
ἐπιδίδωμι, 71
ἐπιζητέω, 77–88
ἐπίθυμα, 195
ἐπωνυμία, 191, 210n.27
ἐπῳδή, 109–37, 188, 189
ἐργαστήριον, 220
ἔρχομαι, 189
ἔρως, 214
εὐαγγέλια θύειν, 95n.19
εὐχά (=εὐχή), 5, 88
εὐχή, 188–213
εὔχομαι, 189

θαψία, 144, 149
θαύματα, 100n.76
θυμίαμα, 191
θυμοκάτοχον, 219
θυμός, 219

ἰατρός, 124n.3
ἱερὰ μαγεία, 248
ἱεροποήμα, 78, 79
ἱεροσυλία, 48
ἱκανοποιεῖν, 85
ἱκετεύω, 6, 25n.28, 66–68

ἵλεως, 45
ἱστορία, 138
ἴυγξ, 218, 241n.84

καθάρσιος, 153
κάθαρμα, 147
κάλλος, 219, 220
καταγράφω, 5, 9–10, 21n.3, 24n.24, 65, 66, 67
καταδεσμεύω, 213
κατάδεσμος, 21n.3, 231
καταδέω, 21n.3, 24n.24, 61
καταδίδωμι, 5, 24n.24
(κατα)κλίνω, 242n.103
κατακλιτικόν, 231
κατατίθημι, 66, 73, 80
κέρδος, 11, 27n.48
κλύω, 189
κοίτη, 220
κολάζω, 41, 66–68, 75, 80, 95n.26
κόλασις, 65–68, 73, 75, 80
κυκεών, 139, 144–45, 166n.37
κύριος, 68
κῦφι, 157, 160, 161

λαικάζω, 241n.96
λόγος, 189, 194
λύω, 26n.38, 76, 79, 86
λύτρον, 79

μαγεία, 248
μαγικός, 248
μάγος, 248
μεταβαίνω, 78
μετέρχομαι, 41, 71, 78, 80
μορφή, 219, 220
μυστήριον, 249–59
  τῆς ἀνομίας, 251, 253
  τῆς βασιλείας τοῦ θεοῦ, 251
  τῆς βασιλείας τῶν οὐρανῶν τοῦ θεοῦ, 251
  τῆς εὐσεβείας, 251
  τοῦ θεοῦ, 249, 251
  θεῖον, 249
  ἱερόν, 249
  τοῦ κανθάρου, 252
  μέγα, 249, 251
  μέγιστον, 249
  τῆς πίστεως, 251
μύστης, 248, 249, 258n.45
μῶλυ, 139, 141, 165n.24
μυσταγωγός, 210n.36, 249, 250
μυστοδόκος, 112

νικητικόν, 111

ὄλλυμι, 26n.38
ὀνειραιτητόν, 176, 182n.2
ὀκειροκρισία, 177
ὀνειροπομπή, 229
ὀνειροπομπία, 180
ὀνειροπομπός, 176, 182n.2
ὀνειροπομπῶν, 179
ὀνείρωξις, 229
ὄνομα, 191, 210n.27
ὁρκίζω, 6, 25n.28, 35
ὄρχις, 148
οὐσία, 14, 224, 228, 240n.65, 75,

παῖς
  ἄφθορος, 185n.33
  μυστοδόκος, 249
πάνακες, 150
παραδέχομαι, 19, 31n.85
παραδίδωμι, 5, 10, 24n.24, 73
παραχωρέω, 79, 80, 83
πάρεδρος, 180, 220, 251
  δαίμων, 251
πάροιστρος, 230
πείθω, 188, 208n.8
πεπρημένος, 73, 84–88
περίαμμον, 107
περιάπτειν, 109, 124n.6, 143
περίαπτον, 107
πιττάκιον, 76
ποτιδέχομαι (προσδέχομαι), 31n.85
προσώπον, 220

ῥιζοτόμος, 124n.3, 138, 141, 144, 149, 150, 151, 152
ῥυστική, 137n.110

σύμβολον, 250
συμμύστης, 249
σύνδικοι, 16
σύνθημα, 250
σύστασις, 177
σῴζω, 120, 137n.110

τέκνα τέκνων, 56n.114
τελέω, 124n.5
τίς (=ὅστις), 97–98

ὑποδιδάσκαλος, 27n.49
ὑπόθεσις, 135n.93
ὑπόκειμαι, 34
ὕποχος, 34

φάρμακον, 109, 125n.12, 139, 140, 142, 143
φαρμακοπώλης, 124n.3, 144, 149, 150

φαρμακός, 146–47, 169n.100
φεύγω, 112, 119
φθισίκηρε, 209n.22
φιλία, 31n.81, 214
φιλτροκατάδεσμος, 194, 220, 231, 242n.97
φίλτρον, 218
φυλακτήριον, 107, 124n.5
φυλάσσειν, 107
φύσις, 139–40, 149
χάρις, 219, 237n.26
χαριτήσιον, 220, 236n.18
χῆρος βίος, 41–42, 54n.87
χολόω, 45, 96n.33
χορηγός, 12, 31n.81

ψυχή, 121, 225

ζητέω, 71, 74, 78, 80

Index of Latin Words

argumentum, 189, 192, 193
amicus, 16

carmen, 113, 239n.58
coercitio, 269
cogere coitus facere, 231
commendare, 80
commonitorium, 87
(con) donare, 90

defigo, 21n.3, 61
defixio, 21n.3, 61, 126n.19
delegare, 83
devotio, 88
domina, 90–91

epaphroditus, 96
excantare, 264
exemplarium, 92

fero (=*affero*), 86, 104n.133
fruges, 264

historiola, 112

indeprehensus, 89
inimicus, 16
invocatio, 189, 191

lamella, 114–37

male facere, 82–83
mandare, 83

NN, 4, 24n.15, 180, 181, 183n.5, 190, 195, 209n.20, 219, 224, 229, 230, 232, 242n.100
negotium, 85

otium, 290n.8

(per)exigere, 89
periculum, 136n.104
persequi, 82, 83
poenicea, 25
preces, 189, 190
pulegium, 145

queri, 88

redimere, 86, 104n.133
regestum (*idem*), 105n.137
repraesentare, 88
reprehendere, 92

sanguine suo, 84–87
sanguinea, 25
satisfacere, 85

tabula ansata, 114

venenum, 264, 272n.27
veneficia, 264
venefica, 239n.58
veneficus, 273n.28
vindicare, 65, 72, 80, 83, 91
vindex, 83
voces magicae, 190, 192, 193

General Index

abortion. *See* conception, contraception, and abortion
abscesses. *See* inflammations
Accadians, 37
Actaeon, 9
actors and actresses, 12
Adonaios, 119
adultery, 235
Aegisthus, 17–18
Aeschylus, 5
Agathocles, 180, 186n.58
Aktiophis, 227, 253
Alcibiades, 262–63
Alexander Trallianus, 118
Alexander-romance, 181, 230
Allecto, 189
altar, 177, 196
amatory curses. *See* erotic magic
Ammianus Marcellinus, 260, 263, 264
Ammon, 72, 157, 160, 181
amoral familist in Mediterranean society, 62
Amphyctionic oath, 37
amulet cases, 110, 111, 114, 115, 158
amulets or phylacteries, 107–37, 218–20, 248, 260
Anaeitis, 44, 46, 58n.158, 77–78, 102
anathemata, 65. *See also* sacrifices and other offerings
Anatolia. *See* Asia Minor and Anatolia
anatomy, 141
Anaxagoras, 264, 265
Anaxilas, 111
angels, 120, 176, 178, 180, 194, 219
anger, restraint of, 219, 236n.22
anonymity and secrecy, 17–18, 62–63, 90
anonymous gods, 45
Antaura, 112–13
Antigone, 236n.20
Antiochus, 223
antipathy, 159. *See also* sympathy and sympathetic magic
Antiphanes, 110
Anubis, 185n.39
aphrodisiacs, 13–15, 145, 148–49, 152, 162, 214–43. *See also* erotic magic
Aphrodite, 189, 198, 223, 231
Apollo, 9, 17–18, 20, 37, 45, 75, 112, 114, 153, 176–82
Apollobex of Coptos, 185n.46
apotropaic rituals and substances, 57, 111, 119, 146
Appian Way, 23
Apuleius, 147, 195, 196, 264
Ara and Arai. *See* Curse personified
Aramaic, 117, 136
Ares or Enyalios, 9, 11, 27n.40, 160, 212n.70, 230
aretologies and miracle stories, 19–20, 74–79, 96, 100. *See also* confessions and confession inscriptions
Aristophanes, 110, 145
Aristophilus of Platea, 149
Aristotle, 141, 142, 162
Artemus, 37, 39, 51n.53, 111, 112, 120, 189. *See also* Hecate; Selene
Artemisia, the curse of, 68–69, 72
Artemus, 38
artisans, 11
Asclepius, 108, 123n.2, 151, 153, 154-56, 179, 184n.19, 268
asebeia. *See* impiety, legal action against
Asia Minor and Anatolia, 18, 33–59 and 60–81 *passim,* 140, 250
aski kataski formula, 121, 137. *See also Ephesia grammata*
assistant/attendant, 180, 193, 211nn.41–42, 220, 251
Astrapsakos, 191, 220
astrology, 118, 134n.84, 151, 154–56, 158, 177–79, 194, 195, 260, 264
Ataecina Turibrigensis, 91
Atargatis, 46
Athena, 37, 72, 179
Athenaeus, 221
Athenian agora, 12
athletic contests or competitions, 11–13, 16–17, 111
atonement, 43, 45, 71–93, 102
Attalus, 113
attitude of performers of ritual. *See* mentality
attraction-spells (*agōgai*), 14, 180, 214–33
Audollent, A., 22n.3, 63, 80
Augustine, 81, 182
Aureliusnomen, 52
Austria, 84
Azande, 8

baboon, blood of, 186n.52

Babrius, 78
Babylonia and Babylonians, 53n.76, 55n.97, 110, 162, 223
Bacchanalia, repression at Rome, 269
Bastet, 115
baths. *See* springs, cisterns, and public baths
bats, 231
beliefs, 176
Bes, 102, 176, 178, 184n.27
binding magic, 3–32, 60–68, 108, 180. *See also defixiones;* voodoo dolls
Bion, 107, 122, 126n.20, 128n.34
Bithynia, 33–59
Björk, G, 63, 80
black magic (*Schadenzauber*), 62, 63, 67, 75, 92, 94n.11, 191
blindness, 37
blood and bleeding, 108, 128n.40, 158, 159, 160, 177, 180, 196, 227, 228, 231
Boeotia, 13–14
bones, 109, 148
bound images. *See* voodoo dolls
boxing, 111
bronze, 3, 7, 9, 11, 26n.35, 73, 74, 118, 135, 144
Brown, P., 16, 268–69
burials. *See* graves, burials, and tombs
Burkert, W., 140
burning, 13, 65, 73, 75, 112, 150. *See also* fire magic and *empyron* spells
business curses. *See* commercial or business curses
Byblos, 38
Byzantine period, 9, 154, 157, 158, 160

Cacus, 83
Caecilia Secundina, 115
Callistratus of Aphidna, 30
Candaules, 222
Caria, 33–59
Carolingian period, 53n.76
Carthage and Punic culture, 13, 113
Cato, 109
Celsus, 141
Chaldaeans. *See* "Babylonia, and Babylonians
chameleon, 16, 21n.3
charioteers, 12–13, 16, 186n.47
charm or charisma, 12, 218–20
chicory, 155–56
children, 42, 43. *See also* untimely dead
Christian and Jewish formulas, 33, 35, 56n.106, 70, 120–21, 122, 131n.50, 133n.68, 189
Christianity and Christian influence, 35, 55n.100, 71, 81, 89, 117, 120, 161, 176, 181–82, 188, 207n.3, 234n.1, 249, 251, 253, 261, 262, 264–65, 266, 269
Chryses, 17–18, 20
chthonic gods or gods of the underworld, 4, 10, 17, 18–19, 24n.15, 34, 35, 45, 46, 64, 66, 92, 195
chthonic sanctuaries, 3, 9, 17, 18–19, 22n.7
Cicero, 15
Cilicia, 33–59
Circe, 139–40, 263
cisterns. *See* springs, cisterns, and public baths
Cnidus and the "Cnidian tablets", 22n.7, 58n.141, 72–75, 86, 89, 90
coercion. *See* magic: coercion and manipulation
coldness or frigidity, 8, 9, 12
commercial or business curses, 11, 62
conception, contraception, and abortion, 145, 148–49, 158–59, 237n.33
confessions and confession inscriptions, 43, 45, 56, 68, 74–81
consecration, dedication, or devotion to a god, 73–93 *passim*. *See also anathemata*
contagion, 73
Coptic texts and magic, 71, 81, 118, 134–35, 242n.106. *See also* Egypt and Egyptian influence
Corinth, 22n.7
corpses, 5, 8, 9, 13, 177, 225, 226, 229
courtroom magic. *See* judicial curses
Crete, 111, 115, 116, 121–22, 250
crimes, 71–93 *passim*. *See also* thieves and theft; stolen property
crocodiles, 8, 159, 221
Croesus, 111
crossroads, 7, 178, 232
Cumont, F., 252
Curse personified (Ara), 42, 49
curse formulas, 5, 10, 41
curse tablets. *See defixiones*
cursing vs. blessing, 42
Cyprus, 23n.11
Cyrene, 221

Dactyls, 121, 127
daemonic animals and monsters, 112–13, 115, 132
daemons, 5, 12, 13, 35, 42, 55, 59, 61, 64, 65, 92, 112, 117, 119, 132n.65, 133n.75, 177, 178, 180, 181, 182, 190, 194, 211n.52, 224, 228, 251
Damnamaneus, 119, 127
dangerous insects and animals, 111, 112–13, 115, 152
debts and securities, 70–71, 75–93
decay of religion, 175
defixiones, 3–32, 60–68, 69–93, 126, 215, 220, 231, 260, 264
Deianeira, 222
Delos, 19, 23n.7, 29, 66
Delphi, 37
Demeter, 16, 22–23n.7, 64, 69–70, 72–73, 78, 144–45, 250
Democritus, 178, 221
Demophon, 252
Demosthenes, 30n.76
diabolai, 23n.9, 196, 227, 253
Dieterich, A., 175–76, 252
Dike, 9, 64, 189, 191
Diocles of Carystos, 151, 152
Dionysus, 181
Diopethes' law of 432/31 B.C., 262
Dioscorides, 153–54, 158, 159, 161

Diotima, 233
Dirae Teiae, 37–38
diseases and illnesses, 71, 75–81, 91, 107–75, 193
divination and prophecy, 175–87, 193–94, 196, 264. *See also* oracles
  from a corpse, 177, 179, 185n.45
  during sleep, 177, 183nn.6 and 8
divinity, union with, 177
divorce or separation curses (*Trennungszauber*), 13, 15, 186n.47, 231
Divus Nodens, 84, 90
doctors or physicians, 107, 124, 138–75, 222–23
doctrine of signatures. *See* sympathy and sympathetic magic
Dodona, 17, 103
dogs and puppies, 21n.3, 112, 113, 147, 191, 196, 212n.70, 228, 230, 231
Doric dialect, 54
dreams, 88, 175–187, 193, 196, 220, 228–30
  spells for inducing or sending, 176, 179, 180, 181, 182n.2, 183nn.4–8
drugs, 109, 138–75, 221–22
drugs-by-degrees classification, 154
drugsellers. *See* rootcutters and drugsellers
duplicity, 217

Earth (Ge), 6, 14, 18–19, 24n.15, 25n.27, 46, 64, 65, 127
economic hardship, 51–52
Edelstein, L., 150–51, 162, 268
Egypt and Egyptian influence, 20, 24n.17, 52, 69, 71, 76, 81, 108, 114, 115, 121, 124, 130, 136–37, 140, 154–62, 176, 191, 248–49, 250–53 *passim*
Eitrem, S., 175–76
Electra, 17
Eleusis and Eleusinian mysteries, 46, 166n.37, 237, 250, 254. *See also* mystery cults and initiation
Empedocles, 142, 238
empiricism, 150
empyron spells. *See* fire magic and *empyron* spells
England, 84–90
enteuxis-formulas, 69, 80, 93
Ephesia grammata, 111, 121–22. *See also aski kataski* formula
Ephesian letters. *See Ephesia grammata*
epigraphic habit, 39–40
epilepsy, 113, 117–18, 119, 120, 133n.75
Epimenides, 141, 147
epistles. *See* letters
epithets, 6, 190–91
erections. *See* male sexual potency and erection
Ereschigal, 121, 178
Erinyes, 42, 46, 55, 64
Eristratos, 225
Eros, 13, 179, 185n.44, 220, 226, 229, 231
erotic magic, 13–15, 183n.10, 195, 214–33. *See also* aphrodisiacs
eruca satira. *See* rocket plant
Eumeneian and Laodiceian formulas, 48n.17
Eusebius, 182
evil eye, 11, 119, 178
evolutionary views of magic and religion, 176
eyeache and eye disease, 114, 117, 128–29n.40, 133, 149, 150, 167n.67

falcon, 181, 186nn.61 and 62
Far Eastern contacts, 156
Fayum mummy protraits, 114
feet, 110, 118, 134
fever, 112, 114, 118
fines and penalties, 34, 40, 52, 54, 74–93
fire magic and *empyron* spells, 224, 225, 250. *See also* burning
fish, 21–22n.3
flax-leaf, 177
flee-formula, 112–13, 116, 119, 122, 128–29
flint, 26n.35
folklore remedies, 113, 146–51
forgiveness, 207n.3
formula, 189
Foucault, M. 218
Fowden, G., 161
frankincense, 177, 178, 179, 191
Frazer, J., 176, 188, 190, 194, 195, 267
Frazerian dichotomy between magic and religion, 188, 190, 194, 195. *See also* magic: and religion as separate categories
Frazerian evolutionism, 176, 244–45
frigidity. *See* coldness or frigidity
frogs, 158, 231, 242
funerary headbands, 116
funerary imprecations or tombstone curses, 8, 20, 33–59, 104n.132
funerary rituals, 40
Furies, 64

Gadamer, H.-G., 268
Galen, 15, 138–39, 151, 154, 155, 159, 223
gemstones, 114–137 *passim,* 180, 186n.53, 231, 252–53
gender and the performance of magical rituals, 161–62, 227–28
Germanicus, the death of, 23n.9, 63
ghosts, 9
gladiators and *venatores,* 13, 121
Gnosticism, 244
goat, 180, 191, 196, 212n.70, 228
God of Abraham, God of Isaac, 118
gods of the underworld. *See* chthonic gods or gods of the underworld
gold 73, 74, 110, 180, 219. *See also* lamellae; stolen property
Gorgon, 119
gossip. *See* slander and gossip
gout, 110, 118–19, 125n.16, 134n.85, 135n.86
graves, burials, and tombs, 3–4, 7, 9, 10, 17, 22n.6, 26n.38, 40, 53, 61, 63, 120, 135, 190, 228, 260
gravestones, 44, 55
Great Mother. *See* Meter

Greek curses vs. Anatolian curses, 36
guilt and innocence, 44, 67, 75–93
gynecology and obstetrics. *See* conception, contraception, and abortion

Hades or Pluto, 26n.38, 45, 46, 64. *See also* underworld
Hadran, 231
hair and nail trimmings, 14, 29, 105, 226
handbooks of recipes and formulas, 23n.11, 54, 91, 112, 121, 124, 127, 209n.19, 218, 220. *See also* professional magicians
hands, 5, 42
Hatti and Hittites, 40
headache, 109, 113, 116, 128n.39, 155–56, 158, 220–21
Hebrew inscriptions. *See* Jews, Judaism, and Jewish influences
Hecate, (see also "Artemis" and "Selene"), 6, 15, 45, 46, 47, 64, 66, 121, 146, 176, 178, 179, 189–90, 212n.70, 226. *See also* Artemis; Selene
Heitsch, E., 197
helebore, 150, 153–54
Helen of Troy, 140, 229
Helios, 34, 35, 41, 42, 45, 46, 47, 59n.155, 70, 71, 155–56, 157, 160, 177–78, 180, 219, 252
heliotropes, 155–56
hell. *See* underworld
Hephaestus, 159, 160
Hera, 157
Heracles, 112, 160, 181, 197, 222, 263
Hermaphrodites, 146
Hermes 3, 4, 5, 6, 9, 14, 15, 18–19, 20, 25n.27, 46, 64, 139–40, 176–82 *passim,* 160, 181, 190–91, 193, 196
Hermes Trismegistus and Hermetism, 154–56, 247
herms, 42
Herodes Atticus, 42, 45
heroes, 43
Herophilus, 138–39
Hesiod, 11, 143
Hestia, 160
hexameters, dactylic, 19, 42, 54, 112–14, 116, 121, 128, 137n.106, 177, 183n.11, 189–215 *passim,* 193
Hippiatrica, 118
Hippocratic writings and practices, 125n.12, 140–41, 143, 145, 154
Hippodameia, 239n.63
hippodromes, 3, 23n.9, 28, 31
Hippolytus of Rome, 161, 182
Hipponax, 146–47, 221
hippopotamus, 180
historiolae, 112, 121–22, 128
homeopathic ritual. *See* sympathy and sympathetic magic
Homer and Homeric hexameters, 17, 59, 108, 111, 116, 117, 110, 132, 139–42, 158–59, 161–62, 179, 188, 219
honor and shame, 216–17
Hosios Dikaios, 59
Hungary, 117
hydrophobia, 119
hymns. *See* prayers, hymns, and liturgies
Hypnos (Sleep), 179
Hypsistos Theos, 70

iambic senarii, 54
iambic trimeters, 41, 54, 116
Iao, 118, 119, 242n.106
Iatros, 112
ibis, 177, 183n.6
illnesses. *See* diseases and illnesses
imperatives, 6, 119–20
impiety, legal action against, 262–63. *See* also sacrilege and impiety
impure air and water, 57
incantations, 11, 15, 19–20, 109, 112–14, 119, 141, 189–90. *See also specific categories (e.g.,* flee-formula; *aski kataski* formula)
incense, 73, 146, 157, 160–61, 177, 178, 179, 180, 183n.10, 185n.31, 191. *See also* frankincense
inefficacy of civil justice, 40, 68
inflammations, 112–14. *See also* swellings and tumors
initiation. *See* mystery cults and initiations
injustice, 65, 68
ink, magical, 177–78, 180, 185n.33
innkeepers, 11
innocence. *See* guilt and innocence
insanity and madness, 154, 158, 225
insomnia, 13, 225–26
Irenaios, 182
iron, 3, 109, 111, 180, 231
irrevocability, 43. *See also* loosening of curses
Isis, 91, 157, 176, 177, 178, 192, 225, 269
Italy, 26n.33, 37, 73, 118
itching, 13
Ixion, 241n.84

Jesus, 183n.6, 229
Jevons, F. B., 20
Jewish formulas. *See* Christian and Jewish formulas
Jews, Judaism, and Jewish influence, 56, 65, 136, 189, 249, 250, 252–53, 265
Jordan, D. A., 61, 117
judicial curses, 15–16, 29n.67
judicial prayers. *See* prayers for justice
Julio-Claudii, 52, 129, 155, 264
Jupiter, 63, 81–82, 90, 104n.124, 269
Justin Martyr, 182

Kagarow, E. G., 6, 80
Kassander, 31
katadesmoi. See defixiones
Kore or Persephone, 4, 22n.7, 25n.7, 46, 64, 72, 78, 91, 189
Kronos, 157, 160, 189, 219
Kuhn, T., 268

Lady goddess Syria, 66–67
lamellae (inscribed gold and silver foil), 114–22, 129–30
lamp magic (*Lampenzauber*), 176–82, 224–26
Latin and Latin texts, 12, 60, 63, 81–93, 117, 121
Latte, K., 36
laurel (sweet bay), 146, 176, 177–78, 183nn.8 and 10
law courts and trials, 5, 15–16, 19–20,
Law of the Twelve Tables at Rome, 63, 264
lead, 3–9, 12, 13, 60–93, 111
leaves, 109, 138–63, 177, 178, 222, 229
legal vocabulary or terminology, 9, 16, 34, 53, 71–106
leges sacrae, 101
legislation against magic and astrology, 261–66
legs, 5
Lemuria, 267
Leto, 37, 45, 46, 75
letters, 4, 65, 81
letters and letterforms, 55, 116
Lex Cornelia de Sicariis et Veneficiis, 264
Libanius, 15–16, 133, 264
libation, 183n.10, 184n.14, 195, 214n.57, 196
linen, 29, 177, 180, 183n.10, 185n.33
lion, 110, 113, 159, 160
literacy, 23n.10
liturgies. *See* prayers, hymns, and liturgies
lizards, 220
Lloyd, G. E. R., 139, 143
logos. *See* formula
loosening of curses, 26n.38, 79
Lords gods Sykonaioi, 66–67
love and lovers, 13, 14, 216–35
Lucian, 109–10, 147, 226
Lycia, 33–59
Lycian curses, 38–39
Lycurgus of Athens, 30
Lydia, 33–59, 75–79,
Lydian curses, 38–39
Lysias, 30–31

MacMullen, R., 39–40
madness. *See* insanity and madness
magic
  and religion as separate categories, vi, 17–20, 36, 61, 92–93, 121–22, 123, 150–51, 188, 190, 191, 194, 244–47, 260–61, 266–67
  and science as separate categories, vi, 247, 260–61, 266–67
  as *Selbsthilfe,* 41, 79
  as shameful or illicit activity, 17, 62–63, 248
  coercion and manipulation in, 92–93, 176, 194–95, 211n.44, 248, 249
  as an art or *technē,* 176
  uses automatic procedures, 61, 122
  as an arcane aspect of religion, 107
  the term as a semantic trap, 188
magical characters (*characteres*), 116, 118
magical herbs and plants, 26n.38, 107, 113, 138–74,
magnets and magnetism, 158–59, 231

Maiistas, 19–20
male sexual potency and erection, 220, 221
Marcellus Empiricus, 117
Marcus Servius Nonianus (consul 35 A.D.), 114
Marett, R. R., 267
marjoram, 184n.14
marriage. *See* weddings and marriage
Mars Silvanus, 87
marsh mallow, 148
Mary Magdalene, 229
maskelli-formula, 184n.13
Medea, 143, 144, 229
Mediterranean family of cultures, 214, 234–35n.5
Megaira, 189
Melampus, 154
Men, 44, 45, 46, 47, 75, 102
mentalities, religious and other, 18–20, 46–47, 57, 63, 91. *See also* non-Greek religious mentality
Mercury, 84, 85, 87, 88
Merkelbach, R., 252
metempsychosis, 114–15
Meter, 46, 75
mid-wives, 145
mind, 15
Minerva (= the Dea Sulis), 85, 87, 90
miniature shrines. *See* shrines and *naiskoi*
miracles. *See* aretologies and miracle stories
misogyny, 215
Mithraic religion, 250
Mithras Liturgy, 197, 251–52, 254
Moira, 189
moon. *See* Selene
Moraux, P., 33
Morgantina, 18–20
Moses, 254
Mother of the Gods, 74
mouth, 177
Mowinckel, S., 246
mummy, 183n.4
Mundus, the affair of, 269
myrtle, 177
Musaeus, 143
Mysia, 33–59
mystery cults and initiation, 112, 121–22, 177, 188, 192, 196–97, 244–59
Mytilene, 39

Nabor, N., 18–19
nail trimmings. *See* hair and nail trimmings
nails and nailing, 3, 9, 24n.24, 61, 62, 90
names and naming, 12, 14, 38, 40, 41–44, 61, 62, 65, 68, 117, 120, 141, 157, 177, 178, 180, 181, 191–94, 219, 223
Near East and Near Eastern rituals, 38, 108, 140, 162, 191
necklaces, 67, 83, 103, 114
necromancy. *See* divination and prophecy: during sleep
Nectanebo, 181, 230
Nemesis, 45, 46, 86, 104
Neoplatonic philosophy, 248

Neptune, 90
Newton, C. T., 80
Newton, I., 247
Nicander of Colophon, 152
night, 176–77
Nike, 112
Nilsson, M., 20, 175
naiskoi. *See* shrines and *naiskoi*
nines, 178
non-Greek religious mentality, 6, 36
normative piety, 18
North Africa, 13, 15, 18
numbers, magical. *See under individual numbers* (*e.g.*, sevens)

oaths, 25n.28, 37, 54, 57, 75–79, 85
obscenity, problems of definition in U.S. Supreme Court, 265
obstetrics. *See* gynecology and obstetrics
Odysseus, 108, 122, 139, 140, 229, 263
oil, 183n.6
olive branches and olivewood, 176, 178, 180, 181, 185n.33
Olympian or heavenly (Ouranian) gods and rituals, 46, 61, 92, 179, 195
Olympias, 181, 230
Olympiodorus of Thebes
Oneiros (Dream), 179
Onians, R. B., 141
onomata. *See* names and naming
opium, 139–40
oracles, 17, 76, 177–79, 193, 196. *See also* divination and prophecy
oracular statue of Hermes, 196
orators, 16
Orestes, 5, 17–18
Orpheus, 126
Orphic beliefs, rituals and texts, 143, 190, 196
"Orphic" lamellae, the so-called, 114–16, 120, 121–22, 130, 197
Osiris, 64, 115, 157, 177, 180, 225, 242n.100
Ostanes, 186n.48
ostraca, 23n.10, 71
ostracism, 23n.10
Ouranos, 157
Ouroboros, 180

Palestine, 26, 54, 116, 121, 270
palindromes, 190
panakes, "All Heal", 150
papyri magical, 108, 111, 114, 115, 118, 121, 151, 156-61, 175–234, 176–234, 248–53
papyrus, as *materia magica*, 177, 180
paralysis, 9, 15, 19–20
parhedros. *See* assistant/attendant
Paris, 229
Parker, R., 147
Paul of Aegina, 154, 160
Paul, the apostle, 209n.19, 251
Paulus Tartareus, 270
payment "with his own blood", 84–90, 104
Pelops, 11, 17–18, 239n.63
penalties. *See* fines and penalties
Penelope, 179
pennyroyal, 144–45
performative utterances, 31n.87, 119
Pericles, 107, 122, 218, 262
perjury, 75, 85
Persephone. *See* Kore or Persephone
Perseus, 119
Persia and Persian influence, 38, 46, 51n.53, 59, 250, 251–52
persistence of rituals, 76, 260
persuasive analogy, 8, 10
persuasion of the gods, 188, 208n.8
Peter, the apostle, 182, 187n.64
Phalasarna, Crete, 111, 115, 116, 121–22
Philinna papyrus, 112–13, 121–22
Philo of Alexandria, 250
philology, 176
Phoenicia and Phoenicians, 38, 120, 124
Phrygia, 33–59, 75–79, 199
physicians. *See* doctors or physicians
physiognomy, 214–15
Physis, 191
Pindar, 11, 108, 143
pirates, 9
Pisidia, 33–59
Pisidian Gods, 46–47
Pithys the Thessalian, 177
plague, 44, 56, 111, 127n.28, 147, 271
plants, Egyptian, 176
Plato and Platonism, 4, 5, 63, 94n.11, 109, 140, 188-89, 195, 220, 233, 263
Pleiades, 178
Pliny the Younger, 151, 159
Pluto. *See* Hades or Pluto
pneumata. *See* daemons, spirits
Poimandres, 155
poison and poisoning, 44, 64, 75–79, 97, 139, 141, 147, 152, 221–22, 264, 271
Polemo, 214–16, 222, 233
political curses, 16
politicians, 16
pollution and purification, 146–47, 153–54, 157, 162, 183n.10
Poseidon, 11, 17
possession by daemon, 181
pottery kilns, 11
Praxidikai, 64
prayer formulas, 6, 10, 18–20, 118, 119–122, 189
prayers for justice, 58, 60–106
prayers, hymns, and liturgies, 6, 18–20, 26n.26, 41, 45, 60–93, 107, 116, 119–22, 150, 153, 157, 179, 180, 188–215, 219, 249–53
prayer, formal structure of, 189, 209n.16
Preisendanz, K., 16, 175, 193
priests and priestesses, 19–20, 44–45, 57n.127, 58n.132, 63, 75–81, 124n.3, 146, 159–60, 163, 237
Proetus, the daughters of, 154
professional magicians, 4, 23

properties (*dynameis*) of herbs, 151–54
prophecy. *See* divination and prophecy
prophets, apparel of, 177, 179, 183n.10
prostitutes, 145
protective spells, 183n.10
Psyche, 231
Ptolemaic period, 69
public display vs. secret burial, 27, 58n.136, 72–74, 80–81. *See also* anonymity and secrecy
pumice ("Stone of Assius"), 159
punishment and torture, 36, 40, 41, 43, 64–93 *passim,* 121. *See also* revenge and vengence
puppies. *See* dogs and puppies
purification. *See* pollution and purification
purity, ritual, 177 (bis), 183nn.8 and 10
Pythagoras and Pythagoreans, 178, 196

Ra, 176
rationalistic supernaturalism, 151
Raven, J. 164
religion. *See also* magic: and religion as separate categories
  popular, 176
  private, 176
  supplication in, 92–93.
  traditional, 176
retrograde inscription, 7–8
revenge and vengeance, 4, 41, 42, 46, 65, 68–93
Rhodes, 37
Riddle, J., 152, 153, 158
rings, 110–11, 114, 231, 237n.26, 252–53
ritual 20, 195–97
ritual purity, 157, 177
ritual, spontaneous development vs. cultural borrowing of, 91
rivalry and competition in Mediterranean society, 10–20, 62–63, 216–20
rocket plant (*eruca sativa*), 223
Rome and Romans, 21, 39, 59n.148, 71, 81, 151, 153, 155, 159, 207n.4, 263–65, 267, 269
Romulus, 117
roosters, 21
rootcutters and drugsellers, 124n.3, 138–75
Rous, 242n.106
rue, 185n.34

Sabaoth, 118, 119, 177, 183n.9, 242n.106
sacred laws or calenders. *See leges sacrae*
sacrifices and other offerings, 43, 45, 56n.106, 75, 146, 147, 150, 157, 183n.10, 190, 191, 195–97, 212nn.58–59, 225, 253
sacrilege and impiety, 33, 56
salt, as offering, 184n.30, 196
salvation, 120–23
Sappho, 189, 198, 223
Sardinia, 115
Sardis, 51
scepters, 43, 44–45, 46, 57, 58, 76, 128n.33, 190
science. *See* magic: and science as separate categories
scruples against homicide, 20
Segal, A., 245–46
Selene, 34, 35, 45, 46, 47, 128, 157, 177, 180, 181, 189–90, 192, 194, 195, 196, 209n.24, 224–25, 226. *See also* Artemis; Hekate
separation spells. *See* divorce or separation curses
Serapis, 19–20, 120, 136
Seth. *See* Typhon and Seth
Sethianorum tabellae 13, 99
sevens and sevenfold repetition, 178, 183nn.10 and 12, 184n.13, 194, 223
Severi, 52, 264, 269
sex and sexual intercourse, 13, 14, 143, 144, 166n.39, 183n.10, 216–35, 253
shrines and *naiskoi,* 51, 179, 180
Sicilian expedition, the, 271
Sicily, 11, 12, 16, 65
silence, ritual, 183n.6
silver. 7, 9, 110, 112, 180. *See also lamellae;* stolen property
similia similibus formula, 6–9, 12. *See* sympathy and sympathetic magic
Simon Magus, 182
slander and gossip, 75–76, 218. *See also diabolai*
slander spells. *See diabolai*
slaves and slavery, 63, 69, 73, 177
sleep. *See* divination and prophecy: during sleep
Smith, W., 142
sociology and sociological methodology, 245–46
sociology of law, 261
Socrates, 31, 109, 122, 218, 262, 263, 271
soil, as offering, 184n.30, 196
Sokhit, 115
soldiers, 56, 91, 121
Sophocles, 144
Sophronius, 9
Soranus, 145, 159
Sosispatra, 223
soul (*psyche*), 5, 14, 15, 40, 120–21
Spain, 60, 91
spirits, evil, divine, and other, 117, 118, 119, 132, 142. *See also* daemons
springs, cisterns, and public baths, 3, 60, 85–88, 90, 224
squill, 146–47, 149, 153
Stewart, P., U.S. Supreme Court Justice, 260
stolen property, 60, 66–93 *passim. See also* thieves and thefts
stomach-ache, 110, 155–56
Strasbourg papyrus, the, 193–94
submissiveness, 45, 69–70, 188, 194. *See also* magic: and religion as separate categories; religion; supplications
suicide, 222
Sulla, 236
Sumerians, 37
sun. *See* Helios
Sunday, 118
superstitions, 151–52, 175
suppliant, 183n.10
supplications, 61, 68, 69–70, 92–93, 211n.43. *See also* submissiveness

surgery, 109
survivals. *See* persistence of rituals
swellings and tumors, 114, 117. *See also* inflammations
sympathy and sympathetic magic, 7, 8, 42, 86, 112, 148–49, 150–51, 224. *See also* antipathy
symposia, 221
syncretism, 176
Syria and Syrians, 13, 15, 46, 112, 117, 119–20
systasis. *See* divinity, union with

talismans, 11, 124, 130. *See also* amulets or phylacteries
Tambiah, S. J., 8, 24
Tartarus. *See* underworld
Tean Curses. *See* Dirae Teiae
temple incubation, 179
temples, 19, 44–45, 58, 72–82, 123n.2, 124n.3, 159–60, 237
Tertullian, 182
testaments. *See* wills and testaments
thapsia ("the deadly carrot"), 144, 149
theatrical contests or competitions, 11, 12, 16–17
Theodosian Code, 265
Theognis, 146
Theophrastus, 138, 146–54
Thessalus of Tralles, 155–56, 157
theurgy, 142, 179, 264
thieves and theft, 27, 44, 64, 66, 67, 75–79. *See also* stolen property
Thoth, 176–77
Thrace, 9
threes and triple repetitions, 41–42, 177, 183n.12, 191–92, 194, 223
throne, 177
Thucydides historicus, 270
Thucydides, son of Melesias, 15
tin, 118, 134, 177
tombs. *See* graves, burials, and tombs
tombstone curses. *See* funerary imprecations or tombstone curses
tongues, 5, 15, 16, 19–20
torture. *See* punishment and torture
touching the earth, 42
tradesmen, 11
tripartite structure of Greek prayer, 189, 193
tripod, 177–78
tumors. *See* swellings and tumors
Typhon and Seth, 64, 180, 231, 242n.100
twelve, 178
Tzetzes, 146–47

uncorrupted youth as medium, 185, 193
underdogs, 20
underworld, 10, 40, 42, 46, 65, 121, 142, 176
union with divinity (*systasis*). *See* divinity, union with
untimely dead (*aoroi* or *ahoroi*), 1, 22, 43, 61, 70, 190, 191, 196, 231
Usener, H., 175–76

vengeance. *See* revenge and vengeance
Venus, 224, 225
verbal vs. written formulas, 4–5
veterinary care and magic. *See Hippiatrica*
violently or unnaturally killed (*biaiothanatoi*), 22, 70, 180, 228, 230, 231
Virtus, 84
voces magicae, 6, 116, 117, 119, 177, 181, 190, 191, 192, 194, 195
voodoo dolls 4, 5, 7, 9, 11, 23, 25, 26, 27, 29, 181, 216, 230, 242n.104
vowels, 116, 219
vows, 61, 65, 77–93

wax, 4, 7, 25, 110, 180, 181, 225, 230
weasel, 110, 221
weddings and marriage, 214, 216–18, 232–34, 235n.11
whispers, 17
widows, 54
Wilamowitz, U. v., 175
wills and testaments, 41
wine, 183n.6
winds, 142
wish formulas, 6
witches, witch fantasies and witchcraft, 223–24, 227–28
wolf, 112, 113, 177, 184n.14, 219
women, 16, 72, 114, 140, 144–45, 152, 158–59, 161–62, 165, 166, 216–33. *See also* gender and the performance of magic rituals
words, the power of, 8, 41–44, 129
workshops, 11
wrath of the gods, 43, 44, 57, 73. *See also* punishments and torture
writ of cession, 10
writing, 5, 23, 24, 124–25. *See also* verbal vs. written formulas
Wünsch, R., 16, 80

youth, as an assistant or medium, 185n.33, 193

Zeus, 23, 34, 35, 45, 46, 112, 120, 128, 136, 157, 179
Zingerle, J., 80–81
Zminis of Tentyra, 180
zodiac, 154–56, 178
Zuntz, G. 114–16